Results and Problems in Cell Differentiation

A Series of Topical Volumes in Developmental Biology

10

Chloroplasts

Edited by J. Reinert

With Contributions by

L.O. Björn   Th. Börner   W. Bottomley
Th. Butterfass   R. Hagemann   F.H. Herrmann
R.G. Herrmann   R.M. Leech   B. Parthier
J.V. Possingham   E. Schnepf   C. Sundqvist
H.I. Virgin   R. Wollgiehn

With 40 Figures

Springer-Verlag Berlin Heidelberg GmbH 1980

Professor Dr. Jakob Reinert
Institut für Pflanzenphysiologie
und Zellbiologie
Königin-Luise-Straße 12–16a
D-1000 Berlin 33

ISBN 978-3-662-21704-7   ISBN 978-3-540-38255-3 (eBook)
DOI 10.1007/978-3-540-38255-3

Library of Congress Cataloging in Publication Data. Main entry under title: Chloroplasts. (Results and problems in cell differentiation; 10). Includes bibliographical references. 1. Chloroplasts. 2. Plant cell differentiation. I. Reinert, J. II. Björn, Lars Olof, 1936–. III. Series. [DNLM: 1. Chloroplasts. 2. Plants – Cytology. W1 RE248X v. 10/QK898.C5 C544] QH607.R4 vol. 10 [QK 725] 574.8761 s 80-15613 [581.87'33].

2131/3130-543210

Preface

Plant cells contain various types of plastid, the best known among which is the chloroplast. Apart from their predominant interest for the work on photosynthesis, however, chloroplasts have attracted considerable attention for other reasons. This pertains to extranuclear inheritance of cell organelles and, particularly important for this series, to the participation of chloroplasts as discrete and partly autonomous cell constituents in the developmental biochemistry of plants.

This volume is composed of articles by investigators who are actively involved in work on various aspects of research on chloroplasts. Each author has independently covered and analyzed as comprehensively as possible the particular aspects assigned to him. This has the advantage of bringing out many different facets of the situation, though some overlapping has to be taken into account. We are sure that this volume will enable the reader to gain a broad theoretical and experimental basis for the understanding of the development of chloroplasts and the relationship between plant cells and these organelles.

Spring 1980

W. BEERMANN
W. J. GEHRING
J. B. GURDON
F. C. KAFATOS
J. REINERT

Contents

Types of Plastids: Their Development and Interconversions

By E. SCHNEPF. With 8 Figures

The Continuity of Plastids and the Differentiation of Plastid Populations

By TH. BUTTERFASS. With 3 Figures

Plastid DNA — The Plastome

By R. G. HERRMANN and J. V. POSSINGHAM. With 8 Figures

RNA and Protein Synthesis in Plastid Differentiation
By R. WOLLGIEHN and B. PARTHIER. With 12 Figures

**Biosynthesis of Thylakoids and the Membrane-Bound Enzyme Systems
of Photosynthesis**
By F. H. HERRMANN, TH. BÖRNER, and R. HAGEMANN. With 3 Figures

Fraction I Protein

By W. BOTTOMLEY

Factors in Chloroplast Differentiation

By C. SUNDQVIST, L.O. BJÖRN, and H. I. VIRGIN. With 4 Figures

The Survival, Division and Differentiation of Higher Plant Plastids Outside the Leaf Cell

By R. M. LEECH. With 2 Figures

Types of Plastids:
Their Development and Interconversions

E. Schnepf
Lehrstuhl für Zellenlehre der Universität
Heidelberg, FRG

I. Characteristics and Distribution of Plastids

Photosynthesis provides both the energy and matter for nearly all biotic processes. Moreover, it has generated all oxygen present in the atmosphere and, thereby, changed the conditions of life and the face of the earth. In photosynthesis light is transformed into chemical energy with the aid of chlorophyll. The overwhelming amount of photosynthetic products are formed by special organelles of eukaryotes, the chloroplasts. In addition to chloroplasts, there are achlorophyllous developmental stages and genetically fixed forms which have other functions in the cell but maintain the potential to develop into chloroplasts or are derived from them. Taken together, this whole family of interrelated organelles is called plastids.

Plastids are bounded by two membranes (Fig. 1). The chlorophyll is localized in an internal membrane system, the thylakoids. The thylakoids form flat sacs which are often arranged into stacks ("grana") (Figs. 1 and 2) and are embedded in the plastidal matrix where DNA and ribosomes as well as starch, lipid droplets

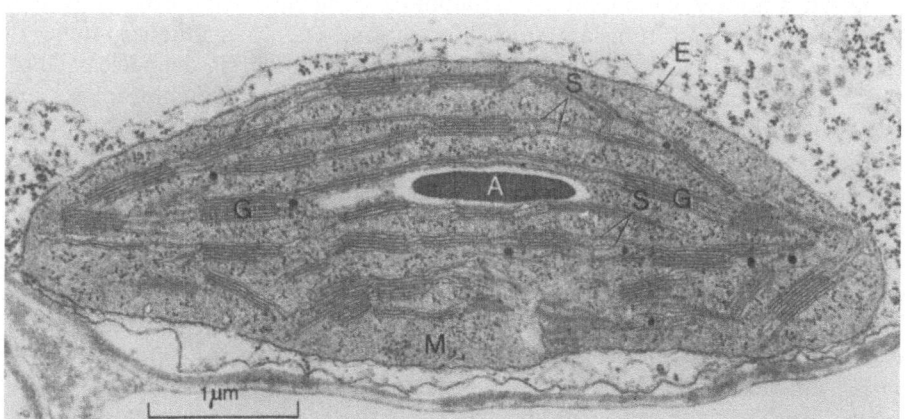

Fig. 1. Chloroplast of *Avena ventricosa. A* starch grain; *E* plastidal envelope; *G* granum; *S* stroma thylakoids. In the matrix *(M)* ribosomes and a few plastoglobuli are visible. (From Gunning and Steer 1975)

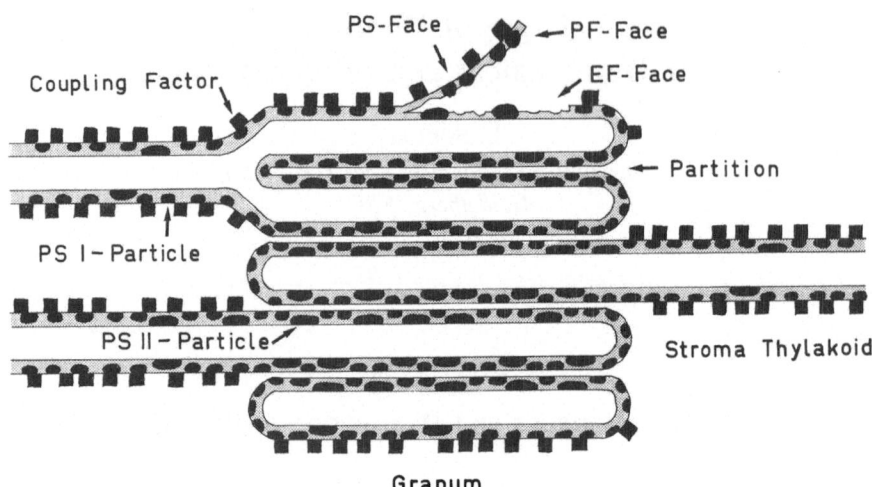

Fig. 2. Diagram of the thylakoidal system of a chloroplast based on thin sectioning and on freeze etching. The particles which represent photosystem I and photosystem II (PS I- and PS II-particles), integrated into the thylakoidal membrane, are exposed in the PF- and EF-face by freeze etching. For details see text

("plastoglobuli") and, occasionally, phytoferritin are found. The compartmentalization is easily explained by the endosymbiosis hypothesis of plastid evolution (Schnepf 1964) but also by other models (Cavalier-Smith 1975; Parthier 1975). Photosynthetically active chlorophyll-containing thylakoids also occur in the prokaryotes, i.e., blue-green algae and some bacteria. In these cells, however, they are not found in a special envelope. Therefore, these prokaryotic thylakoids are not homologous with chloroplasts but, instead, represent special differentiations and invaginations of the plasmalemma, even though they have often lost their connection with the latter, especially in the blue-green algae.

Plastids are semiautonomous: *auto*nomous in having their own DNA, and they transcribe and translate this information, i.e., to synthesize some of their proteins; *semi*autonomous in being strongly dependent on the nuclear DNA and the cytoplasmic translation system. In higher plants the chloroplasts are uniformly shaped like a lense with a diameter of $3 - 10\mu m$. In 75% of the 200 plants studied by Möbius (1920) they measure $4 - 6\mu m$. Other plastids are, as a rule, smaller and tend to be spherical or irregular in shape. In algae the variability of chloroplast shape is greater. Although they frequently resemble those of higher plants, many are cup-, band-, or star-shaped.

Plastids are integral parts of the cell. In higher plants and in most groups of algae, they are found, with very few exceptions, in every living cell. They are, for example, lacking in some highly reduced male sexual cells (e.g., *Stephanopyxis turris*; von Stosch and Drebes 1964, cf. Kirk and Tilney-Bassett 1967), and they do not occur in the highly specialized hair cells of the green algae, *Bulbochaete hiloensis* (Fraser and Gunning 1973). In extremely rare cases, single cells of mosses and angiosperms were found to be free of plastids (Bauer 1942/1943; Butterfass 1969). These apoplastidic cells do not divide further.

Apoplastidic algae occur, e.g., in the Euglenophyceae, Chrysophyceae, and Dinophyceae. However, in most groups of algae (Pringsheim 1963), as well as in higher plants, achlorophyllous organisms and cells contain achlorophyllous plastids. They are apochlorotic (see Sect.III.A.2). All fungi are apoplastidic, just as are all proto- and metazoa. There are, however, marine mollusks which have acquired plastids by feeding on siphonalean green algae and incorporating the algal chloroplasts into special cells. Such chloroplasts remain photosynthetically active for several weeks but cannot multiply in the new host cell (Muscatine and Greene 1973). Obviously, the host cannot supply them with sufficient proteins and other substances needed for their development, e.g., DNA polymerase and factors for chlorophyll biosynthesis.

It is well established that plastids never arise de novo but by division. It is known from *Chlamydomonas* that after gamete fusion not only the nuclei but also the chloroplasts fuse (Blank et al. 1978). Even the leucoplasts of the apochlorotic flagellate *Polytoma* undergo "plastogamy" (Gaffal 1978). But plastogamy seems to be an exception.

Detailed descriptions of plastid division and its biochemical basis are given in the following chapters. Other authors report on the formation of the photosynthetic system and on the differentiation of chloroplasts. The purpose of this article is to present an introductory account of the different types of plastids, their distribution and modes of development and interconversions.

II. Classification of Plastids

A. Factors Which Effect the Diversity of Plastids

The diversity of plastids is due to three main factors. It is sometimes difficult to separate their effects on plastidal differentiation but only an independent analysis can elucidate the relationships.

The first is the systematic position of the plant which reflects its general genetic background, as represented by its genome and plastome. On the one hand, the genetic information determines the complexity of the organism — whether it is a unicellular or a multicellular one, and, if multicellular, how many different cell types can be formed. On the other hand, it codes for the principal constituents of the plastids (i.e., different species of chlorophyll and other substances) and, thereby, also influences plastid structure. This factor is less important in higher plants (for exceptions see Behnke 1975a) than in the algae, but special mutants and apochlorotic parasites and saprophytes should also be grouped here.

The second factor affecting plastid diversity is cell differentiation which depends on the *active* genes only. Gene activation and expression is governed by environmental factors (light, temperature) and, in multicellular plants, by the location of the cell within the organism, and within the sequence of developmental steps. In higher organisms, nearly every cell type possesses its special type of plastid (Hohl 1960). The effect of light on cell and plastid differentiation is mediated by the phytochrome system (Gehring et al. 1977; Girnth et al. 1978) provided that phytochrome does not directly act on plastid development by changing membrane properties of the plastids (Schmidt and Hampp 1977).

The third factor is the environment, especially light and nutrition, affecting directly the organelles, e.g., the photoconversion of protochlorophyllide into chlorophyll (for survey see Treffry 1978) and the membrane transformations connected therewith in etioplasts (see Sect.III.B.2). The temperature-sensitive plastids of several higher plants are deficient in chloroplast rRNA at high temperature. As a consequence, the developing leaves are chlorotic (Schäfers and Feierabend 1976; Feierabend and Mikus 1977; for *Euglena*, see Brandt and Wiessner 1977). The mechanism by which water stress affects plastid development (Duysen and Freeman 1978) is unknown.

If it is correct that these three factors exclusively determine the development of plastids, the population of plastids within *one* cell should be composed of identical members, provided that the plastomes do not differ. Indeed this is generally the case. The few exceptions are restricted to polyenergid "cells" as, e.g., in siphonalean algae (Caulerpales) (Borowitzka 1976; Colombo 1978). In these cells starch-accumulating plastids are found in the endoplasm close to chloroplasts in the ectoplasm. The mechanisms for the differentiation of these plastids and the conditions which allow their coexistence are not known. Monoenergid mixed cells with genetically different plastids (formed by the fusion of unequal gametes or by mutation in the plastome) form instable cell lines. After a rather short sequence of cell divisions the descendants have a uniform population (Michaelis 1962). In such mixed cells the genetically different plastids may influence each other in an unknown way (Röbbelen 1966).

It is obvious that the diversity of plastids within one plant is related to the diversity of its cells. If there are developmental gradients within a cell, the polarity in differentiation can include the plastids and lead to different plastids within a single, normal cell (Bopp 1955: moss protonema; Wada and O'Brien 1975: fern protonema; Puiseux-Dao et al. 1977: *Acetabularia*). These observations, as well as those in the Caulerpales, suggest that the local microenvironment can influence the development of the plastids to a certain degree and that plastids can be captured in certain regions of the cell.

Partial darkening of the giant cells of *Nitella* results in differential development of the plastids into two phenotypically distinct chloroplasts within one and the same cell (Craig and Gibor 1971).

B. Principles of Classification

A classification system for plastids which follows the systematic position of the plants and, hence, the general pigment content, although valid, is of little help in classifying the diversity of plastidal stages in higher plants. Attempts to classify the latter were often confusing since, unfortunately, stages from different categories were sometimes grouped together (Newcomb 1967). The discriminating features usually are color, developmental stage, and function. In principle, classification systems which are based on these criteria are parallel to and independent of each other. Clearly in all systems, intermediate stages occur which cannot be easily classified.

1. Color

The first principle used to classify the plastids of higher plants was their color (Schimper 1885): white or colorless types are "leucoplasts", green types are "chloroplasts" and yellow or red plastids (because of their relatively high content in carotenoids) are referred to as "chromoplasts". Some authors include the chloroplasts within the chromoplast group because of their color; usually the term is applied in the narrow sense. Within such a system based on color, only the chloroplasts are a natural unit. In contrast the leucoplasts, occurring in meristematic cells (as proplastids), in storage tissues (as amyloplasts etc.), in roots, white petals, and in many other cell types, form a very nonhomogeneous group. Nevertheless, this terminology is still widely employed because of its simplicity.

2. Developmental Stage

According to their developmental stage distinctions can be made between juvenile ("proplastids"), differentiating, fully developed, and senescent plastids. Strictly speaking, proplastids are restricted to embryonic and meristematic, i.e., juvenile cells. Fully developed plastids occur in fully differentiated cells and senescent plastids in aging cells. As will be shown later, redifferentiation of a cell involves a redifferentiation of their plastids. The developmental stage of the plastids directly depends on the developmental stage of the cell.

3. Function

Functional stages are: photosynthesizing plastids ("chloroplasts"), storage plastids ("amyloplasts", "proteinoplasts", "elaioplasts"), plastids with secondary functions (participating in the production of secondary plant substances as well as chromoplasts and leucoplasts in flowers and fruits), and plastids with the exclusive function to divide and to grow ("proplastids"). The term proplastid, therefore, designates plastids of a special function as well as of a special developmental stage. Many cells contain leucoplasts of uncertain or unknown function. Indeed many leucoplasts appear to have no function whatsoever, except to maintain themselves; these are usually poorly developed and have been called proplastids by some authors even though they occur in differentiated, nondividing cells. The choice of the most appropriate classification system depends on the special problem. Each of the three can be used with advantage. It can be confusing, however, if the categories of the different systems are mixed together.

III. Forms and Stages of Plastids

In a survey on the forms and stages of plastids, it is reasonable to treat separately the algal plastids from those of higher plants. The diversity is mainly based on the fact that the general genetic information is important for the former because cell differentiation is of minor importance, while, on the contrary, cell differentiation via differential gene activation dominates plastid modification in the latter.

A. Plastids of Algae

1. Chloroplasts

The systematics of algae is partly based on their photosynthetic pigments and hence on the structure of the plastids (Dodge 1973; Dodge and Kowallik 1974).

2. Leucoplasts

The low degree of differentiation in algal organisms implies a relatively high uniformity of plastids within a thallus. Only in large Phaeophyceae and Rhodophyceae with meristem-like tissues do plastids occur with a poorly developed thylakoidal system (Borowitzka 1978). They are comparable to the proplastids of higher plants. In contrast to most higher plants, algae frequently form chloroplasts in the absence of light, but light can influence pigmentation and the development of the thylakoidal system (Sheath et al. 1977; Vesk and Jeffrey 1977). Moreover, there are light-dependent species (*Ochromonas*: Gibbs 1962) or mutants, for instance of *Chlamydomonas reinhardii*, which develop only leucoplasts in the dark. They are especially suitable tools for studying the development and function of the thylakoidal system (see Herrmann et al., this vol.). Several flagellates, e.g., *Euglena gracilis*, can be grown heterotrophically in the dark. Under these conditions these organisms do not synthesize chlorophyll but develop leucoplasts which can re-green in the light; this bleaching is reversible (Klein et al. 1972; Ophir et al. 1975). The addition of antibiotics to the culture medium also produces leucoplasts (e.g. Kronestedt and Walles 1975) which then are irreversibly bleached; however, some authors believe these bleaching treatments do not produce apochlorotic but rather apoplastidic strains (see Nigon and Heizmann 1978). The irreversible bleaching of *Euglena gracilis*, strain Z, by high temperature was explained recently on molecular basis. The plastidal RNA-polymerase of this strain is inactive at $34°-35°$ C, in contrast to the nuclear RNA-polymerases (Brandt and Wiessner 1977).

In *Chlorella protothecoides* the differentiation of the plastids is controlled by nutrition and light (Osafune and Hase 1975). Green cells develop leucoplasts in a medium containing a high concentration of glucose or another organic carbon source, either in the presence or absence of light. They regenerate normal green plastids in the presence of light if they are transferred into a nitrogen-containing medium without organic carbon supply. When transferred to the dark, they become pale-green, because the chloroplasts develop poorly in the dark. In general, the leucoplasts of naturally occurring apochlorotic algae seem to have retained the ability to synthesize reserve polysaccharides. The leucoplast DNA of *Polytoma* has anomalous physical characteristics but also considerable affinities to the chloroplast DNA of its green relative, *Chlamydomonas* (Siu et al. 1975).

B. Plastids of Higher Plants

1. Proplastids

The embryonic and meristematic cells of higher plants have undeveloped proplastids. They are spherical or irregular in shape, usually colorless, and smaller

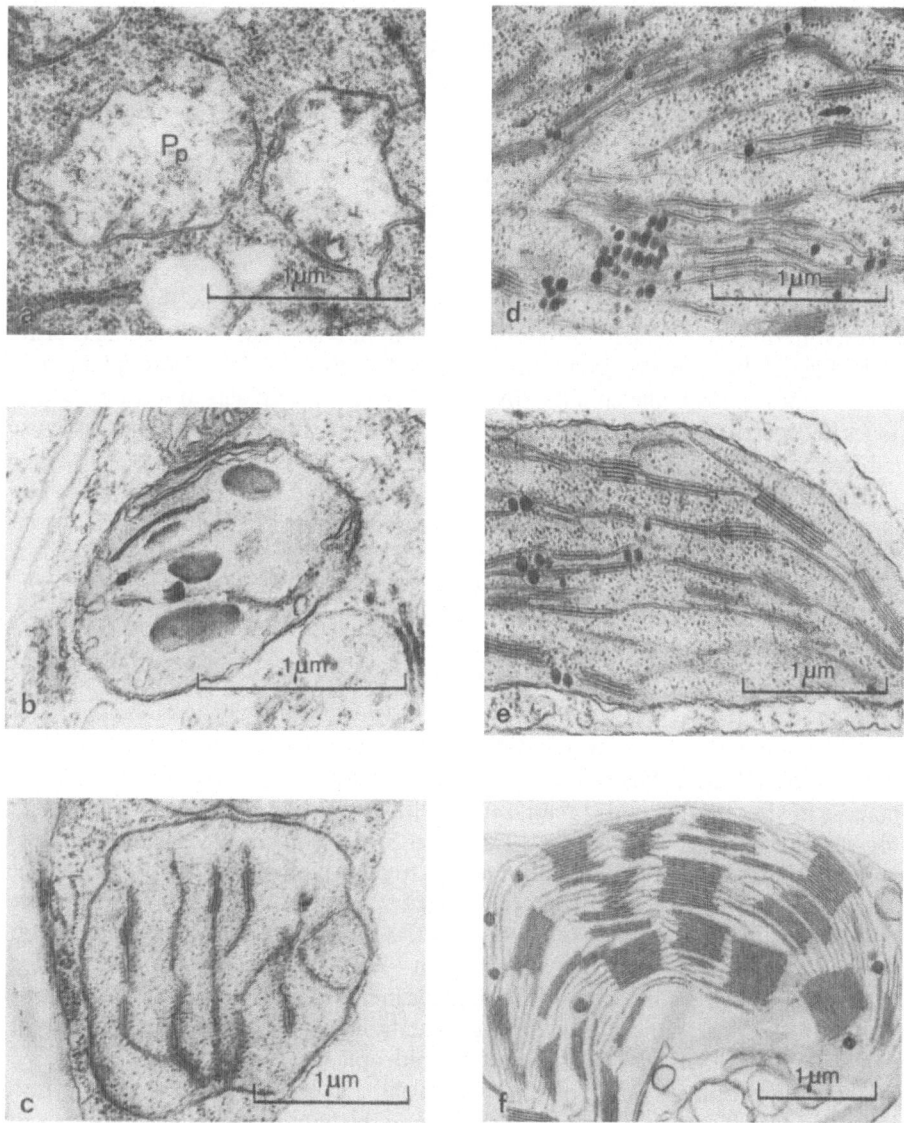

Fig. 3a-f. Development of chloroplasts in embryos and primary leaves of *Hordeum vulgare* during continuous illumination (5 klux). Age: **a** 0 h; **b** 48 h; **c** 72 h; **d** 96 h; **e** 120 h; **f** 168 h. **a** represents proplastids *(Pp)*, **f** a chloroplast. (From Sprey 1973)

than chloroplasts (generally about 1–1.5μm in diameter). They have a dense matrix with many ribosomes and contain very few small, single thylakoids, tubules and vesicular structures, often connected with the inner plastidal membrane (Figs. 3a and 5a). In addition, small plastoglobuli and, less frequently, small starch grains are found. The proplastids of the shoot promeristem do not differ in structure from those of the root (cf. Hinchman 1972), but begin to develop into chloroplasts in the postmeristem so that the proplastids sensu stricto are restricted to a small group of cells. In embryogenesis diversification of plastids begins early (Nagl and Kühner 1976). In leaf meristems plastids may (Menke 1960) or may not (Brossard 1972) begin to transform into chloroplasts, and also in the shoot cambium they are structurally modified (Robards and Kidwai 1969).

Even in cotyledons of dry seeds chloroplasts can occur (Marin and Dengler 1972). It is unlikely that the so-called plastid initials indeed exist. They were postulated by Mühlethaler and Frei-Wyssling (1959) and assumed to be more reduced precursors of proplastids.

Proplastids divide in coordination with cell division (see Butterfass and Herrmann, this vol.), usually by constriction but in a few cases by budding (Pettitt 1976). Plastid division, however, is not restricted to proplastids. In lower plants and mosses, but also in higher plants, division of both chloroplasts and intermediate stages between proplastids and chloroplasts has been observed directly or, at least, deduced (e.g., Possingham and Saurer 1969).

Usually roots do not become photosynthetically active even in the light. The plastids remain proplastid-like, functionless, and do not divide further in normal development. Somewhat confusingly, they have often been called proplastids. In the root cap amyloplasts (see Sect.III.B.4) are formed.

2. Etioplasts

In the shoot and in leaves, the differentiation of chloroplasts from proplastids requires light. With sufficient light intensities chlorophyll is synthesized, thylakoids increase in size, branch and spread out, and finally form the grana-stroma-system (for survey see e.g., Wehrmeyer 1972) (Fig. 3). In contrast to *Chlamydomonas* the growth of thylakoidal membranes in higher plants does not seem to be controlled mainly by feed-back mechanisms but by parameters which are independent of the developmental stage of the thylakoids and their composition (Akoyunoglou and Argyroudi-Akoyunoglou 1978). The thylakoids of developing chloroplasts may be connected with the inner membrane of the plastidal envelope (Fig. 4). In darkness these processes are partly inhibited, and a special stage of plastid, the etioplast, develops instead of chloroplasts (Fig. 5).

Typical etioplasts are found exclusively in cells which develop from meristematic cells directly and would be green in the light. In dark-grown roots and calli (e.g., Laetsch and Stetler 1965; Sjolund and Weier 1971; but see also Blackwell et al. 1969), as well as in the promeristem itself, etioplasts are generally lacking. Their development largely depends on the age of the tissue (Robertson and Laetsch 1974). Seedlings of some gymnosperms (e. g., *Pinus*) can synthesize chlorophyll and form chloroplasts with stacked thylakoids in darkness, but in addition develop prolamellar bodies in the prolonged absence of light (Camefort 1963); however, these latter structures do not accumulate protochlorophyllide or chlorophyllide in correlative amounts (Michel-Wolwertz and Bronchart 1974). Other gymnosperms (e.g., *Larix*) develop only an immature thylakoidal system with tiny grana (Walles and Hudák 1975). Presumably, the differences are related to the time during which the embryo is in contact with the primary endosperm (Laudi and Fanelli 1964).

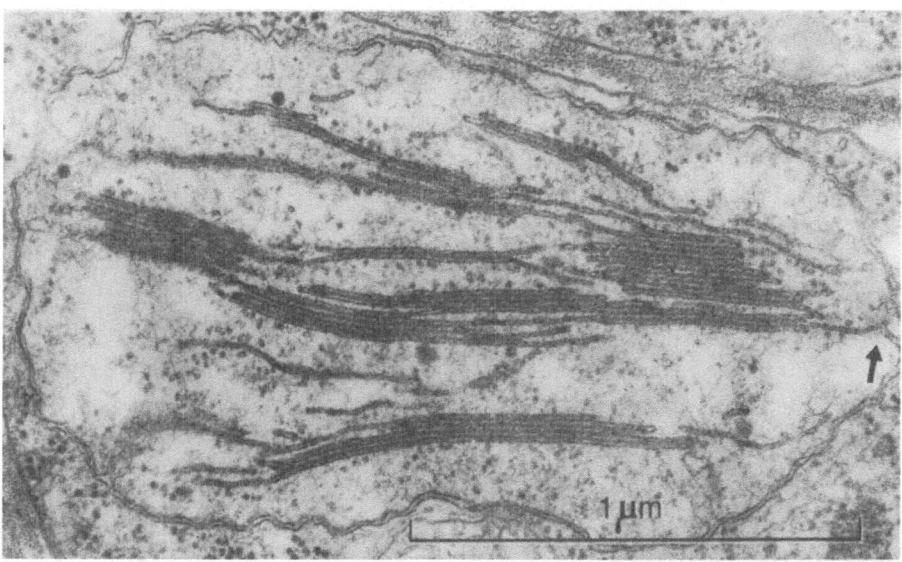

Fig. 4. Chloroplast in a young *Sphagnum* leaflet. The thylakoidal system is not yet fully developed and is in connection with the inner membrane of the plastidal envelope *(arrow)*

Etioplasts seem to contain all of the morphological and biochemical precursors necessary to become photochemically active chloroplasts after a few hours' illumination (Alberte et al. 1972): the prolamellar body and protochlorophyllide which both are sources for the thylakoidal system (e.g., Henningsen and Boynton 1974). However, Girnth et al. (1978) could not find a correlation between the rate of grana and stroma thylakoid formation and the extent of the prolamellar body disintegration. The conversion processes will be treated in detail in other chapters of this volume. The prolamellar body serves as a storage of membranes (Sprey 1973) in the form of tubular, branched, and interconnected units (diameter: 18–20 nm) which form a crystalline lattice (further chemical characterization: Lütz 1975a, b; Lütz and Klein 1979).

Extensive studies especially of Gunning and Wehrmeyer (for survey see Wehrmeyer 1972; Gunning and Steer 1975) revealed that the lattice can occur in different types, even within a single etioplast. The "units" may be tetrahedrally four-armed or, more rarely, cubically six-armed. The former are mainly arranged in "zinc blend-like" or "wurtzite-like" lattices, but five-sided rings also occur with these units, and various combinations are also possible (Fig. 6).

In addition to the prolamellar body and often in connection with it, the etioplasts also contain single thylakoids which usually are fenestrated rather regularly. Naturally occurring etioplasts, in darkened parts of buds and bases of monocotyledons etc., contain more or less extended stacks of thylakoids, i.e., the developing grana. In these cases the membranes of the prolamellar body and of the developing grana seem to be readily interconverted. The equilibrium is influenced not only by the light intensity and quality which reaches the plastids, but also by the

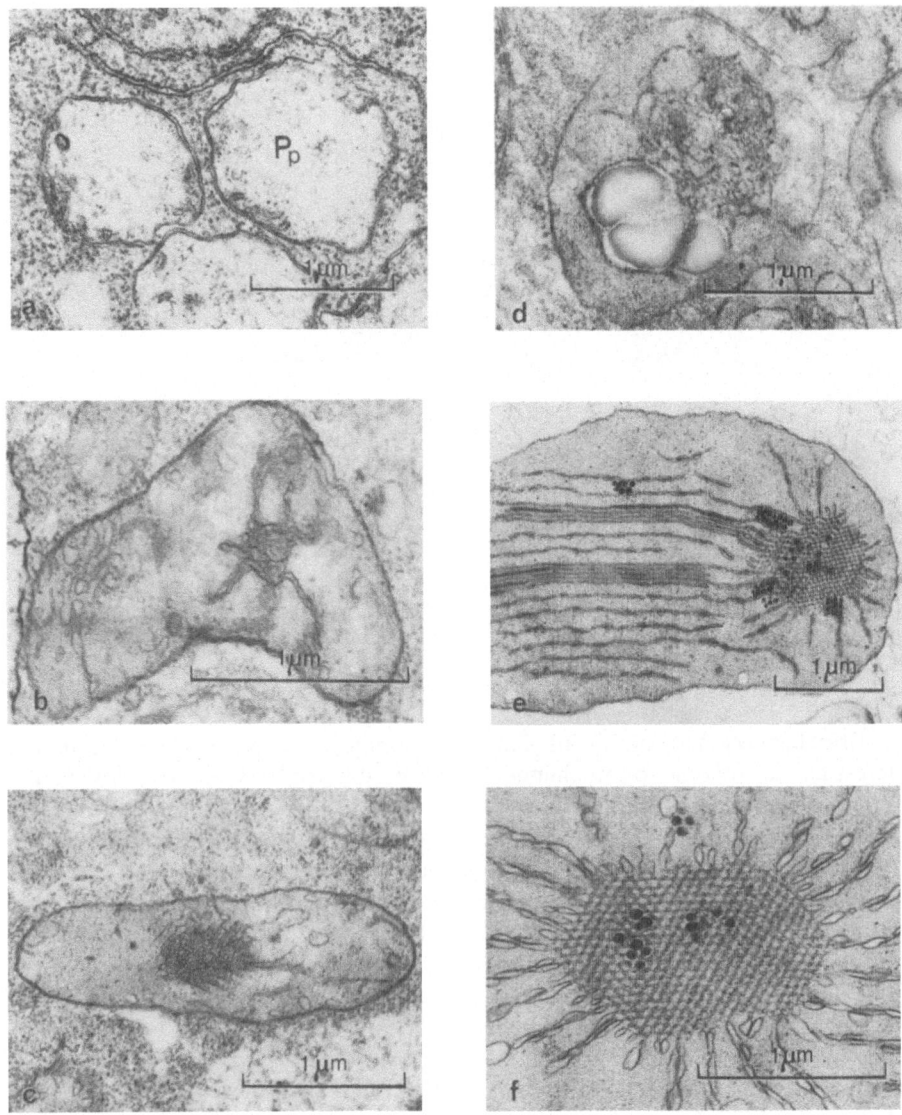

Fig. 5a-f. Development of etioplasts in embryos and primary leaves of *Hordeum vulgare* in darkness. Age: **a** 0 h; **b** 48 h; **c** 72 h; **d** 96 h; **e** 120 h; **f** 168 h; **a** represents proplastids *(Pp)*; **e, f** fully developed etioplasts. (From Sprey 1973)

"pretreatment" (Wrischer 1966) and the temperature (Ikeda 1971; for effects on pigment accumulation see Treffry 1973). The same intensity gives rise to more grana thylakoids if it is higher than the preillumination and, vice versa, to more extended prolamellar bodies if the preillumination itself was more intense. Prolamellar bodies are formed obviously when the rate of membrane production exceeds the rate of membrane conversion into thylakoids (Ikeda 1971).

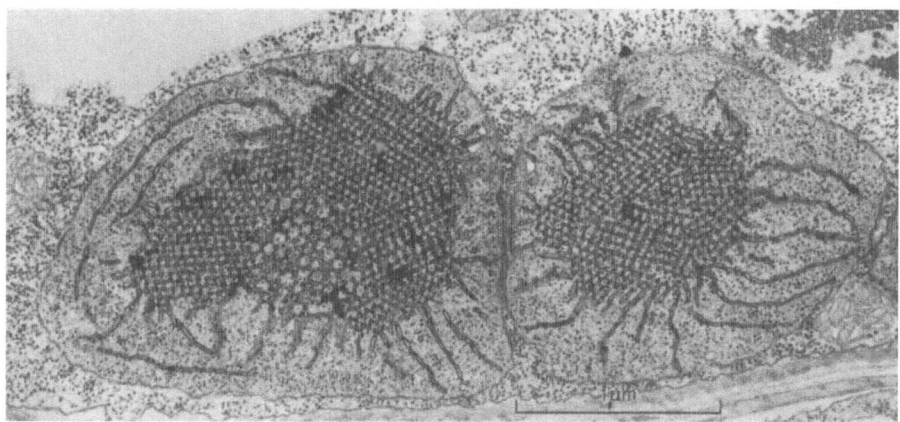

Fig. 6. Etioplasts from a leaf of a seedling of *Avena sativa* with semicrystalline prolamellar bodies. (From Gunning and Steer 1975)

3. Chloroplasts

Photosynthesis (see also Herrmann et al. and Bottomley, this vol.) consists of two sequential processes, the photochemical light reactions and the biochemical dark reactions. The two light reactions are localized in the thylakoids. The thylakoidal membranes contain two different photosystems (I and II), the function of which is to produce both ATP and NADPH at the expense of H_2O as the ultimate electron donor. Photosystem I on its own can generate ATP in a cyclic photophosphorylation. In C_3-plants CO_2 is fixed in the dark reactions of the matrix by a sequence of reactions known as the Calvin cycle. An additional mechanism to fix CO_2 more efficiently is used by C_4-plants. Among other unique features, they produce C_4-acids in the initial fixation steps.

The light microscope shows that chloroplasts contain tiny green cylindrical bodies with a diameter of about 0.2–0.4μm, the grana, embedded in a less dense "stroma". Electron microscopy (Fig. 1) reveals that the grana are made up of stacks of closely appressed circular thylakoids (granum thylakoids or discs) which, in places, extend into the stroma to form stroma thylakoids or frets (Fig. 2). They consist of tubules or flat sacs which often are irregularly perforated. The stroma thylakoids connect, at least partly, single grana but also grana thylakoids within one stack (Wehrmeyer 1964) thus forming one continuous compartment for the thylakoidal system (Paolillo 1970). As a rule, the grana and stroma thylakoids are oriented more or less parallel to the long diameter of the chloroplast, but often their arrangement is rather irregular. Pale-green cells, as e.g., in the shoot cortex or in the leaf epidermis of many plants, contain plastids with a few small grana but show no other peculiarities. The chloroplasts of guard cells resemble those of the mesophyll.

Chloroplasts in long-lived tissues may undergo seasonal changes in the structure of the thylakoidal system and in the function to store starch (Senser et al. 1975). Unusual pigments result in abnormal thylakoid arrangement: the green plastids of the *Cucurbita maxima* seed coat contain protochlorophyll which is not

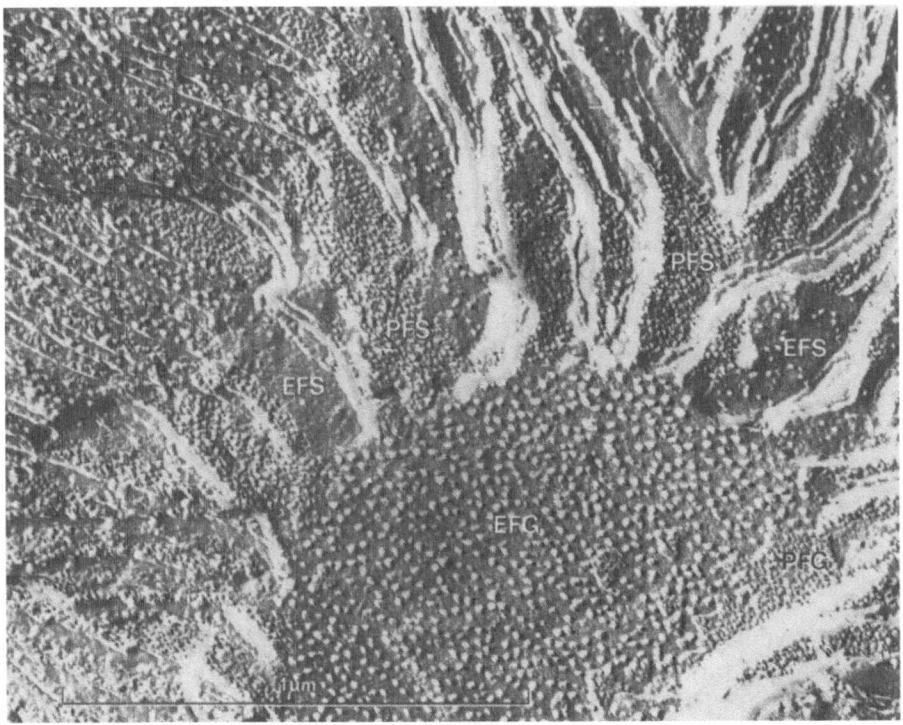

Fig. 7. Freeze-etched chloroplasts of *Lomandra longifolia*. Granum with large particles in the EF-face *(EFG)* and small particles in the PF-face *(PFG)*; in the stroma thylakoids the frequency of the large particles is much lower *(EFS*-area) whereas the number of the smaller particles (in the *PFS*-area) is as high as in the granum (further explanation in Fig. 2). Electron micrograph from Dr. D. Goodchild, CSIRO Division of Plant Industry, Canberra. (From Gunning and Steer 1975)

converted into chlorophyll by light; the thylakoids consist of branching tubules (Lott and Darley 1973). The number of discs per granum in chloroplasts may range from two up to some dozens. The number as well as the size depends on the cell type but also on light adaptation (Hariri and Brangeon 1977, for survey see Boardman 1977). The lipid composition in plastids of sun and shadow plants is also different (Guillot-Salomon et al. 1978). Because of the close contacts (partitions) of adjacent thylakoids within a granum, only the frets and the outer parts of the uppermost and lowest disc of a granum are in immediate contact with the matrix (Fig. 2).

In general, the partition zone seems to contain photosystem I and II activity. Here the ratio chlorophyll a:b is lower than in stroma thylakoids. As revealed by freeze etching (Figs. 2 and 7), two types of particles (9–12 nm and 13–17 nm in diameter respectively) are localized within the membranes of these partition zones. The largest particles are found in the EF (exoplasmic fracture) face of the membranes and seem to represent a central photosystem II reaction center plus associated light-harvesting protein (Miller et al. 1976; Armond and Arntzen 1977). In the stroma region predominantly photosystem I, visualized by the 11-nm

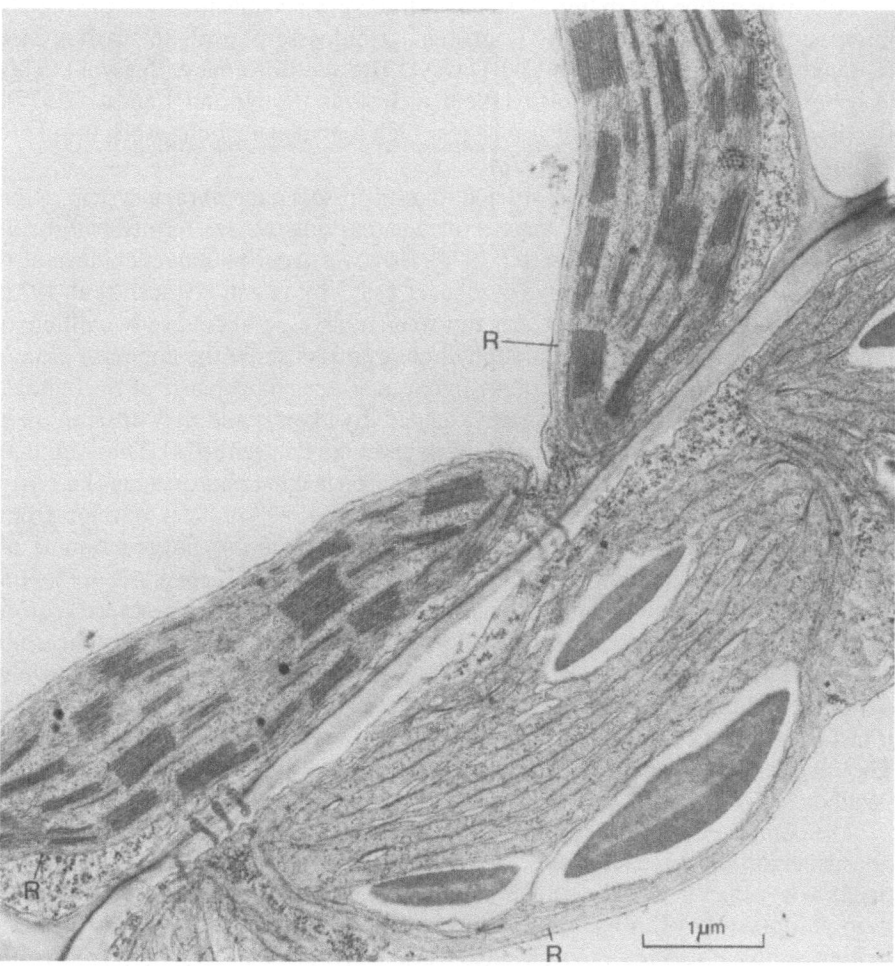

Fig. 8. Dimorphic chloroplasts in the C_4-plant, *Zea mays*. The bundle sheath chloroplasts *(lower right)* contain starch grains but no grana; the mesophyll chloroplasts *(upper left)* grana, but no starch. Both types of chloroplasts contain a peripheral reticulum *(R)*. The chloroplasts had not completed their development, as indicated by the presence of a small prolamellar body in the upper plastid. (From Gunning and Steer 1975)

particles has been found (Gasanov and French 1973). The size of the particles changes during thylakoid differentiation (Armond et al. 1977), perhaps due to an increasing aggregation with the light-harvesting pigment protein. The smaller particles are observed in the PF-(matrix) fracture face of the membranes. Attached to the true PS (matrix) surface particles with a diameter of 10–11 nm occur (Wellburn 1977). They seem to represent the coupling factor for photophosphorylation (Miller and Staehelin 1976) which is presumably identical with the ATPase. The concentration of photosystem II activity in the grana region was shown also by cytochemical methods (Marty 1977).

Spherules of fibrillar material (stroma centers) or crystals found in the matrix obviously represent Fraction I protein, ribulosebisphosphate carboxylase (Gunning et al. 1968; Willison and Davey 1976; see Bottomley, this vol.). This enzyme also forms crystalline thylakoid inclusions (Sprey and Lambert 1977). Pyrenoids of algal plastids likewise have a high activity of ribulose-bisphosphate carboxylase (Kerby and Evans 1978).

In C_4-plants all chloroplasts studied to date possess a membrane system in the periphery of the plastids which consists of anastomosing tubules, termed peripheral reticulum (Fig. 8) (Sprey and Laetsch 1978). However, a peripheral reticulum is also found in some C_3-plant mesophyll and guard cell chloroplasts (Gracen et al. 1972; for a survey see Laetsch 1974). Corn and some other C_4-plants have two different types of leaf chloroplast. The mesophyll chloroplasts show the common grana-stroma pattern but lack starch grains. In contrast, the chloroplasts of the bundle-sheath cells contain starch and single extended thylakoids, and they differ in their chlorophyll a:b ratio (Woo et al. 1971; for a survey see Laetsch 1974). These agranal chloroplasts often are said to have a reduced or even lacking photosystem II activity (see however Kirchanski and Park 1974; Jupin et al. 1975). They develop from chloroplasts which have small grana and accumulate initially large amounts of starch (Miyake and Maeda 1976). Occasionally, remnants of these grana are found even in fully differentiated bundle sheath cells (for quantitative analyses see Woo et al. 1971). The transitory granal stage is formed either directly (in the light) or from etioplasts with crystalline prolamellar bodies (during illumination following darkness) (Suzuki 1974). Farineau et al. (1978), however, observed a transformation of etioplasts in bundle-sheath cells without passing the grana stage. The bundle sheath and the mesophyll chloroplasts contain identical genomes (Walbot 1977).

The comparison of the results on corn bundle-sheath chloroplasts and other unusual chloroplast types (Jupin et al. 1975: phloem chloroplasts, poor in stroma thylakoids) suggests that the stacking of the thylakoids into grana is related to the ratio photosystem I:II. Generally even in these extreme cases no photosystem is lacking completely. A few examples are known where plastids are not able to evolve oxygen in light but can perform photosystem I-dependent cyclic photophosphorylation (Schmid et al. 1976; Menke and Schmid 1976: *Neottia*, labellum of *Aceras*). These plastids contain single, unstacked thylakoids.

4. Storage Plastids

Amyloplasts are specialized in synthesizing and accumulating starch. This reserve carbohydrate is produced and deposited in the matrix in the form of grains whose size, shape, and structure vary among different plants and among different cells of a plant (Badenhuizen 1973).

Amyloplasts of sieve tubes produce starch grains with a relatively high content in amylopectin (Palevitz and Newcomb 1970). The endosperm of some corn varieties contains amyloplasts which predominantly produce amylose, others nearly exclusively synthesize amylopectin ("waxy" varieties), and others the even more branched glycogen.

Glycogen, the common storage carbohydrate of animals and fungi, is found only rarely in higher plants. It occurs in small aggregates rather than in distinct grains in the plastids of the Müllerian bodies of *Cecropia* (Rickson 1976), but not in the other parts of this plant. The Müllerian bodies are cellular structures on the abaxial junction of the petiole and the stem which are harvested and eaten by ants. Moline and Jensen (1975) observed glycogen formation induced by virus infections. These examples show that the synthetic capacity of amyloplasts can be switched by mutations (corn), by cell differentiation (*Cecropia*, sieve tubes), and by exogenous factors (virus infections).

Most amyloplasts are achlorophyllous. Particularly the "long-term" amyloplasts of storage tissues contain only a poorly developed thylakoidal system. In some cases (*Pellionia:* Salema and Badenhuizen 1967) chloroplasts serve also as amyloplasts, and accumulate starch in unusually high amounts. Proplastids can contain smaller starch grains. In the spadix appendix of some Araceae, starch is stored in chromoplasts (Schnepf and Czygan 1966).

Frequently an accumulation of starch (in transitory amyloplasts, Gunning and Steer 1975) seems to be nothing more than a response to the actual supplying of the cell with sugars, and their disappearance indicates sugar consumption.

Proteinoplasts store proteins, *elaioplasts* accumulate lipids. These processes play a minor role in the life of plants and are only occasionally observed, mostly in leucoplasts. The intrathylakoidal inclusions of juvenile chloroplasts of different plants (Casadoro et al. 1977) seem to store protein used later in lamellar formation (Hurkman and Kennedy 1977).

5. Plastids Participating in Elimination Processes

In many glandular cells, especially in those which produce and export lipophilic materials such as essential oils and resins, achlorophyllous plastids are found which are associated with and often nearly completely ensheathed by a cisterna of the endoplasmic reticulum (ER) (Vasiliev 1970; Dumas 1974). This conformation is frequently considered to indicate a special function. A participation of plastids in the elimination of secondary plant products is suggested by the occurrence of e.g. shikimate dehydrogenase (Feierabend and Brassel 1977), phenylalanine ammonia lyase and of flavonoids (Saunders and McClure 1975; Weissenböck et al. 1976; Charrière-Ladreix 1979) and visualized by the detection of lipophilic masses in or at plastids of some gland cells (Hammond and Mahlberg 1978; Charrière-Ladreix 1979) but also of tapetum cells where they are thought to produce "pollenkitt" (Hesse 1978).

6. Plastids as Optical Attractants

In flowers and fruits, plastids serve as attractants for pollinators and distributors of seeds. Formerly it was assumed that they were involved in odor production of *Platanthera*, but this was disproved by Heinrich (1973). The optical signal function of red and yellow chromoplasts and of leucoplasts in flowers and fruits is, however, not controversial.

The leucoplasts of white petals and similar organs contain only a few single thylakoids. In some cases they remain undeveloped during cell differentiation although they are generally formed from chloroplasts (Schürmann and Villiger 1967: bracts of *Davidia involucra*; Orsenigo 1963: *Hyacinthus*) and thus represent regressive stages. These leucoplasts develop at a very early stage of anthesis and exist even several weeks long since there are some very long-lived flowers (e.g., in orchids). Thus, they should not be thought of as senescent stages (see Sect.III.B.8) though, admittedly, a distinct separation cannot be made.

The chromoplasts in flowers and in fruits are equivalent to the just described leucoplasts and occur in equivalent cells. A well-known example is represented by the flowers of *Narcissus poeticus* which have leucoplasts in the perianth tube and chromoplasts in the corona. Another example concerns the yellow or white spatha of *Zantedeschia* (Araceae) in which both the chromoplasts (Grönegress 1974) as well as the leucoplasts (Pais 1972) re-green during fruit development. In addition to chromoplasts with an attractant function and those which represent true senescent stages, there are others which appear to have no useful purpose (Kirk and Tilney-Bassett 1967; Sitte 1974). Examples are the chromoplasts of carrot roots and the calyx epidermis of *Chrysosplenium* after anthesis. The former are probably the result of a mutation which has no selective advantage for the plant but was propagated by man; indeed roots of wild carrots do not contain chromoplasts.

Chromoplasts occur as different structural types. A contemporary classification was given by Sitte (1974). He distinguished between globulous, membraneous, tubulous, reticulo-tubulous, and crystallous chromoplasts. The morphological differences between the carotinoid-containing structures obviously are related to the surface:volume ratio which is correlated to the ratio of their semipolar and nonpolar components (Sitte 1978).

It is important for the developmental potency of the chromoplasts with attractant function that they or at least some of them — like other plastids — contain DNA (Falk et al. 1974; Wuttke 1976).

7. Apparently Functionless Plastids

Many tissues contain leucoplasts which apparently have no actual function in the cell. They are achlorophyllous, although the cells often receive sufficient light as, for example, in the shoot and leaf epidermis and their derivatives such as hairs and glandular epithelia. However, epidermal plastids are often pale-green. In guard cells they are true chloroplasts. Root plastids — of all tissues — often remain colorless in the light. As an exception, such plastids green upon extended illumination (e.g. Heltne and Bonnett 1970; Mesquita 1971).

Leucoplasts with no obvious function are round or ameboid in shape and possess a poorly developed thylakoidal system. Occasionally they have single dilated thylakoids which contain electron-dense substances, obviously proteins, and in these instances, can be considered to be proteinoplasts (Newcomb 1967). In several cases tubular complexes resembling prolamellar bodies are found in such plastids. Sieve-element plastids also contain protein aggregates (Behnke 1975b) and, in addition or exclusively, starch grains, depending on the systematic position of the plant (Behnke 1975a). Nevertheless, these plastids seem to have no actual (storage or other) function because, as far as is known, the accumulated products are never re-metabolized in the living cell. They develop from proplastids but in cases of abnormal phloem formation even from chloroplasts (Esau and Hoefert 1978).

The existence of even rudimentary plastids is a prerequisite for a cell to divide and to produce new cell lines with plastids of different functions. Cells with poorly developed, nonspecialized leucoplasts (e.g., epidermal cells of *Begonia* leaves) often seem to initiate regeneration processes. The dormant buds of potato tubers also contain such nonspecialized leucoplasts (Marinos 1967).

Most chromoplasts and leucoplasts of apochlorotic parasites and saprophytes (*Orobanche* chromoplasts: Kollmann et al. 1969) also seem to be apparently functionless. They have, however, retained their ability to form starch (see Sect.III.B.4). The above-mentioned carrot root chromoplasts also belong to this group, as well as, e.g., the plastids in the xylem transfer cells of *Hieracium floribundum* rhizomes which lack a defined thylakoidal system and develop regressively from chloroplasts (Yeung and Peterson 1975).

8. Plastids in Senescent and Mineral Deficient Tissues

As mentioned in Sect.III.B.5, chromoplasts of the globulous type develop during senescence of leaf chloroplasts (Ljubešić 1968; Dodge 1970; for degeneration processes in detached leaves or in leaf discs see, e.g., Młodzianowski and Ponitka 1973; Cran and Possingham 1974). Their plastoglobuli increase in size and in number, whereas their thylakoids become irregular and are reduced. The change in color form green to red or yellow in senescent cells is a consequence of the shifted ratio of chlorophyll to carotenoids, the former being degraded faster than the latter (Egger 1965). In contrast, the carotenoid content usually increases during the formation of chromoplasts with attractant function. The ultrastructural changes in senescing plastids are related to changes in the lipid composition (Huber and Newman 1976). Chloroplast breakdown in detached senescent leaves is not obligatorily connected with a rise in plastoglobuli (Shaw and Manocha 1965). Photosystem I seems to function even in plastids with a rather degraded thylakoidal system. For citrus fruits, ethylene is used to promote senescence of plastids transforming them from chloroplasts to chromoplasts and, thereby, to "improve" the color quality (Shimokawa et al. 1978).

Nitrogen deficiency in algae induces the formation of similar "chromoplasts" (more exactly: chloro-chromoplasts) of the globulous type as in senescent leaves (Kleinig and Wrischer 1968). Nutrient deficiencies in higher plants often result in "chlorotic" tissues with abnormal plastids. The variation in plastidal structure depends on the particular mineral which is deficient, on the plant, and even on the tissue (Thomson and Weier 1962; Marinos 1963; Vesk et al. 1966; Whatley 1971; Hall et al. 1972). Under conditions of nutrient deficiency, chloroplast formation is inhibited in general; their size is reduced and the thylakoidal system is poorly developed. Occasionally, relatively high amounts of starch are stored at the same time. Frequently the number and size of plastoglobuli is also increased in these plastids. Under conditions of severe deficiencies the thylakoids and the plastids tend to swell and to break down completely.

Another way of plastidal degeneration is found in cells which lose their cytoplasm and die after differentiation. The plastids diminish in size and simply disappear, obviously without participation of lysosomal activities (Schnepf 1973: hyalocytes of *Sphagnum*).

In connection with the senescence of plastids, it is interesting to note that the symbiotic chloroplasts of marine gastropods (see Sect.I), which remain photosynthetically active for several weeks, still incorporate photosynthetically fixed carbon-14 into carotenoids but not into chlorophyll (Trench and Smith 1970).

IV. Plastid Interconversions

During the more than 20 years following Frey-Wyssling's postulation of the "monotropic plastid metamorphosis" (Frey-Wyssling et al. 1955), in contrast to Schimper's (1885) classical concept of reversal transformations, plastidal interconversions have been discussed extensively. It should be mentioned, however, that many questions resulted from misunderstandings due to an inadequate terminology.

Schimper (1885) suggested that leucoplasts, chloroplasts, and chromoplasts can develop into each other while Frey-Wyssling et al. (1955) postulated that the development proceeds monotropically:

$$\text{proplastid} \rightarrow \text{leucoplast} \rightarrow \text{chloroplast} \rightarrow \text{chromoplast.}$$

As explained in Sect.II.B color and developmental stages are two different, independent categories though, admittedly, "young" plastids frequently are colorless and "old" ones yellow or red. In higher plants, the metamorphosis of plastids proceeds by two different routes, namely the true differentiation of a single cell on the one hand and its remeristematization or reembryonalization in connection with cell divisions on the other.

Remeristematization is connected with the formation of proplastid-like stages. They develop from other plastids, in general after a series of cell divisions, e.g., when a wound cambium is formed or, in culture, a callus and, subsequently, an embryo. Their progeny then undergoes a new differentiation phase during the development of a new plant. Many cells can generate calli and a few of them even new plants. In the latter case the plastids must interconvert. This is a consequence of the totipotency of a normal plant cell. Improved culture techniques will probably provide cell lines and plant regeneration even from more highly differentiated cells.

The stage of plastids in a callus depends, at least in older cultures, on external factors (light, growth substances) but not on the initial cell type. These plastids are chloroplasts or proplastid-like leucoplasts which are easily interconverted (e.g., Laetsch and Stetler 1965; Sjolund and Weier 1971). Relative to metamorphosis in leaf development, these interconversions are slow processes (Bergmann and Berger 1966). In this connection it is noteworthy that, in general (for an exception see

Blackwell et al. 1969), etioplasts with crystalline prolamellar bodies are not found in dark-grown callus cultures and that Stetler (1973) has detected protochlorophyllide which is not photoconvertible in such cultures. Having in mind the often-neglected fact that every cell type has its own type of plastid, it is not possible to give a generally valid scheme for plastid interconversions during the differentiation of a single cell. Frequently the development begins with proplastids. It can end in all stages of plastids found in cells which die as a consequence of the development of the plant. Especially it must be emphasized that plastids, even chloroplasts, do not necessarily end as "chromoplasts".

The fact that the cell differentiation does govern the plastid differentiation is described in a legion of papers. Only a few, containing detailed analyses, can be mentioned here: Hinchman (1972): oat coleoptile; Cran and Possingham (1972): spinach leaves; Robertson and Laetsch (1974): barley leaves; Lerbs and Eicke (1974): *Selaginella* shoots.

One example showing how the developmental steps are controlled and correlated is given by Feierabend in the development of seedlings after germination. He determined the effects of environmental factors (light, temperature), and endogenous factors (growth substances, protein synthesis) on the development of plastidal enzymes (Feierabend 1966, 1969) and the morphogenesis of the plastids (Berger and Feierabend 1967) and the interrelation of these processes with the development of other organelles, especially the microbodies (Feierabend 1972). Some of the following chapters in this volume deal with and are restricted to the problem of chloroplast formation in normal development. The full developmental capacity, however, is evident in some special cases.

As a result of the thesis on plastid metamorphosis of Frey-Wyssling et al. (1955), the reversion of chromoplasts into chloroplasts has been the subject of several studies. It has now been proven that the reversion occurs in chromoplasts with attractant function of flowers (e.g., Grönegress 1974; Sitte 1974) and fruits (Devidé and Ljubešić 1974: pumpkins) as well as of senescent, yellowed *Nicotiana* leaves (Ljubešić 1968). The greening of carrot chromoplasts was described by Grönegress (1971) and by Wrischer (1972). In general the carotenoid-containing plastoglobuli and crystals are reduced during the conversion; simultaneously the thylakoidal system with grana and frets develops. The development of chromoplasts instead of chloroamyloplasts in detached cucumber cotyledons after treatment with the growth retardant, 2-chloroethyl-trimethylammonium chloride (CCC), is described by Mikulska et al. (1973). As mentioned in Sect.III.B.4, leucoplasts in the flower region have the same status; they can likewise be converted into chloroplasts (Pais 1972). When root leucoplasts become green they follow the same pathway without the intermediacy of etioplasts (for an exception see Murakami 1962). This process thus resembles plastid transformations in callus cultures (Heltne and Bonnett 1970; Mesquita 1971). Chlorophyll formation requires blue light (Björn and Odhelius 1966) which seems to stimulate the synthesis of specific chloroplast rRNA (Dirks and Richter 1975). The interconversion is a rather slow process as is the case in callus cultures. With respect to the light requirements as well as to the morphology of the plastid metamorphosis and its kinetics, the transformation of amyloplasts in storage tissues into chloroplasts follows the same rules (Berger and Bergmann 1967; Anstis and Northcote 1973).

Obviously this mode of chloroplast formation differs considerably from the way in which leaf chloroplasts develop from true proplastids of meristematic cells. In the latter case there are several synthetic activities which are light-independent and result in the formation of protochlorophyllide and a prolamellar body in darkness. Hence the appearance of a prolamellar body might be an indication for a remeristematization (Schnepf and Pross 1976).

Finally it remains to be mentioned that pathogens — though they generally induce chlorosis or, in other words, a transformation of chloroplasts into leucoplasts (e.g., Sun 1965: virus — *Abutilon*; Syrop 1975: *Taphrina* — almond) — also can effect a conversion of leucoplasts into chloroplasts. This was observed, e.g., in petals of *Cheiranthus* infected by mycoplasms (Gourret and Le Normand 1971) and in galls on *Cuscuta australis* (Laudi 1968).

V. Conclusions

At first glance the questions regarding the biosynthesis of plastids and their interconversion seem to be resolved: (1) Plastids are not formed de novo but arise by division and (2) all different stages of plastids in higher plants are readily interconverted. Careful consideration of all the data, however, reveals many unsolved problems. Among them the most important are those which concern the interrelationship between plastidal development and cell differentiation.

Acknowledgments. I thank Drs. B. Sprey and M.W. Steer for providing me with electron micrographs and Drs. H. Bothe and T.J. Mabry for their constructive criticism of the manuscript. Supported in part by the Deutsche Forschungsgemeinschaft.

References

Akoyunoglou, G., Argyroudi-Akoyunoglou, J.H.: Control of thylakoid growth in *Phaseolus vulgaris*. Plant Physiol. 61, 834-837 (1978)

Alberte, R.S., Thornber, J.P., Naylor, A.W.: Time of appearance of photosystem I and II in chloroplasts of greening jack bean leaves. J. Exp. Bot. 23, 1060-1069 (1972)

Anstis, P.J.P., Northcote, D.H.: Development of chloroplasts from amyloplasts in potato tuber discs. New Phytol. 72, 449-463 (1973)

Armond, P.A., Arntzen, C.J.: Localization and characterization of photosystem II in grana and stroma lamellae. Plant Physiol. 59, 398-404 (1977)

Armond, P.A., Staehelin, L.A., Arntzen, C.J.: Spatial relationship of photosystem I, photosystem II, and the light-harvesting complex in chloroplast membranes. J. Cell Biol. 73, 400-418 (1977)

Badenhuizen, N.P.: Fundamental problems in the biosynthesis of starch granules. Ann. N.Y. Acad. Sci. 210, 11-16 (1973)

Bauer, L.: Untersuchungen zur Entwicklungsgeschichte und Physiologie der Plastiden von Laubmoosen. Flora 136, 30-84 (1942/43)

Behnke, H.-D.: Mikromorphologische Merkmale der Siebelement-Plastiden als ein Beitrag der Transmissionselektronenmikroskopie zur Systematik der Samenpflanzen. Ber. Deutsch. Bot. Ges. 88, 361-368 (1975a)

Behnke, H.-D.: P-type sieve-element plastids: a correlative ultrastructural and ultrahistochemical study on the diversity and uniformity of a new reliable character in seed plant systematics. Protoplasma 83, 91-101 (1975b)

Berger, C., Bergmann, L.: Farblicht und Plastidendifferenzierung im Speichergewebe von *Solanum tuberosum* L. Z. Pflanzenphysiol. 56, 439-445 (1967)

Berger, C., Feierabend, J.: Plastidenentwicklung und Bildung von Photosynthese-Enzymen in etiolierten Roggenkeimlingen. Physiol. Vég. 5, 109-122 (1967)

Bergmann, L., Berger C.: Farblicht und Plastidendifferenzierung in Zellkulturen von *Nicotiana tabacum* var. Samsun. Planta 69, 58-69 (1966)

Björn, L.O., Odhelius, I.: Chlorophyll formation in excised roots of cucumber and pea. Physiol. Plant. 19, 60-62 (1966)

Blackwell, S.J., Laetsch, W.M., Hyde, B.B.: Development of chloroplast fine structure in aspen tissue culture. Am. J. Bot. 56, 457-463 (1969)

Blank, R., Grobe, B., Arnold, C.G.: Time-sequence of nuclear and chloroplast fusions in the zygote of *Chlamydomonas reinhardii*. Planta 138, 63-64 (1978)

Boardman, N.K.: Comparative photosynthesis of sun and shade plants. Ann. Rev. Plant Physiol. 28, 355-377 (1977)

Bopp, M.: Die Entwicklung von Zelle und Kern im Protonema von *Funaria hygrometrica* Sibth. Planta 45, 573-590 (1955)

Borowitzka, M.A.: Some unusual features of the ultrastructure of the chloroplasts of the green algal order Caulerpales and their development. Protoplasma 89, 129-147 (1976)

Borowitzka, M.A.: Plastid development and floridean starch grain formation during carposporogenesis in the coralline red algae *Lithothrix aspergillum* Gray. Protoplasma 95, 217-228 (1978)

Brandt, P., Wiessner, W.: Unterschiedliche Temperaturoptima der DNA-abhängigen RNA-Polymerasen von *Euglena gracilis*, Stamm Z und ihre Bedeutung für die experimentelle Erzeugung der permanenten Apochlorose durch höhere Temperatur. Z. Pflanzenphysiol. 85, 53-60 (1977)

Brossard, D.: Ultrastructure comparée des "méristèmes foliaires", actifs ou inhibés, et des points végétatifs de bulbilles en formation ou achevées, chez le *Bryophyllum daigremontianum* Berg. C. R. Acad. Sci. Paris Ser. D 274, 2639-2642 (1972)

Butterfass, T.: Die Plastidenverteilung bei der Mitose der Schließzellenmutterzellen von haploidem Schwedenklee (*Trifolium hybridum* L.). Planta 84, 230-234 (1969)

Camefort, H.: La différenciation et l'organisation des chloroplastes dans le mésophylle des cotylédons du pin pignon (*Pinus pinea* L.). C. R. Acad. Sci. Paris 257, 2876-2879 (1963)

Casadoro, G., Rascio, N., Paganelli Cappelletti, E.M.: Membrane-bound plastidial inclusions in belladonna (*Atropa belladonna* L.). Biol. Cell. 29, 61-66 (1977)

Cavalier-Smith, T.: The origin of nuclei and of eukaryotic cells. Nature (London) 256, 463-468 (1975)

Charrière-Ladreix, Y.: La sécrétion flavonique chez *Populus nigra* L. et autres ligneux: un exemple de compartimentation métabolique. Thèse, Grenoble (1979)

Colombo, P.M.: An ultrastructural study of thallus organization in *Udotea petiolata*. Phycologia 17, 227-235 (1978)

Craig, I.W., Gibor, A.: Chloroplast development in Charophyceae. J. Cell Biol. 49, 950-953 (1971)

Cran, D.G., Possingham, J.V.: Variation of plastid types in spinach. Protoplasma 74, 345-356 (1972)

Cran, D.G., Possingham, J.V.: The effect of cell age on chloroplast structure and chlorophyll in cultured spinach leaf discs. Protoplasma 79, 197-213 (1974)

Devidé, Z., Ljubešić, N.: The reversion of chromoplasts to chloroplasts in pumpkin fruits. Z. Pflanzenphysiol. 73, 296-306 (1974)

Dirks, W., Richter, G.: Bedeutung von Blaulicht für die Synthese von RNA-Komponenten bei der Chloroplastendifferenzierung in isolierten Wurzeln *(Pisum sativum)*. Biochem. Physiol. Pflanz. 168, 157-166 (1975)

Dodge, J.D.: Changes in chloroplast fine structure during the autumnal senescence of *Betula* leaves. Ann. Bot. (London) 34, 817-824 (1970)

Dodge, J.D.: The Fine Structure of Algal Cells. London-New York: Academic Press 1973

Dodge, J.D., Kowallik, K.V.: Die Algen. In: Ultrastruktur der pflanzlichen Zelle; Robards, A.W. (ed.); pp. 181-224. Stuttgart: Thieme 1974

Dumas, C.: Contribution à l'étude cyto-physiologique du stigmate. VIII. Les associations réticulum endoplasmique — plaste et la sécrétion stigmatique. Botaniste 56, 81-102 (1974)

Duysen, M.E., Freeman, T.P.: Thylakoid development and chlorophyll distribution in water-stressed and hormone-treated wheat leaves. Can. J. Bot. 56, 1941-1945 (1978)

Egger, K.: Die Verbreitung von Vitamin K_1 und Plastochinon in Pflanzen. Planta 64, 41-61 (1965)

Esau, K., Hoefert, L.L.: Hyperplastic phloem in sugarbeet leaves infected with the beet curly top virus. Am. J. Bot. 65, 772-783 (1978)

Falk, H., Liedvogel, B., Sitte, P.: Circular DNA in isolated chromoplasts. Z. Naturforsch. 29c, 541-544 (1974)

Farineau, J., Guillot-Salomon, T., Popovic, R.: Étude de la structure et des activités photochimiques des deux types d'etiochloroplastes formés chez la feuille de Zea Mays L. au cours d'un verdissement en régime d'éclairs brefs. Biol. Cell. 32, 307-316 (1978)

Feierabend, J.: Enzymbildung in Roggenkeimlingen während der Umstellung von heterotrophem auf autotrophes Wachstum. Planta 71, 326-355 (1966)

Feierabend, J.: Der Einfluß von Cytokininen auf die Bildung von Photosyntheseenzymen in Roggenkeimlingen. Planta 84, 11-29 (1969)

Feierabend, J.: Regulation der Entwicklung von Zellorganellen bei höheren Pflanzen. Ber. Deutsch. Bot. Ges. 85, 601-613 (1972)

Feierabend, J., Brassel, D.: Subcellular localization of shikimate dehydrogenase in higher plants. Z. Pflanzenphysiol. 82, 334-346 (1977)

Feierabend, J., Mikus, M.: Occurrence of a high temperature sensitivity of chloroplast ribosome formation in several higher plants. Plant Physiol. 59, 863-867 (1977)

Fraser, T.W., Gunning, B.E.S.: Ultrastructure of the hairs of the filamentous green algae Bulbochaete hiloensis (Nordst.) Tiffany: an apoplastidic plant cell with a well-developed Golgi apparatus. Planta 113, 1-19 (1973)

Frey-Wyssling, A., Ruch, F., Berger, X.: Monotrope Plastiden-Metamorphose. Protoplasma 45, 97-114 (1955)

Gaffal, K.P.: Configural changes in the plastidome of Polytoma papillatum after completion of cytokinesis and during fusion of the gametes. Protoplasma 94, 175-191 (1978)

Gasanov, R.A., French, C.S.: Chlorophyll composition and photochemical activity of photosystems detached from chloroplast grana and stroma lamellae. Proc. Natl. Acad. Sci. USA 70, 2082-2085 (1973)

Gehring, H., Kasemir, H., Mohr, H.: The capacity of chlorophyll-a biosynthesis in the mustard seedling cotyledons as modulated by phytochrome and circadian rhythmicity. Planta 133, 295-302 (1977)

Gibbs, S.P.: Chloroplast development in Ochromonas danica. J. Cell Biol. 15, 343-361 (1962)

Girnth, C., Bergfeld, R., Kasemir, H.: Phytochrome-mediated control of grana and stroma thylakoid formation in plastids of mustard cotyledons. Planta 141, 191-198

Gourret, J.-P., Le Normand, M.: Les plastes des pétales virescents de la giroflée infectée par des mycoplasmes. J. Microsc. 12, 151-156 (1971)

Gracen, V.E., Jr., Hilliard, J.H., Brown, R.H., West, S.H.: Peripheral reticulum in chloroplasts of plants differing in CO_2 fixation pathways and photorespiration. Planta 107, 189-204 (1972)

Grönegress, P.: The greening of chromoplasts in Daucus carota L. Planta 98, 274-278 (1971)

Grönegress, P.: The structure of chromoplasts and their conversion to chloroplasts. J. Microsc. 19, 183-192 (1974)

Guillot-Salomon, T., Tuquet, C., De Lubac, M., Hallais, M.-F., Signol, M.: Analyse comparative de l'ultrastructure et de la composition lipidique des chloroplastes de plantes d'ombre et de soleil. Cytobiology 17, 442-452 (1978)

Gunning, B.E.S., Steer, M.W.: Ultrastructure and the Biology of Plant Cells. London: Arnold 1975

Gunning, B.E.S., Steer, M.W., Cochrane, M.P.: Occurrence, molecular structure, and induced formation of the "stromacentre" in plastids. J. Cell Sci. 3, 445-456 (1968)

Hall, J.D., Barr, R., Al-Abbas, A.H., Crane, F.L.: The ultrastructure of chloroplasts in mineral-deficient maize leaves. Plant Physiol. 50, 404-409 (1972)

Hammond, C.T., Mahlberg, P.G.: Ultrastructural development of capitate glandular hairs of *Cannabis sativa* L. (Cannabaceae). Am. J. Bot. 65, 140-151 (1978)

Hariri, M., Brangeon, J.: Light-induced adaptive responses under greenhouse and controlled conditions in the fern *Pteris cretica* var. *ouvrardii*. I. Structural and infrastructural features. Physiol. Plant. 41, 280-288 (1977)

Heinrich, G.: Die "Duftplastiden" von *Platanthera bifolia* und *Platanthera chlorantha*. Cytobiology 7, 138-144 (1973)

Heltne, J., Bonnett, H.T.: Chloroplast development in isolated roots of *Convolvulus arvensis* (L.). Planta 92, 1-12 (1970)

Henningsen, K.W., Boynton, J.E.: Macromolecular physiology of plastids. IX. Development of plastid membranes during greening of dark-grown barley seedlings. J. Cell Sci. 15, 31-55 (1974)

Hesse, M.: Entwicklungsgeschichte und Ultrastruktur des Pollenkitts bei *Tilia* (Tiliaceae). Plant Syst. Evol. 129, 13-30 (1978)

Hinchman, R.R.: The ultrastructural morphology and ontogeny of oat coleoptile plastids. Am. J. Bot. 59, 805-817 (1972)

Hohl, H.-R.: Über die submikroskopische Struktur normaler und hyperplastischer Gewebe von *Datura stramonium* L. I. Teil: Normalgewebe. Ber. Schweiz. Bot. Ges. 70, 395-439 (1960)

Huber, D.J., Newman, D.W.: Relationships between lipid changes and plastid ultrastructural changes in senescing and regreening soybean cotyledons. J. Exp. Bot. 27, 490-511 (1976)

Hurkman, W.J., Kennedy, G.S.: Development and cytochemistry of the thylakoidal body in tobacco chloroplasts. Am. J. Bot. 64, 86-95 (1977)

Ikeda, T.: Prolamellar body formation under different light and temperature conditions. Bot. Mag. 84, 363-375 (1971)

Jupin, H., Catesson, A.M., Giraud, G., Hauswirth, N.: Chloroplasts à empilements granaires anormaux, appauvris en photosystème I, dans le phloème de *Robinia pseudoacacia* et de *Acer pseudoplatanus*. Z. Pflanzenphysiol. 75, 95-106 (1975)

Kerby, N.W., Evans, L.V.: Isolation and partial characterization of pyrenoids from the brown alga *Pilayella littoralis* (L.) Kjellm. Planta 142, 91-95 (1978)

Kirchanski, S.J., Park, R.B.: Thylakoid proteins from mesophyll and bundle sheath chloroplasts of *Zea mays*. Plant Physiol. 54, Suppl. 29 (1974)

Kirk, J.T.O., Tilney-Bassett, R.A.E.: The Plastids Their Chemistry, Structure, Growth and Inheritance. London-San Francisco: Freeman 1967

Klein, S., Schiff, J.A., Holowinsky, A.W.: Events surrounding the early development of *Euglena* chloroplasts. II. Normal development of fine structure and the consequences of preillumination. Dev. Biol. 28, 253-273 (1972)

Kleinig, H., Wrischer, M.: Die Feinstruktur von *Acetabularia*-Chloroplasten bei Sekundärcarotinoid-Bildung. Z. Pflanzenphysiol. 58, 248-251 (1968)

Kollmann, R., Kleinig, H., Dörr, I.: Fine structure and pigments of plastids in *Orobanche*. Cytobiology 1, 152-158 (1969)

Kronestedt, E., Walles, B.: On the presence of plastids and the eyespot apparatus in a porfiromycin-bleached strain of *Euglena gracilis*. Protoplasma 84, 75-82 (1975)

Laetsch, W.M.: The C₄ syndrome: a structural analysis. Annu. Rev. Plant Physiol. 25, 27-52 (1974)

Laetsch, W.M., Stetler, D.A.: Chloroplast structure and function in cultured tobacco tissue. Am. J. Bot. 52, 798-804 (1965)

Laudi, G.: Ultrastructural researches on the plastids of parasitic plants. IV. Galls of *Cuscuta australis*. Giorn. Bot. Ital. 102, 1-19 (1968)

Laudi, G., Fanelli, L.: Ricerche comparate sulla morfologia e sulla fisiologia di *Larix* e di *Picea*. Contenuto in clorofilla di semi e plantule durante la germinazione in condizioni di oscurità. Giorn. Bot. Ital. 71, 171-176 (1964)

Leedale, G.F.: Euglenoid Flagellates. Englewood Cliffs, NJ: Prentice-Hall 1967

Lerbs, V., Eicke, R.: Vergleichende elektronenmikroskopische Untersuchungen im Vegetationskegel von *Selaginella martensii*. Entwicklungsstadien der Chloroplasten in Meristemen. Ber. Dtsch. Bot. Ges. 87, 303-315, (1974)

Ljubešić, N.: Feinbau der Chloroplasten während der Vergilbung und Wiederergrünung der Blätter. Protoplasma 66, 369-379 (1968)

Lott, J.N.A., Darley, J.J.: Unusual thylakoid arrangements in plastids from *Cucurbita maxima* seed coat cells. Cytobiology 8, 55-60 (1973)

Lütz, C.: Biochemische und cytologische Untersuchungen zur Chloroplastenentwicklung. I. Die chemische Charakterisierung der Prolamellarkörper aus Etioplasten von *Avena sativa* L. Z. Pflanzenphysiol. 75, 346-359 (1975a)

Lütz, C.: Biochemische und cytologische Untersuchungen zur Chloroplastenentwicklung. II. Isolierung und Charakterisierung eines Glykoproteins aus den Prolamellarkörpern der Etioplasten von *Avena sativa* L. Z. Pflanzenphysiol. 76, 130-142 (1975b)

Lütz, C., Klein, S.: Biochemical and cytological observations on chloroplast development. VI. Chloroplasts and saponins in polamellar bodies and prothylakoids separated from etioplasts of etiolated *Arena sativa* L. leaves. Z. Pflanzenphysiol. 95, 227-237 (1979)

Marin, L., Dengler, R.E.: Granal plastids in the cotyledons of the dry embryo of *Kochia childsii*. Can. J. Bot. 50, 2049-2052 (1972)

Marinos, N.G.: Studies on submicroscopic aspects of mineral deficiencies. II. Nitrogen, potassium, sulfur, phosphorus, and magnesium deficiencies in the shoot apex of barley. Am. J. Bot. 50, 998-1005 (1963)

Marinos, N.G.: Multifunctional plastids in the meristematic region of potato tuber buds. J. Ultrastruct. Res. 17, 91-113 (1967)

Marty, D.: Localisation ultra-structurale des sites d'activité des photosystèmes I et II dans les chloroplastes in situ. C. R. Acad. Sci. Paris Ser. D 285, 27-30 (1977)

Menke, W., Schmid, G.H.: Cyclic photophosphorylation in the mykotrophic orchid *Neottia nidus-avis*. Plant Physiol. 57, 716-719 (1976)

Mesquita, J.F.: Altérations ultrastructurales des plastes au cours du verdissement expérimental des racines du *Lupinus albus* L. Port. Acta Biol. Ser. A 12, 33-52 (1971)

Michaelis, P.: Über Zahlengesetzmäßigkeiten plasmatischer Erbträger, insbesondere der Plastiden. Protoplasma 55, 177-231 (1962)

Michel-Wolwertz, M.R., Bronchart, R.: Formation of prolamellar bodies without correlative accumulation of protochlorophyllide or chlorophyllide in pine cotyledons. Plant Sci. Lett. 2, 45-54 (1974)

Mikulska, E., Zołnierowicz, H., Narolewska, B.: Ultrastructure of chromoplasts in detached cotyledons of cucumber treated with growth retardant (2-chloroetyhl-trimethyl-ammonium chloride). Biochem. Physiol. Pflanz. 164, 514-521 (1973)

Miller, K.R., Staehelin, L.A.: Analysis of the thylakoid outer surface. J. Cell Biol. 68, 30-47 (1976)

Miller, K.R., Miller, G.J., McIntyre, K.R.: The light-harvesting chlorophyll-protein complex of photosystem II. Its location in the photosynthetic membrane. J. Cell Biol. 71, 624-638 (1976)

Miyake, H., Maeda, E.: Development of bundle sheath chloroplasts in rice seedlings. Can. J. Bot. 54, 556-565 (1976)

Młodzianowski, F., Ponitka, A.: Ultrastructural changes of chloroplasts in detached parsley leaves yellowing in darkness and the influence of kinetin on that process. Z. Pflanzenphysiol. 69, 13-25 (1973)

Möbius, M.: Über die Größe der Chloroplasten. Ber. Dtsch. Bot. Ges. 38, 224-232 (1920)

Moline, H.E., Jensen, S.G.: Histochemical evidence for glycogen-like deposits in barley yellow dwarf virus-infected barley leaf chloroplasts. J. Ultrastruct. Res. 53, 217-221 (1975)

Mühlethaler, K., Frey-Wyssling, A.: Entwicklung und Struktur der Proplastiden. J. Biophys. Biochem. Cytol. 6, 507-512 (1959)

Murakami, S.: Prolamellar body of the proplastids in barley root cells. Experienta 18, 168-169 (1962)

Muscatine, L., Greene, R.W.: Chloroplasts and algae as symbionts in molluscs. Int. Rev. Cytol. 36, 137-169 (1973)

Nagl, W., Kühner, S.: Early embryogenesis in *Tropaeolum majus* L.: Diversification of plastids. Planta 133, 15-19 (1976)

Newcomb, E.H.: Fine structure of protein-storing plastids in bean root tips. J. Cell Biol. 33, 143-163 (1967)

Nigon, V., Heizmann, P.: Morphology, biochemistry, and genetics of plastid development in *Euglena gracilis*. Int. Rev. Cytol. 53, 211-290 (1978)

Ophir, I., Talmon, A., Polak-Charcon, S., Ben-Shaul, Y.: Aspects of structure and photosynthetic competence of *Euglena* plastids under conditions of greening and degreening. Protoplasma 84, 283-295 (1975)

Orsenigo, M.: Sul naturale imbianchimento dei tepali di giacinto in precedenza verdi e sulla irreversibilità dei cloroplasti. G. Bot. Ital. 70, 467-475 (1963)

Osafune, T., Hase, E.: Some structural characteristics of the chloroplast in the "glucose-bleaching" and re-greening cells of *Chlorella protothecoides*. Biochem. Physiol. Pflanz. 168, 533-542 (1975)

Pais, M.S.S.: Sur le reverdissement de la spathe de *Zantedeschia aethiopica* au cours de la frutification. Étude en microscopie électronique de l'ontogénie des chloroplastes. Isolement de substances du type kinine dans les fruits et dans la spathe frutifère. Port. Acta Biol. Ser. A 12, 101-124 (1972)

Palevitz, B.A., Newcomb, E.H.: A study of sieve element starch using enzymatic digestion and electron microscopy. J. Cell Biol. 45, 383-398 (1970)

Paolillo, D.J., Jr.: The three-dimensional arrangement of intergranal lamellae in chloroplasts. J. Cell Sci. 6, 243-255 (1970)

Parthier, B.: Zur Evolution von Chloroplasten und Mitochondrien. Nova Acta Leopold. N. F. 42, 223-239 (1975)

Pettitt, J.M.: Developmental mechanisms in heterospory. III. The plastid cycle during megasporogenesis in *Isoetes*. J. Cell Sci. 20, 671-685 (1976)

Possingham, J.V., Saurer, W.: Changes in chloroplast number per cell during leaf development in spinach. Planta 86, 186-194 (1969)

Pringsheim, E.G.: Farblose Algen. Ein Beitrag zur Evolutionsforschung. Stuttgart: Fischer 1963

Puiseux-Dao, S., Hoursiangou-Neubrun, D., Dubacq, J.P., Oblin, S., Borghi, H., Dazy, A.C.: New information on the plastid heterogeneity in *Acetabularia mediterranea* cells. Protoplasma 91, 226-227 (1977)

Rickson, F.R.: Ultrastructural differentiation of the Müllerian body glycogen plastid of *Cecropia peltata* L. Am. J. Bot. 63, 1272-1279 (1976)

Robards, A.W., Kidwai, P.: A comparative study of the ultrastructure of resting and active cambium of *Salix fragilis*, L. Planta 84, 239-249 (1969)

Robertson, D., Laetsch, W.M.: Structure and function of developing barley plastids. Plant Physiol. 54, 148-159 (1974)

Röbbelen, G.: Chloroplastendifferenzierung nach geninduzierter Plastommutation bei *Arabidopsis thaliana* (L.) Heynh. Z. Pflanzenphysiol. 55, 387-403 (1966)

Salema, R., Badenhuizen, N.P.: The production of reserve starch granules in the amyloplasts of *Pellionia daveauana* N. E. Br. J. Ultrastruct. Res. 20, 383-399 (1967)

Saunders, J.A., McClure, J.W.: Phytochrome controlled phenylalanine ammonia lyase in *Hordeum vulgare* plastids. Phytochemistry 14, 1285-1289 (1975)

Schäfers, H.-A., Feierabend, J.: Ultrastructural differentiation of plastids and other organelles in rye leaves with a high-temperature-induced deficiency of plastid ribosomes. Cytobiology 14, 75-90 (1976)

Schimper, A.F.W.: Untersuchungen über die Chlorophyllkörner und die ihnen homologen Gebilde. Jahrb. Wiss. Bot. 16, 1-27 (1885)

Schmid, G.H., Jankowicz, M., Menke, W.: Cyclic photophosphorylation and chloroplast structure in the labellum of the orchid *Aceras anthropophorum*. J. Microsc. Biol. Cell. 26, 25-28 (1976)

Schmidt, H.-W., Hampp, R.: Regulation of membrane properties of mitochondria and plastids during chloroplast development. II. The action of phytochrome in a cell-free system. Z. Pflanzenphysiol. 82, 428-434 (1977)

Schnepf, E.: Zur Feinstruktur von *Geosiphon pyriforme*. Ein Versuch zur Deutung cytoplasmatischer Membranen und Kompartimente. Arch. Mikrobiol. 49, 112-131 (1964)

Schnepf, E.: Mikrotubulus-Anordnung und -Umordnung, Wandbildung und Zellmorphogenese in jungen *Sphagnum*-Blättchen. Protoplasma 78, 145-173 (1973)

Schnepf, E., Czygan, F.-C.: Feinbau und Carotinoide von Chromoplasten im Spadix-Appendix von *Typhonium* und *Arum*. Z. Pflanzenphysiol. 54, 345-355 (1966)

Schnepf, E., Pross, E.: Differentiation und redifferentiation of a transfer cell: development of septal nectaries in *Aloe* and *Gasteria*. Protoplasma 89, 105-115 (1976)

Schürmann, P., Villiger, W.: Veränderungen des Chlorophyllgehaltes und der Chloroplastenfeinstruktur in den Hochblättern von *Davidia involucrata* Baill. Planta 76, 335-347 (1967)

Senser, M., Schötz, F., Beck, E.: Seasonal changes in structure and function of spruce chloroplasts. Planta 126, 1-10 (1975)

Shaw, M., Manocha, M.S.: Fine structure in detached, senescing wheat leaves. Can. J. Bot. 43, 747-755 (1965)

Sheath, R.G., Hellebust, J.A., Sawa, T.: Changes in plastid structure, pigmentation and photosynthesis of the conchocelis stage of *Porphyra leucosticta* (Rhodophyta, Bangiophyceae) in response to low light and darkness. Phycologia 16, 265-276 (1977)

Shimokawa, K., Sakanoshita, A., Horiba, K.: Ethylene-induced changes of chloroplast structure in Satsuma mandarin (*Citrus unshiu* Marc.). Plant Cell Physiol. 19, 229-236 (1978)

Sitte, P.: Plastiden-Metamorphose und Chromoplasten bei *Chrysosplenium*. Z. Pflanzenphysiol. 73, 243-265 (1974)

Sitte, P.: Die lebende Zelle als System, Systemelement und Übersystem. Nova Acta Leopold. 47, Nr. 226, 195-216 (1978)

Siu, C.-H., Chiang, D.-S., Swift, H.: Characterization of cytoplasmic and nuclear genomes in the colorless alga *Polytoma*. V. Molecular structure and heterogeneity of leucoplast DNA. J. Mol. Biol. 98, 369-391 (1975)

Sjolund, R.D., Weier, T.E.: An ultrastructural study of chloroplast structure and dedifferentiation in tissue cultures of *Streptanthus tortuosus* (Cruciferae). Am. J. Bot. 58, 172-181 (1971)

Sprey, B.: Lichtinduzierte Entwicklung von Etioplasten zu Chloroplasten: Induktion und Regulation der Membranbildung. Ber. Kernforschungsanlage Jülich Nr. 1019 (1973)

Sprey, B., Laetsch, W.M.: Structural studies of peripheral reticulum in C_4 plant chloroplasts of *Portulacca oleracea* L. Z. Pflanzenphysiol. 87, 37-53 (1978)

Sprey, B., Lambert, C.: Lamellae-bound inclusions in isolated spinach chloroplasts. II. Identification and composition. Z. Pflanzenphysiol. 83, 227-247 (1977)

Stetler, D.A.: Nonphotoconvertible protochlorophyllide in etiolated tissue lacking prolamellar bodies. Bot. Gaz. 134, 290-295 (1973)

Stosch, H.A. von, Drebes, G.: Entwicklungsgeschichtliche Untersuchungen an zentrischen Diatomeen IV. Die Planktondiatomee *Stephanopyxis turris* - ihre Behandlung und Entwicklungsgeschichte. Helgol. Wiss. Meeresunters. 11, 209-257 (1964)

Sun, C.N.: Structural alterations of chloroplasts induced by virus in *Abutilon striatum* v. Thompson. Protoplasma 60, 426-434 (1965)

Suzuki, S.: Electron microscope studies on the morphogenesis of plastids in C_4-plants. II. Development of etioplasts under illumination in *Zea mays* L. Sci. Rep. Tokyo Kyoiku Daigaku Sect. B 15, 255-263 (1974)

Syrop, M.: Leaf curl disease of almond caused by *Taphrina deformans* (Berk.) Tul. II. An electron microscope study of the host/parasite relationship. Protoplasma 85, 57-69 (1975)

Thomson, W.W., Weier, T.E.: The fine structure of chloroplasts from mineral-deficient leaves of *Phaseolus vulgaris*. Am. J. Bot. 49, 1047-1055 (1962)

Treffry, T.: Chloroplast development in etiolated peas: reformation of prolamellar bodies in red light without accumulation of protochlorophyllide. J. Exp. Bot. 24, 185-195 (1973)

Treffry, T.: Biogenesis of the photochemical apparatus. Int. Rev. Cytol. 52, 159-196 (1978)

Trench, R.K., Smith, D.C.: Synthesis of pigment in symbiotic chloroplasts. Nature (London) 227, 196-197 (1970)

Vasiliev, A.E.: On the localization of terpenoid synthesis in plant cells (Russian). Akad. Nauk SSSR 5, 29-45 (1970)

Vesk, M., Jeffrey, S.W.: Effect of blue-green light on photosynthetic pigments and chloroplast structure in unicellular marine algae from six classes. J. Phycol. 13, 280-288 (1977)

Vesk, M., Possingham, J.V., Mercer, F.V.: The effect of mineral nutrient deficiencies on the structure of the leaf cells of tomato, spinach, and maize. Aust. J. Bot. 14, 1-18 (1966)

Wada, M., O'Brien, T.P.: Observations on the structure of the protonema of *Adiantum capillus-veneris* L. undergoing cell division following white-light irradiation. Planta 126, 213-227 (1975)

Walbot, V.: The dimorphic chloroplasts of the C_4 plant *Panicum maximum* contain identical genomes. Cell 11, 729-737 (1977)

Walles, B., Hudák, J.: A comparative study of chloroplast morphogenesis in seedlings of some conifers (*Larix decidua, Pinus sylvestris* and *Picea abies*). Stud. For. Suec. 127, 1-22 (1975)

Wehrmeyer, W.: Über Membranbildungsprozesse im Chloroplasten. II. Mitteilung. Zur Entstehung der Grana durch Membranüberschiebung. Planta 63, 13-30 (1964)

Wehrmeyer, W.: Entwicklung, Bau und Funktion der Grana in Chloroplasten. Biol. unserer Zeit 2, 112-119 (1972)

Weissenböck, G., Plesser, A., Trinks, K.: Flavonoidgehalt und Enzymaktivitäten isolierter Haferchloroplasten (*Avena sativa*). Ber. Dtsch. Bot. Ges. 89, 457–472 (1976)

Wellburn, A.R.: Distribution of chloroplast coupling factor (CF_1) particles on plastid membranes during development. Planta 135, 191–198 (1977)

Whatley, J.M.: Ultrastructural changes in chloroplasts of *Phaseolus vulgaris* during development under conditions of nutrient deficiency. New Phytol. 70, 725–742 (1971)

Willison, J.H.M., Davey, M.R.: Fraction 1 protein crystals in chloroplasts of isolated tobacco leaf protoplasts: A thin-section and freeze-etch morphological study. J. Ultrastruct. Res. 55, 303–311 (1976)

Woo, K.C., Pyliotis, N.A., Downton, W.J.S.: Thylakoid aggregation and chlorophyll a/chlorophyll b ratio in C_4-plants. Z. Pflanzenphysiol. 64, 400–413 (1971)

Wrischer, M.: Neubildung von Prolamellarkörpern in Chloroplasten. Z. Pflanzenphysiol. 55, 296–299 (1966)

Wrischer, M.: Transformation of plastids in young carrot callus. Acta Bot. Croat. 31, 41–46 (1972)

Wrischer, M.: Ultrastructural localization of photosystem I in plastids of senescent spinach leaves. Acta Bot. Croat. 36, 57–61 (1977)

Wuttke, H.-G.: Circular DNA in chromoplasts of *Tulipa gesneriana*. Planta 132, 317–319 (1976)

Yeung, E.C., Peterson, R.L.: Fine structure during ontogeny of xylem transfer cells in the rhizome of *Hieracium floribundum*. Can. J. Bot. 53, 432–438 (1975)

The Continuity of Plastids and the Differentiation of Plastid Populations

TH. BUTTERFASS

Fachbereich Biologie der Johann Wolfgang Goethe-Universität
Frankfurt/Main, FRG

I. Introduction

The development of a proplastid into a chloroplast is only one aspect of the development of the photosynthetic system. We have to ask further questions about (1) the continuity of plastids from cell to cell, and (2) how the number of plastids in a cell is regulated. If we accept that proplastids are transmitted from cell to cell and multiply by division, then the central problem of plastid continuity becomes one of how cell divisions and plastid divisions are coordinated. The coordination is so efficient that for hundreds of millions of years all ancestors of our recent plants have received plastids. As to the size of the population of plastids in a cell, the mean number of chloroplasts in a guard cell may be quite different from the mean number in a palisade cell, and indeed, great differences are found in this and in similar comparisons. Thus, the pattern of different plastid numbers reflects the pattern of cell-specificity. The pattern of chloroplast numbers originates from cell-specific differences in the regulation of plastid divisions; these divisions are not coupled to cell divisions, and hence are not essential for the continuity of plastids. They are

---

Terminology. *Basic number of plastids:* A number not changing with differences in nuclear ploidy. The term cannot be defined more exactly so far, and is used rather loosely. *Endopolyploidy:* This is shown by a cell in whose nucleus the deoxyribonucleic acid (DNA) has passed at least one accessory replication beyond the 4C-state, as occurring in certain cells of many plants and animals during normal differentiation. *Monoplastidic:* A cell containing one single plastid. *Pattern:* Non-random distribution of elements of the same kind (here chloroplasts) in a plant in space and time. *Plastidome:* All plastids of a cell taken together and considered from the morphological point of view (Dangeard 1919), *Plastome:* The entire genetic system of all plastids of a cell (Renner 1934). *Polyplastidic:* A cell containing more than two plastids. *Prepattern:* Subsidiary pattern contributing to the formation of the terminal pattern (Stern 1954). Prepatterns may follow one another epigenetically, or they may be produced independently and be superimposed in forming the terminal pattern. *Terminal pattern:* A pattern in its final form, produced by the combined effects of all prepatterns (Stern 1954). Here, a pattern of chloroplast numbers in different cells. *Trisomic plants:* Diploid plants containing one extra chromosome in all diploid nuclei, i.e., one particular chromosome occurs in triplicate.

regulated by a variety of different underlying factors, only some of which can at present be analyzed. To each of these factors a characteristic pattern of its own can be ascribed, viz. a partial pattern or a prepattern (Stern 1954) which contributes to the terminal pattern of chloroplast numbers that is seen by direct counting when the mature state has been reached.

II. Do Plastids Arise Only by Division?

A wealth of observations shows that division is the predominant form of plastid reproduction, and no evidence exists showing that plastids arise de novo, i.e., otherwise than from pre-existing plastids. In addition, observations also on plants with polyplastidic cells, strongly suggest that a reproduction other than from pre-existing plastids is excluded. Three examples will be mentioned here.

1. The chlorocytes of stem leaflets of submerged *Sphagnum cuspidatum* plants contain exactly 1, 2, 4, 8, 16, or about 32 chloroplasts. This series is seen in neighboring cells, but it reflects the sequence that occurs in each single cell, providing direct visual evidence that all of the plastids multiply by division (Butterfass 1971).

2. Most *Selaginella* species contain exactly four chloroplasts in mature guard cells of stomata. Very young guard cells of *Selaginella* contain one plastid each, as do the meristematic cells, and slightly older guard cells contain two plastids. If juvenile guard cells contain three plastids, two of them are smaller, and adjacent to each other. Again, a reproduction other than from pre-existing plastids appears improbable.

3. In a haploid plant of *Trifolium hybridum*, the plastid number per guard cell mother cell is reduced to two. After mitosis, many guard cells devoid of plastids are found (Butterfass 1969), and it is significant that no plastids are formed anew in these cells. Obviously, a reproduction of plastids other than from pre-existing plastids does not occur.

III. The Coordination of Cell Division and Plastid Division as the Central Problem

A. Coordination is Indirect

Since all cells of green plants possess plastids (rare exceptions are not to be discussed here) a general coordination between cell division and plastid division is indicated. The nature of this coordination is problematical: it might be a direct one, cell division inducing plastid division or vice versa; or it might be an indirect one, both divisions being controlled by a third process. There are some indications that the latter is the case. Thus, during cell division, the plastids may behave in either of two ways (Butterfass 1963, 1973). At one extreme, plastid division precedes cell division, in which case the daughter cells each receive their final set of plastids from the mother cell. At the other extreme, plastid division follows upon cell division. The former sequence is observed in guard cell mother cells of diploid sugar beets, the latter in those of haploid sugar beets and of alsike clover, *Trifolium hybridum*. Obviously, between these extremes there are many examples of less clear-cut behavior. Baker (1926) stated that whether chloroplasts of *Euglena* divide before or after the cell depends on the growth conditions. Since both sequences are observed in nature, it may be argued that neither cell division nor plastid division directly initiates each other; hence a third process may control both.

B. Nuclear DNA and Plastid Numbers

Small tetraploid cells of the sugar beet contain many more chloroplasts than large diploid cells of the same size (Fig.1). Thus, whereas cell size and plastid reproduction may well be closely correlated, cell size is not the trigger for the division of plastids. Rather, the dosage of nuclear DNA may have a more direct influence upon plastid division (Butterfass 1963, 1973). According to this hypothesis, internal and external conditions determine whether, after nuclear DNA replication, cells or plastids divide first. In polyplastidic cells the order of events appears of little relevance, provided that the number of chloroplasts is not so small that, as in the guard cell mother cells of the haploid *Trifolium hybridum* referred to in Sect.II.C above, there is a risk that one daughter cell may not receive any. It

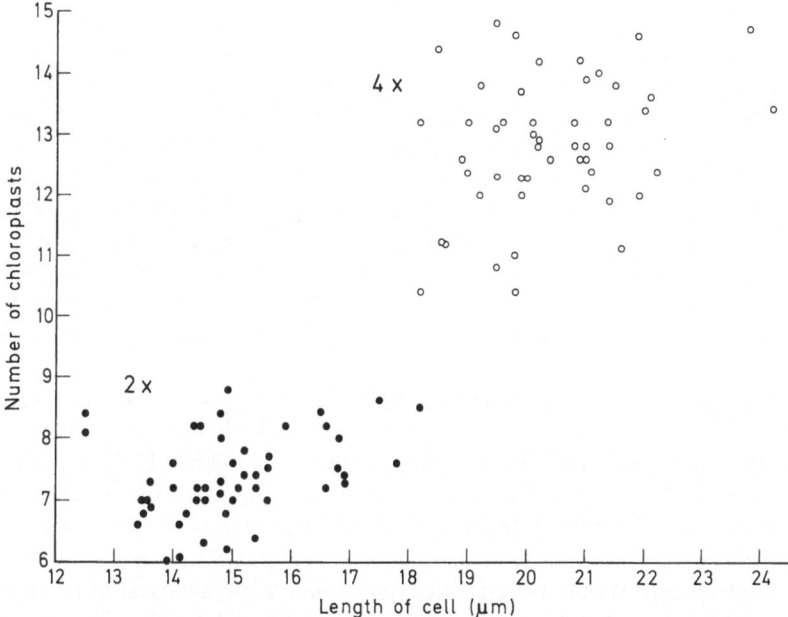

Fig. 1. The mean lengths of guard cells of stomata plotted against the mean numbers of chloroplasts in 50 diploid and 50 tetraploid sugar beet plants grown together

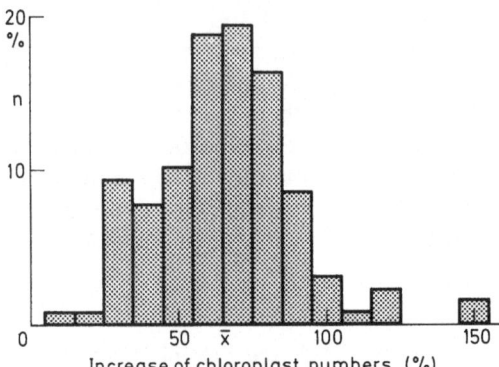

Fig. 2. Increase percentages of chloroplast numbers in guard cells of stomata after the duplication of the chromosome numbers (autopolyploidy). "Aged" polyploids from nature are excluded. 120 species

should be noted that, as the dosage of nuclear DNA in the two G_1 daughter cells taken together is the same as in the parental G_2 cell, the ultimate total plastid number between two S-phases is about the same whether it is attained before or after cell division, if no differentiation occurs.

If correct, the hypothesis that the nuclear DNA dosage is the trigger for plastid division must meet one more requirement, i.e., a change of ploidy should result in a change of chloroplast number per cell. As green plants have a long history, the duplication of plastids in meristematic cells is obviously tightly controlled. In general, cell doubling in meristems must be accompanied by plastid doubling. In differentiated polyploid cells, however, the increase of plastid numbers need not reflect exactly the increase of the dosage of nuclear DNA because the control might have been made less tight. If this control system, possibly modified, is working at all after the polyploidization of differentiating cells, the inference is that there is no strong selection pressure against it. Figures 1 and 2 show that, in fact, there are about 60 to 80% more chloroplasts in guard cells with duplicated ploidy. (As chloroplast numbers depend on genotype, cell type, and growing conditions, these parameters must not be varied when making such comparisons.) There is no well-documented exception to the rule that higher ploidy occurs along with higher chloroplast numbers (Butterfass 1973), endopolyploid cells and haploid higher plants included.

As stated above, the hypothesis that the dosage of nuclear DNA controls plastid replication in meristematic cells requires that polyploid cells contain more chloroplasts. Since this is shown to occur, the amount of nuclear DNA or, rather, of some part of this DNA (see Sect. V), may indeed be the controlling factor in the coordination of cell division and plastid division in meristems. At present there appears to be no evidence for any other factor. The replication of the genome is part of different sequences of events. In meristems, one sequence leads to mitosis, cell division, and, perhaps, cell growth, another one to plastid division. In contrast, in nonmeristematic cells, mitosis and cell division are inhibited at a time when plastid division is still proceeding. Because of its enormous adaptive value, the flexible mechanism is common to dividing cell lineages throughout the world of green plants.

The idea that the plastid population in a sequence of meristematic cells will oscillate always between a basal number and a number twice the basal number is correct only inasmuch as the population is approximately halved during each mitosis; but it is wrong to claim that a constant basic figure is involved. A final number of, e.g., 8 could be attained by plastid division either before mitosis, requiring 16 plastids per dividing mother cell, or after mitosis, requiring only 8 plastids per dividing mother cell. The ranges would be 16–8 or 8–4, respectively. If some plastids divide prior to nuclear prophase and others after nuclear telophase, the sharing ratio of plastid divisions between these periods will determine the numbers produced, the smaller number lying between 4 and 8. As the sharing ratio is variable, so are the ranges and the means. The final G_1 figure may be constant to a certain degree, but other figures are not.

C. Plastid Number and Plastid Size

Plastids become smaller by division, and therefore a negative correlation is usually found between plastid number and plastid size. This correlation may last until the final stage of differentiation, showing that a small plastid need not grow and that a large one need not divide. A size difference similar to that found between chloroplasts of sun and shade leaves (see Sect.IV.C) can, for instance, be observed in *Cleome spinosa:* guard cells of stomata in the pericarp epidermis on average contained 9.5 chloroplasts with a diameter of 11.5 arbitrary units, but in a foliage leaf the values were 5.2 and 14.0, respectively; thus the plastidome volume remained virtually unchanged.

Enlarged mature plastids occur nearly always at the expense of number. Genome-induced giant chloroplasts were observed repeatedly since Eyster (1929) in plants of different taxa. As a rule, all such chloroplasts grow uniformly, genetic factors inhibiting or postponing their division. When two extra chromosomes, each of which changes the chloroplast numbers (see Sect. V), are combined in one cell (double trisomic plant), the combined effect of both may result in enlarged chloroplasts rather than increased numbers, provided that the two effects of the two chromosomes do not cancel each other completely. An example is seen in the double trisomics $2x + III + VIII$ of sugar beets (Butterfass 1965). However, in many cells of the double trisomic of the same species, only one of the chloroplasts grows excessively (the length may become eightfold!) while the other chloroplasts remain unchanged. This result demands further investigation.

Plastid growth and plastid division can be separated from each other even more easily than can cell growth and cell division, and just as nuclear DNA replication is separable from cell division (polyploidy), so is plastid DNA replication from plastid division. Whether a plastid obeys an external signal to divide does not depend primarily on plastid size or plastid DNA content, but on other conditions. Furthermore, as nuclear DNA replication is directly correlated with nuclear and cellular growth, so presumably is plastid DNA replication with plastid growth. At least in sugar beets, the size of the plastids is indicative of the relative amount of plastid DNA (Herrmann 1968, 1970; Herrmann and Kowallik 1970; Kowallik and Herrmann 1972).

IV. Accessory Plastid Divisions Unrelated to Cell Division

A. Patterns of Plastid Reproduction

The simplest situation is one in which all cells of a plant contain about the same number of plastids. This is so in many algae where the vegetative cells belong to a single cell type, and the number of plastids per cell can be constant. In higher plants, however, cells of different kinds usually contain plastids in different numbers. A radish (*Raphanus sativus* var. *sativus*), for instance, contains 4–5 chloroplasts in its guard cells and at least 30, or many more, chloroplasts in its spongy parenchyma cells. The species-specific numbers can be very different, but it is probable that all flowering plants show cell-specificity patterns of plastid numbers, reflecting patterns of plastid reproduction. Unequal allotment of plastids also occurs during cell division in higher plants, but there is no evidence that it might account for permanent differences (see e.g., Anton-Lamprecht 1967).

Cell-specificity of chloroplast numbers can arise by at least two processes (Butterfass 1968). One is the dosage effect of the genome (ploidy effect). The other, which is the first in the course of events, can only be defined so far as being independent of nuclear ploidy. The role of both mechanisms will be evaluated below.

Michaelis (1962) assumed that the pattern of chloroplast numbers in a leaf is developed out of a basic number typical for meristematic cells, by waves of duplication. Bartels (1964) contested this idea. Figure 3, however, presents new evidence in the form of a compilation of all known ratios of chloroplast numbers between epidermal and guard cells and between mesophyll and guard cells. All data were plotted in one frequency distribution because the character of both distributions as to the location of the peaks is the same. For one pair of cells (e.g., epidermal/guard cells) only one figure per species (viz., the mean ratio, if more ratios were available) is included.

The distribution is expected to be skewed because the ratios plotted include the results of endopolyploidy that works by doubling. Therefore logarithmic

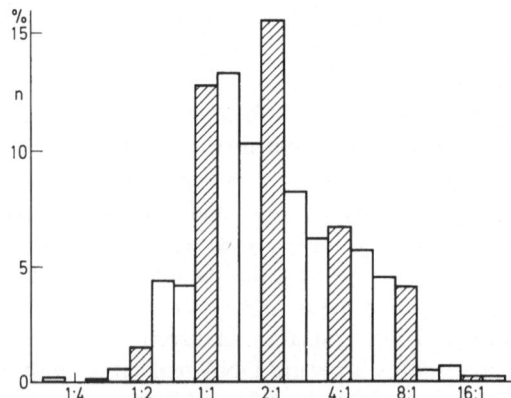

Fig. 3. The ratios of chloroplast numbers between epidermis or mesophyll cells and guard cells of stomata. Hatched bars indicate ratios required by theory. 827 entries

transformation is appropriate. The resulting distribution might reflect random variation (statistical 'noise') and hence be Gaussian, which is not because $\chi^2 = 92$ is three times as large as required for significance (30.6 for P = 1% and 15 d.f.). There is no other homogeneous distribution to be expected for random variation of means of this kind than a Gaussian distribution. Hence the distribution is heterogeneous. Then two possibilities remain. Either the partial distributions differ only in standard deviations, sharing the same mean. There is not the least indication for that, and the assumption of unimodality would contradict the well-established experience that endopolyploidy works on plastid numbers in approximate proportion to the dosage of nuclear DNA. The alternative is, as taken for granted from the beginning, that partial distributions differ by their means, and the combined distribution is multimodal. The question thus reduces to where the peaks are located. Figure 3 shows that main peaks occur at the ratios indicating just one or (perhaps) two doublings of the chloroplast numbers in epidermal and in mesophyll cells as related to that of guard cells. Hence Michaelis (1962) was right in principle. Because one may argue that all ratios of about 1:1 should be excluded as not contributing to the series of doublings, more weight should be attributed to the other peaks.

A peak may also be present in the neighborhood of the ratio 1:2, indicating that in some species the reference number itself can be the result of an increase, not far from a doubling, as related to numbers in epidermal or mesophyll cells. The numbers of chloroplasts in guard cells are used as references only for lack of better ones. The regularity of the peaks is all the more striking since many of the epidermal and mesophyll figures used come from cell populations of mixed ploidy and thus of very different chloroplast numbers. In addition, inevitable sampling and counting errors are involved.

The relative frequencies among the peak ratios may be heavily biased and, therefore, cannot be evaluated from Fig. 3. Herbs, for instance, are over-represented; so are ratios between epidermal cells and guard cells, because they are more easy to attain. Hence it is not known whether the ratio 1:1 occurs less frequently than the ratio 2:1, and if so, how much. But this is hardly interesting and is not the point. Figure 3 shows at which ratios peaks are found, and that they fit approximately into a series of doublings.

The pattern produced by duplication of plastid numbers combines with, and is partly concealed or modified by, other patterns of different kinds and origins in forming the terminal pattern of plastid numbers.

B. Prepatterns

Most higher plants show endopolyploidy in some or many of their cells. This differentiation is common in mesophyll cells, less frequent in epidermal cells, and almost never found in the guard cells of stomata. It is brought about by one or more accessory replications of the nuclear DNA, and leads to an endopolyploidy pattern in the plant. Normally endopolyploidy is restricted to cells which will not divide again; in the affected cells the plastids increase in number according to the level of ploidy attained (for details see Butterfass 1968). One clear-cut example is the

Table 1. The cooperation of two prepatterns in producing a terminal pattern of chloroplast numbers. Diploid sugar beet cotyledons. The three figures characterizing a pattern give the chloroplast numbers or ploidy levels in (in this order) guard, epidermal, and spongy parenchyma cells. Diagrammatic; mean plastid numbers and ploidies vary to some extent

Prepattern of diploid basic numbers 7/5/25 Prepattern of actual ploidy 2x (4x and 8x) 16x	Terminal pattern of counted numbers 7/14/120

oxalate cells in cotyledons of sugar beets, possessing up to 650 chloroplasts and being highly endopolyploid at levels of about 64 to 128 ploid; the high chloroplast number is a result of the high polyploidy. Thus a prepattern for chloroplast numbers, viz. the endopolyploidy pattern, is recognized contributing to the terminal pattern of plastid reproduction. The peaks shown in Fig. 3 are in part due to influences from this prepattern.

Differentiation of chloroplast numbers in cells of different type is, however, not due to endopolyploidy alone. In secondary leaves of *Oenothera hookeri*, for example, the mesophyll cells contain about 30, and the guard cells about 8 chloroplasts, endopolyploidy being completely absent. Therefore, the pattern found cannot result from endopolyploidy; instead, there must be a pattern of a different kind. Table 1 shows a more complex example, with a prepattern due to endopolyploidy superimposed upon the other prepattern. Such combinations are rather the rule than the exception. Each prepattern again may be the result of other prepatterns, i.e., of a series of homeogenetic or heterogenetic inductions (Bünning 1965).

If endopolyploidy effects could be eliminated from the ratios plotted in Fig. 3, probably the peaks of the distribution would decrease in number at the expense of higher ratios, but the remaining peaks would be sharper, because the mean figures used as numerators would be calculated from less heterogeneous cell populations.

Chloroplasts increasing in number by an increase of ploidy or by other reasons of normal leaf differentiation usually grow to normal size, thereby increasing the plastidome volume. An increase in numbers, however, can be attained also by separation of the plastidome into smaller plastids, for instance in sun leaves (see Sect.IV.C). This process is not accompanied by growth and does not occur in cells of all kinds. In this case, an environmentally induced prepattern of plastidome parceling superimposes upon other prepatterns in forming the terminal pattern. The effects of other external factors are considered below.

C. Influences of Some External Factors

Most changes in the growing conditions result in a shift of chloroplast numbers. In analyzing these shifts, care should be taken to evaluate which prepattern is affected, although this might be rather difficult. An old example (Butterfass 1964b) will

Table 2. The water supply as affecting the cell-specificity pattern of chloroplast numbers in *Beta vulgaris*. Adapted from Butterfass (1964b). The shifts of endopolyploidy were confirmed by counting chromosomes in cells induced to divide; the result is qualitative, not quantitative, as shown by Butterfass and Schlayer (1965)

	Mean number of chloroplasts			
	Water supply		Difference	Endopolyploidy
	Low	High		
Guard cells	8.1	7.5	− 0.6 [a]	Absent
Epidermal cells	8	11	+ 3 [a]	Increased
Spongy parenchyma cells	43	59	+ 16 [a]	Increased

[a] Significant at the $P = 5\%$ level

stress the requirement: if fodder beets *(Beta vulgaris)* are grown under different water regimes, the chloroplast numbers in cells of different types may shift to different degrees, and even in opposite directions (Table 2). Probably a decrease in numbers would be found in epidermal and in mesophyll cells as well as it is in guard cells, though it is masked there by an increase due to changes in the ploidy level. Without information on endopolyploidy (its occurrence, response to changes in the environment, and its effect on chloroplast numbers) observations on shifts in chloroplast numbers can be highly perplexing, and it is essential that the effect of the factor that is being investigated upon ploidy level is monitored.

Light has obvious effects on plastid reproduction, but the prepattern involved is usually not known. For this only a few examples will be mentioned. In plants kept in the dark and then exposed to light, the plastids will nearly always resume division (Boasson and Laetsch 1968; Boasson et al. 1972; Bennett and Radcliffe 1975; Possingham et al. 1975; Rose et al. 1975; and other workers), at least if the leaves investigated are not fully mature. Rose et al. (1975) found the dosage of nuclear DNA also increased. The size of the plastids may also be increased (Robertson and Laetsch 1974). Sucrose appears to be able to replace light in inducing immature chloroplasts to divide again (Ch. Lorey, unpublished); proplastids, of course, can divide in the absence of light. Sun leaves contain more, but smaller, chloroplasts in their cells (Hanson 1917, and many other workers); endopolyploidy is not involved here. Phytochrome may trigger plastid division (Hahn and Miller 1966; Possingham 1973b; Bradbeer et al. 1974; Kass and Paolillo 1974a, b), but this point is controversial. Strong light of different wave lengths did not lead to striking differences of chloroplast numbers in the experiments of Kakhnovich and Klimovich (1971). In other experiments, green light allowed growth and greening, but no division of plastids (Possingham 1973a; Possingham and Rose 1976). In *Acetabularia,* blue light favored division (Schmid and Clauss 1974, 1975).

There is controversy about the effects of other factors. It will suffice here to present a few selected reports, omitting the bulk of references. Chloroplasts may be induced to multiply via endopolyploidy by providing them better growing conditions (mineral salt supply, water conditions, light; Schlayer 1971). They may multiply independently of ploidy if environmental conditions are unfavorable; the chloroplasts then usually remain smaller. However, reduced size combined with increased number of chloroplasts need not indicate poor growth; chloroplasts from sun leaves are an example of such behavior. In *Nicotiana* leaves, fluorodeoxyuridine (10^{-4}M) prevents further growth of chloroplasts but does not inhibit chloroplast division (Verbeek-Boasson 1969). This indicates that plastid division does not need to be preceded or accompanied by plastid DNA replication. There are many more reports confirming this result.

Generally, it is easier to increase than to reduce plastid numbers per cell. They could be increased, for example, in epidermal cells of cotyledons of *Helianthus annuus* by application in physiological or at least not lethal doses of cyclic adenosine-3'-5'-monophosphate, adenosine, adenine, cytokinins, gibberellic acid, ethylene, carbon dioxide, cysteine, glucose, fructose, cycloheximide, chloramphenicol, and ethanol (Macchini 1975). It is surprising that inhibitors of protein synthesis should be included in this heterogeneous list. The effect of chloramphenicol, which inhibits plastid protein synthesis, is especially noteworthy and deserves further investigation.

V. The Genetic Analysis of the Motive Powers

"The dosage of nuclear DNA" is a convenient term for a complex gene group initiating or controlling plastid reproduction, or for other factors involved (see at the end of this chapter). In sugar beets, selection continued for three generations for low or for high chloroplast numbers in guard cells was enough to produce strains with five and ten chloroplasts, respectively, instead of about seven (Butterfass 1968). The control system thus seems to be amenable to alteration. It is therefore remarkable that about every second species of all higher plants investigated so far, natural polyploids included, shows six to eight chloroplasts in its guard cells. What may be the advantage of such a mean? Would smaller figures increase the chance of apoplastidic cells arising? These questions have yet to be answered.

Each extra chromosome in sugar beet trisomics affecting chloroplast numbers does so specifically (Butterfass 1964a, 1967). In the trisomic state, five out of nine chromosomes of the sugar beet have been found to induce changes of the basic numbers, i.e., changes not brought about by differences in ploidy. All chromosomes of the sugar beet with the possible exception of one (no. IX: Butterfass 1973, p.184) may change the endopolyploidy pattern. As the measure of endopolyploidization in triploids does not exceed that in diploids, interactions between chromosomes of the third complement obviously prevent all surplus endopolyploidy. The interaction of different chromosomes can be demonstrated with double trisomics where increases of basic or of endopolyploidy-dependent numbers vanish in certain combinations (Butterfass 1965). Whether a given chromosome is involved in the control of plastid numbers of normal eudiploid plants cannot be stated unequivocally because the action of an extra chromosome upon the multiplication of chloroplasts depends upon the genetic background of the cell and upon the growing conditions (cf. the results of Butterfass 1964a with those of 1967). The chromosomes cooperate in the promotion and control of plastid numbers, and they do so cell-specifically.

It would be premature, however, to identify the chromosomal factors with genes. Different activities of different chromosomes as well as the response of plastid numbers to selection are no proof for genes being involved in a more direct way. Instead, effects of the nucleotype (Bennett 1972, 1973) might prevail, viz. effects of the amount of DNA per se, of volume and of heterochromatin (see Nagl 1976). For a discussion of chloroplast numbers along these lines, however, our knowledge is still too restricted.

VI. Further Evolutionary Significance

One may speculate that the increase of chloroplast numbers in polyploid cells is not only the inescapable by-product of a relationship absolutely necessary for dividing plant cells, but has an evolutionary benefit in itself. If polyploid cells were only larger than diploids, but did not contain more plastids, the internal cell functioning might be upset. Thus, without concomitant adjustment of plastid numbers endopolyploidy and polyploidy of entire plants might have been ruled out as means of evolution. Furthermore, plants with polyploid meristems, like all green plants, can exist only if plastids are transmitted reliably to daughter cells. If only the size of the cells, but not the number of plastids were increased, the plastids would be more loosely scattered; therefore, they might not interact in the original degree, and thus might not spread as regularly as in smaller cells. (A spatially random distribution of plastids is far from a distribution with similar distances between neighboring plastids.) Therefore, the more randomly the plastids are scattered, the higher is the danger of apoplastidic daughter cells to occur, or, the more plastids must be present to avoid this danger. This relationship is significant because the plastid numbers of young diploid cells are not far from minimum numbers required for the safe transmission of plastids, and are well below the numbers that are tolerable with absence of interactions among plastids which regulate their intracellular distribution (Butterfass 1979). Thus, the nuclear system with a dosage effect on the multiplication of plastids may be regarded as a prerequisite for the evolutionary utilization of polyploidy.

As to the multiplication of plastids during differentiation, two kinds have been distinguished: either the multiplication is the result of a preceding endopolyploidization, or it is not. Functionally, the two mechanisms are not fully interchangeable; there is a special benefit for each of them, and they may be adaptations to different requirements. Endopolyploidy appears to be far more reactive to environmental changes, enabling plants to respond within a few days to, for instance, an improvement of water relations by expanding cells of susceptible tissues and endowing them with more chloroplasts. In contrast, the prepattern of chloroplast numbers produced by differentiation steps other than endomitoses is

Table 3. The occurrence of monoplastidy in different plant groups on different developmental stages. Compiled from extended literature

	Monoplastidy known to occur during				
	Micro-gameto-genesis	Macro-gameto-genesis	Sporo-genesis	Meris-tematic activity	Photosyn-thesis of mature cells
Many algae, *Anthoceros*	+	+	+	+	+
Selaginella, Isoetes	+	+	+	+	− [a]
Lycopodiales	+	?	+	+	−
Musci	+	−	+	−	−
Hepaticae except *Anthoceros*	+	−	−	−	−
Filicatae, Equisetum, Spermatophyta	− [b]	−	−	−	−

+ Members of the plant group are known to show monoplastidy on this stage. Other members may lack it
− Evidence is known for the absence of monoplastidy on this stage, but not for its presence
[a] Except few *Selaginella* species with single giant plastids in some cells
[b] The report of Hagemann (1976) according to which higher plants exist whose sperm cells contain only one or no plastid is not well documented in the cited literature

less responsive to external factors, and changes may take more time. Some plant species are almost unable to use endopolyploidy to adapt ecologically, among them *Oenothera hookeri, Helianthus annuus,* and *Crepis capillaris.* On the other hand, they have the benefit that their leaves do not have to live with an endopolyploidy irreversibly acquired under transient conditions. One may speculate that for plants growing under conditions changing repeatedly, endopolyploidy might be a burden rather than an advantage. The ecological approach to chloroplast numbers as initiated by Mokronosov and his co-workers (Mokronosov et al. 1973; Bagautdinova et al. 1975) should be extended to the role and the adaptive value of endopolyploidy.

There is an additional point which has to be discussed in the context of evolution (Table 3). In protonemata, stems, and sporophytes of mosses, the tip cells are polyplastidic and divide like cells of higher plants, not showing any indication that a mechanism specifically related to the distribution of the plastids to the daughter cells might be at work; during the formation of sperm cells and of spore mother cells, however, diplastidic cells show a regular allotment of plastids to daughter cells, and mature spore mother cells containing four plastids distribute them one into each of the four resulting spores. These long-known facts (Dyer 1976), taken together, pose questions that have been neglected so far.

It seems clear that lower plants have retained a mechanism which regularly allots plastids to the daughter cells at division, whereas higher plants probably have not. The point is exemplified by comparing the moss *Sphagnum cuspidatum* with the flowering plant *Trifolium hybridum.* It was described in Sect.II how plants of the moss, when grown submerged, can become monoplastidic in certain leaflet meristems in contrast to the normal polyplastidic condition, but nevertheless do not become apoplastidic. In the *Trifolium*, however, apoplastidic cells do arise if the number of plastids per mother cell drops below a critical number (Sect. II).

As higher plants are lacking monoplastidic stages (Table 3), monoplastidy is an ancient character (Proskauer 1969) and, in the long run, would appear to have been of smaller adaptive value in higher plants than polyplastidy. Why, then, has monoplastidy not been replaced completely also in lower plants by polyplastidy (Table 3)? Is monoplastidy an adaptation to a requirement that does not occur in higher plants, or in certain dividing cells of other plants, and if so, what is this requirement?

VII. Conclusion

The central problem in considerations of plastid continuity is the question of the nature of the extremely efficient homeostatic system that ensues the safe transmission of plastids. It works independently of the general physiological efficiency of the cell if the genetics is heavily disturbed, as in multiple trisomic or in hypoploid plants. Part of an explanation might be that the plastids as semiautonomous organelles are always ready to divide if conditions allow that. Even in plants with an unbalanced genotype, the population of plastids expands to its limits, is halved in mitosis, and then expands again. The limits depend mainly upon the relative content of the cell of nuclear DNA as oscillating between G_1 and G_2 phase and as changing with ploidy. In addition, the quality of the genotype, other internal conditions and the external conditions all act as modifiers of the plastid division control system.

As compared with the importance of safe transmission of plastids to progeny cells and plants, the differentiations of chloroplast numbers by endopolyploidy or otherwise may appear as mere adornments of minor adaptive value. This is not to say that they are unimportant. But the first-mentioned process is a prerequisite for the very existence of green plants, whereas the latter may well improve the efficiency of the green plants in metabolism and reproduction. As a result, also the agricultural performance as measured by yield may depend upon patterns of chloroplast numbers (Butterfass 1972).

The terminal pattern arises from the cooperation of several prepatterns, which are open to many influences from the genotype and the environment. Different prepatterns may be modified with variable speeds and intensities, and they may contribute in variable measure to the terminal pattern. (The prepattern of endopolyploidy, for instance, may be absent completely.) The internal buffering and flexibility of the entire system of prepatterns producing the terminal pattern can be expected to reflect ecological capacity and adaptability. However, nothing is known about that.

Studies on the differentiation of plastid populations may well lead to general insights into the nature, and the limitations, of the nucleoplasmic ratio, and of pattern formation. If we really knew how plastids are made to divide, and why in certain cells they divide more often than in others, we would understand much more about differentiation in general.

Acknowledgments. During the preparation of this article I had several thorough and valuable discussions on content and formulation with Professor R.G. Herrmann, Düsseldorf. He discussed the first drafts of a few parts with several English-speaking colleagues whose remarks and suggestions contributed significantly to clearness and linguistic correctness of these parts at that early stage. My investigations have been supported by the Deutsche Forschungsgemeinschaft. All help is gratefully acknowledged.

Note Added in Proof. For a detailed presentation of the subject see Butterfass, Th.: Patterns of Chloroplast Reproduction. Cell Biology Monographs, Vol. VI. Wien-New York: Springer 1979.

References

Anton-Lamprecht, I.: Anzahl und Vermehrung der Zellorganellen im Scheitelmeristem von *Epilobium*. Ber. Dtsch. Bot. Ges. 80, 747-754 (1967)

Bagautdinova, R.I., Fedoseeva, G.P., Mokronosov, A.T.: Struktura i funktional naya aktivnost' assimiliruyushchikh organov u rasteniy raznykh sistematicheskikh i ekologicheskikh grupp. Abstr. XII Int. Bot. Congr. No. 418 (1975)

Baker, W.B.: Studies in the life history of *Euglena*. I. *Euglena agilis*, Carter. Biol. Bull. 51, 321-362 (1926)

Bartels, F.: Plastidenzählungen bei *Epilobium hirsutum*. I. Mitt. Zählungen in Zellen aus unterschiedlich differenzierten Geweben des Laubblattes. Planta 60, 434-452 (1964)

Bennett, J., Radcliffe, C.: Plastid DNA replication and plastid division in the garden pea. FEBS Let. 56, 222-225 (1975)

Bennett, M.D.: Nuclear DNA content and minimum generation time in herbaceous plants. Proc. R. Soc. London Ser. B 181, 109-135 (1972)

Bennett, M.D.: Nuclear characters in plants. Brookhaven Symp. Quant. Biol. 25, 344-366 (1973)

Boasson, R., Laetsch, W.M.: Effects of inhibitors of DNA synthesis on chloroplast growth and development. Plant Physiol. 43, Suppl., S-7, Abstract (1968)

Boasson, R., Laetsch, W.M., Price, J.: The etioplast-chloroplast transformation in tobacco: Correlation of ultrastructure, replication, and chlorophyll synthesis. Am. J. Bot. 59, 217-223 (1972)

Bradbeer, J.W., Gyldenholm, A.O., Smith, J.W., Rest, J., Edge, H.J.W.: Plastid development in primary leaves of *Phaseolus vulgaris*. IX. The effects of short light treatment on plastid development. New Phytol. 73, 281-290 (1974)

Bünning, E.: Die Entstehung von Mustern in der Entwicklung der Pflanzen. In: Handbuch der Pflanzenphysiologie, Ruhland, W. (ed.), Vol. XV/1, pp. 383-408. Berlin-Heidelberg-New York: Springer 1965

Butterfass, Th.: Die Abhängigkeit der Plastidenvermehrung von der Reproduktion der Erbsubstanz im Kern. Ber. Dtsch. Bot. Ges. 76, 123-134 (1963)

Butterfass, Th.: Die Chloroplastenzahlen in verschiedenartigen Zellen trisomer Zuckerrüben *(Beta vulgaris* L.). Z. Bot. 52, 46-77 (1964a)

Butterfass, Th.: Die Steigerung des Endopolyploidiegrads in Blättern von *Beta vulgaris* L. durch bessere Wasserversorgung. Ber. Dtsch. Bot. Ges. 77, 285-290 (1964b)

Butterfass, Th.: Verschiedenartige Ursachen der Plastidenvermehrung in verschiedenen Zellen. Ber. Dtsch. Bot. Ges. 78, (105)-(110) (1965)

Butterfass, Th.: Endopolyploidie und Chloroplastenzahlen in verschiedenartigen Zellen trisomer Zuckerrüben. Planta 76, 75-84 (1967)

Butterfass, Th.: Das Muster aus zellspezifischen Chloroplastenzahlen und seine Ursachen. Naturwiss. Rundsch. 21, 466-469 (1968)

Butterfass, Th.: Die Plastidenverteilung bei der Mitose der Schließzellenmutterzellen von haploidem Schwedenklee (*Trifolium hybridum* L.). Planta 84, 230–234 (1969)

Butterfass, Th.: Fourfold exact duplication of chloroplasts in cells of *Sphagnum*. Naturwissenschaften 58, 420 (1971)

Butterfass, Th.: Endopolyploidie und Ertrag bei diploiden und tetraploiden Zuckerrüben. III. Ergebnisse zur Methodik. Theor. Appl. Genet. 42, 41-43 (1972)

Butterfass, Th.: Control of plastid division by means of nuclear DNA amount. Protoplasma 76, 167-195 (1973)

Butterfass, Th., Schlayer, G.: Widersprüchliche Ergebnisse bei der Bestimmung des Endopolyploidiegrads von Zuckerrüben durch Auslösen von Mitosen. Chromosoma 17, 303-308 (1965)

Dangeard, P.-A.: Sur la distinction du chondriome des auteurs en vacuome, plastidome et sphérome. C. R. Acad. Sci. 169, 1005-1010 (1919)

Dyer, A.F.: The visible events of mitotic cell division. In: Cell Division in Higher Plants, Yeoman, M.M. (ed.), pp. 50-110. London-New York: Academic Press 1976

Eyster, W.H.: Variation in size of plastids in genetic strains of Zea mays. Science 69, 48 (1929)

Hagemann, R.: Plastid distribution and plastid competition in higher plants and the induction of plastom mutations by nitroso-urea-compounds. In: Genetics and Biogenesis of Chloroplasts and Mitochondria, Bücher, Th. et al. (eds.), pp. 331-338. Amsterdam: North-Holland 1976

Hahn, L.W., Miller, J.H.: Light dependence of chloroplast replication and starch metabolism in the moss Polytrichum commune. Physiol. Plant. 19, 134-141 (1966)

Hanson, H.C.: Leaf-structure as related to environment. Am. J. Bot. 4, 533-560 (1917)

Herrmann, R.G.: Chloroplastengröße und inkorporierte ^{3}H-Thymidinmenge. Autoradiographische Studien zur Frage: Gibt es genetisch mehrwertige Plastiden? Ber. Dtsch. Bot. Ges. 81, 332 (1968)

Herrmann, R.G.: Multiple amounts of DNA related to the size of chloroplasts. I. An autoradiographic study. Planta 90, 80-96 (1970)

Herrmann, R.G., Kowallik, K.V.: Multiple amounts of DNA related to the size of chloroplasts. II. Comparison of electron-microscopic and autoradiographic data. Protoplasma 69, 365-372 (1970)

Kakhnovich, L.V., Klimovich, A.S.: Fotosinteticheskiy apparat v zavisimosti ot intensivnosti sveta. Fiziol. Rast. 18, 893-897 (1971)

Kass, L.B., Paolillo, D.J., Jr.: The effect of darkness and inhibitors of protein synthesis on the replication of chloroplasts in the moss, Polytrichum. Z. Pflanzenphysiol. 73, 198-207 (1974a)

Kass, L.B., Paolillo, D.J., Jr.: On the light requirement for replication of plastids in Polytrichum. Plant Sci. Lett. 3, 81-85 (1974b)

Kowallik, K.V., Herrmann, R.G.: Variable amounts of DNA related to the size of chloroplasts. IV. Three-dimensional arrangement of DNA in fully differentiated chloroplasts of Beta vulgaris L. J. Cell Sci. 11, 357-377 (1972)

Macchini, L.: Die Wirkung von stofflichen Außeneinflüssen auf die ploidieunabhängige Chloroplastenvermehrung. Diss. Heidelberg (1975)

Michaelis, P.: Über Zahlengesetzmäßigkeiten plasmatischer Erbträger, insbesondere der Plastiden. Protoplasma 55, 177-231 (1962)

Mokronosov, A.T., Bagautdinova, R.I., Bubnova, E.A, Kobeleva, I.V.: Fotosinteticheskiy metabolizm v palisadnoy i gubchatnoy tkanyakh lista. Fiziol. Rast. 20, 1191-1197 (1973)

Nagl, W.: Zellkern und Zellzyklen. Stuttgart: Ulmer 1976

Possingham, J.V.: Chloroplast growth and division during the greening of spinach leaf discs. Nature New Biol. 245, 93-94 (1973a)

Possingham, J.V.: Effect of light quality on chloroplast replication in spinach. J. Exp. Bot. 24, 1247-1258 (1973b)

Possingham, J.V., Rose, R.J.: Chloroplast replication and chloroplast DNA synthesis in spinach leaves. Proc. R. Soc. London Ser. B 193, 295-305 (1976)

Possingham, J.V., Cran, D.G., Rose, R.J., Loveys, B.R.: Effects of green light on the chloroplasts of spinach leaf discs. J. Exp. Bot. 26, 33-42 (1975)

Proskauer, J.: The probable cell structure of the original green land plant. XI Int. Bot. Congr. Abstr., p. 174 (1969)

Renner, O.: Die pflanzlichen Plastiden als selbständige Elemente der genetischen Konstitution. Ber. Sächs. Akad. Wiss. Math. Phys. Kl. 86, 241-266 (1934)

Robertson, D., Laetsch, W.M.: Structure and function of developing barley plastids. Plant Physiol. 54, 148-159 (1974)

Rose, R.J., Cran, D.G., Possingham, J.V.: Changes in DNA synthesis during cell growth and
   chloroplast replication in greening spinach leaf disks. J. Cell Sci. 17, 27-41 (1975)
Schlayer, G.: Modifikationen des DNS-Gehalts in Zuckerrübenzellen. Planta 98, 294-299
   (1971)
Schmid, R., Clauss, H.: Die Vermehrung der Chloroplasten von *Acetabularia* im Rot- und
   Blaulicht. Protoplasma 82, 283-287 (1974)
Schmid, R., Clauss, H.: Multiplication and protein content of chloroplasts of *Acetabularia
   mediterranea* in blue light after prolonged irradiation with red light. Protoplasma 85, 315-
   325 (1975)
Stern, C.: Genes and developmental patterns. Caryologia 6, Suppl., 355-369 (1954)
Verbeek-Boasson, R.: Chloroplast replication and growth in tobacco. Diss. Groningen
   (1969)

Plastid DNA — The Plastome

R.G. HERRMANN[1] and J.V. POSSINGHAM[2]

[1] *Botanisches Institut der Universität Düsseldorf, FRG*
[2] *C.S.I.R.O., Division of Horticultural Research,*
Box 350, Adelaide, Australia 5001

I. Introduction

The development of autotrophic eukaryotic organisms results from a close co-operation between three distinct cellular compartments (using the compartment definition of Schnepf 1966), of nucleus/cytosol, plastids, and mitochondria, each of which contains its own genetic machinery. Understanding the nature of this cooperation and its evolution is a fundamental problem in biology. This chapter focuses primarily on genetic elements of plastids.

Plastids have a particularly wide range of structure and function and are involved in cellular energy metabolism, storage, and reproduction. Present knowledge suggests that the DNA of the organelle is involved in the genetic anchoring of at least some of these processes, such as photoautotrophy. Specialized modifications of this organelle are proplastids, etioplasts, leucoplasts, amyloplasts, elaioplasts, chloroplasts (rhodoplasts, phaeoplasts), sieve-tube plastids, and chromoplasts (Schnepf, this vol.). It is not known whether the organelle DNA is involved in the biogenesis and function of all modifications.

The genetic machinery of plastids is organized in a prokaryotic manner. Existing evidence indicates that a single, double-stranded circular DNA molecule, the chromosome, contains the coding potential of the organelle. Plastids usually contain many copies of this chromosome which is sufficiently large to code for several hundred polypeptides of 20 kd M.W. and it is now possible to map genes physically on this chromosome. The genetic system of the organelle is more complex than was originally assumed. Apart from the enzymatic equipment to duplicate (Kolodner and Tewari 1975a), transcribe (Hartley and Ellis 1973), and translate (Blair and Ellis 1973; Bottomley et al. 1974) genetic information, the organelle contains enzymes for posttranscriptional modification of RNA (Bohnert et al. 1974) and for posttranslational modification of polypeptides both imported from the cytosol (Dobberstein et al. 1977) or synthesized internally (Edelman and Reisfeld 1978; Grebanier et al. 1978). Information about DNA recombination in

---

Abbreviations. *bp:* base pairs; *kbp:* kilo base pairs; *kd:* kilodalton; *Md:* megadalton; $G+C$: mole-% guanine plus cytosine; *M.W.:* molecular weight; *UV:* ultraviolet; *ptDNA:* is used for plastid DNA regardless of the organelle modification; *nucDNA:* nuclear DNA; *mtDNA:* mitochondrial DNA.

plastids is restricted to *Chlamydomonas* (Sager 1977). For the same organism restriction/modification is discussed as one means to achieve uniparental inheritance (Sager and Lane 1972). Virtually no information is available about DNA repair in chloroplasts.

II. Terminology

It is difficult to formulate an unequivocal terminology to describe the genetic make-up of eukaryotic cells because of recent discoveries of genetic components such as plasmids (Schell et al. 1977), viruses (Shepherd 1976), viroids (Diener 1974), or pro- and eukaryotic endosymbionts (Schnepf 1966; Tomas and Cox 1973; Jeffrey and Vesk 1976) which may be associated with different parts of the cell. We have used the terms *genome* (Winkler 1920) and *plastome* (Renner 1934) to describe the genetic elements in nuclei and plastids, respectively, since these terms underline the interdependence of these compartments. Alteration of the genetic material in any one of the compartments leads to an impairment of cellular activities and influences the continuity of the whole cell.

The term chromosome is used for prokaryotic as well as eukaryotic organization of genetic material and a distinction is made between *polyploid* and *polyenergidic* arrangement of multiple chromosome sets. Both expressions describe two states of disposition with probably different functional and evolutionary potentialities. Polyploidy implies that several chromosome sets are concentrated in one place, nucleus or nucleoid, whereas in polyenergidic organization they are distributed in several places, such as in *Physarum* or *Vaucheria*. Each nucleus or nucleoid of a polyenergidic cell or organelle may in turn be polyploid.

Whenever possible, expressions like *extrachromosomal* or *extrakaryotic* (*extranuclear*) have been avoided. Extrachromosomal DNA is not necessarily extranuclear and extranuclear DNA may be situated in the cytosol (e.g., Guerineau et al. 1976; Hollenberg et al. 1976) rather than in plastids and mitochondria. It is also conceivable that facultative extrachromosomal genetic elements may reside in plastids and mitochondria.

III. Characteristics of Plastid DNA

The existence of DNA in the multitude of organelles defined as plastids is probably universal. It has been demonstrated cytologically in all major plant phyla as well as in most developmental modifications of the organelle. Some genetic evidence is available especially from experiments with *Oenothera* indicating that ptDNA is ontogenetically distinct and transmitted independently of nucDNA (see Sect.V).

Research on ptDNA has received impetus from recent technical advances including those concerned with improved methods of subcellular fractionation which yield organelles of high purity and quality (Ellis 1975; Morgenthaler et al. 1975; Schmitt and Herrmann 1977), the discovery and application of a wide variety of site-specific restriction endonucleases and subsequent developments of the recombinant DNA technology (for review see Hollenberg 1978) and of the Southern method (Southern 1975), the development of DNA- and RNA-programmed cell-free translation systems (e.g., Zubay et al. 1970; Marcu and Dudock 1974; Pelham and Jackson 1976), the separation (e.g., Shapiro et al. 1967; Weber and Osborn 1969; Tzagoloff and Meagher 1971) and characterization of plastid polypeptides (e.g., Chua and Schmidt 1979), the selection of plastome mutants with impaired plastids (for review see Kirk and Tilney-Bassett 1978) and the use of the highly sensitive fluorochromes such as DAPI which make it possible to visualize with the light microscope the small amounts of DNA present in plastids (Dann et al. 1971; Coleman 1979; James and Jope 1979).

A. Identification, Purity, Quality

An important aspect in any discusion on ptDNA is the nature of criteria used to describe its identity and purity.

Several properties have been used to identify ptDNA. These include determination of average base composition, lack of detectable methylated bases, intramolecular heterogeneity in base composition, ease of reassociation and circularity. Although the physicochemical characterization of ptDNA is an essential element of the identification procedure, none of these properties is unique to ptDNA and is capable of establishing that a DNA is in fact derived from a plastid. Collectively, however, they can be used to distinguish ptDNA from mt- and nucDNA. PtDNA shares ease of reassociation, compositional heterogeneity, as well as lack of detectable methylated bases, with mtDNA (Borst 1972) and some nuclear DNA fractions (Skinner and Beattle 1974; Sinclair et al. 1975). Circularity is shared with mtDNA (Borst 1972; Kolodner and Tewari 1972a), nuc-rDNA (Hourcade et al. 1974), plant viral DNA's (Shepherd 1976) and plasmid-like DNA of probably cytosolic origin (Guerineau et al. 1976; Hollenberg et al. 1976).

Proof that a DNA species is an integral plastid constituent should be based on the integrity of the isolated organelle and on the purity of the organelle preparation from which it is obtained. Some of the past confusion about the identity and physicochemical properties of ptDNA (for review see Kirk 1971) can be attributed to problems with the isolation of plastids

A major step forward has been the isolation of unbroken plastids in isotonic media by short-time centrifugation (see Walker 1971) followed by DNAase treatment to remove contaminating nucDNA (Wells and Birnstiel 1969). Alternative possibilities are centrifugation in osmotically balanced isopycnic gradients with or without DNAase treatment (Schmitt and Herrmann 1977), or phase partition in binary aqueous systems (Larsson et al. 1971; Blomquist et al. 1975). Intact plastids prepared in this way can meet high standards of quality and purity and can yield pure and largely undegraded ptDNA.

The possibility of identifying ptDNA by nonaqueous subcellular fractionation offers the advantage of minimizing redistribution of soluble components which generally attends aqueous techniques, but the technique has not yet been fully explored. For example DNA isolated from *Antirrhinum* chloroplasts in the mid-sixties by this means was one of the rare exceptions of correct identification of ptDNA in higher plants (Ruppel 1967). However, Bird et al. (1973) emphasize that even nonaqeuous fractionation fails to yield pure plastid preparations.

The assessment of DNA purity[1] is straightforward if there is a sufficiently large difference in base composition between pt- and nucDNA to allow their pycographic separation by equilibrium (Meselson et al. 1957) or relaxation (Anet and Strayer 1969) centrifugation in salt gradients (CsCl, NaI, KI; Blin et al. 1975). With closely banding DNA components, improvement of the sensitivity of detection may be achieved if marginal peak fractions from preparative gradients are recentrifuged analytically (Bard and Gordon 1969).

With kryptically banding ptDNA which applies to many higher plants, advantage is usually taken of differences in reassociation rates of pt- and nucDNA (Tewari and Wildman 1966, 1970). Single-stranded ptDNA reassociates readily when held at annealing temperature, in contrast to nucDNA, and this can be demonstrated qualitatively by density shifts in CsCl gradients. This criterion which is conceptually identical to that used to assess purity of mtDNA's (see Borst et al. 1967) has since found widespread application, but while there can be no question of its relevance for diagnostic purposes, there are problems associated with using it as a primary tool for the estimation of DNA purity. Several authors claim that pt- and nucDNA form common reassociation complexes since the appearance of unimodal bands at densities intermediate between denatured and native DNA, or bimodal banding, depending on the ratio of the two components have been observed (Richards 1967; Spencer and Whitfeld 1969; Kung et al. 1972; Herrmann et al. 1974). When bimodal banding occurs separation may not be stoichiometric (Herrmann et al. 1974). The reaction appears to be a widespread phenomenon and to be nonspecific, since heterologous pt- and nucDNA may cross-react. No detailed investigation has yet been performed delineating the nature of the cross-reaction, but control experiments have excluded the possibility that such complexes can be due to physical interlocking of the DNA strands (Herrmann et al. 1974). Irrespectively of whether this reaction is due to sequence similarity, sequence homology or even preparation artifact, its practical consequence is that it sheds considerable doubt on the interpretation of many experimental data on higher plant ptDNA. The purity of plastid DNA can be severely overestimated and a pronounced effect on reassociation and molecular hybridization processes may be expected. Indications of significant nuclear contamination on ptDNA preparations can be found in the literature, e.g., linear DNA fibrils 120–160μm in length (Woodcock and Fernandez-Moran 1968; Tewari and Wildman 1970; Manning et al. 1972; Herrmann et al. 1975) which usually can be attributed to the presence of nucDNA, although it is possible that these molecules could be derived from tri- or tetramer unicircular ptDNA. Likewise, if both DNA's differ slightly in composition, a discrepancy between the average base composition of ptDNA calculated from buoyant densities and that from thermal denaturation (Kolodner and Tewari 1975b) can be indicative of the presence of nucDNA. No deviations would be expected in the absence of methylated bases (Kirk 1967).

Alternative approaches based on the lack of detection of methyl cytosine (Baxter and Kirk 1969) or electron microscopy (Herrmann et al. 1975) have not enjoyed a popularity comparable to the reassociation procedure. They require

---

1   The term purity is used here to describe the absence of another DNA species from ptDNA. It also has a second connotation describing the absence of non-DNA material.

relatively large quantities of DNA and are relatively insensitive because methyl cytosine itself is only a minor constituent of nucDNA. The methods involving electron microscopy at best are semiquantitative, tedious, and require that the DNA be largely undegraded.

Although each of the methods has problems, taken together they can provide at least a rough estimate of an upper limit of contamination. The importance of careful consideration of the methods used to prepare plastid fractions, the possible artifacts these methods can produce and the need for a more critical interpretation of experiments on isolated plastids and their nucleic acids cannot be overemphasized.

B. Average Base Composition

For Chlorophyta ptDNA's G + C contents are usually between 36 and 38 mol% calculated from buoyant density determination in CsCl gradients, from thermal dissociation (Tm) analysis or from direct biochemical analysis (Wells and Ingle 1970; Kirk 1971). Some deviations, however, are known as DNA from plastid fractions of the liverwort *Sphaerocarpos donnellii* has been found to have a lower average G + C content (31.6%) than the other Chlorophycean species studied (Herrmann et al. 1980) while it is well known that *Euglena gracilis* (Euglenophyta) has the low average G + C content of 25% (Stutz et al. 1976). Table 1 includes the average G + C contents of those ptDNA's whose circularity has been established.

PtDNA in higher plants and algae may be higher, lower, or about equal in average G + C content than nucDNA (Kirk and Tilney-Bassett 1978). In *Acetabularia* ptDNA may be more dense than nucDNA (Green et al. 1967), while ptDNA of the Xanthophycean species *Vaucheria sessilis* might be an example that ptDNA of algae can band within the density distribution of nucDNA (Kowallik and Hennig, pers. commun.).

C. The Plastid Chromosome

1. Structure, Complexity

Circular DNA molecules of plastids were discovered after those of mitochondria (Borst and Kroon 1969; Manning et al., 1971). Their considerable size, equivalent to a mass weight of 80–130 Md depending on the species, and the resulting tendency to break, made their isolation difficult. Circular molecules have been demonstrated in about 20 species including representatives of Chlorophyta, Euglenophyta, and Xanthophyta. They have been isolated from chloroplasts and chromoplasts but not from other organelle modifications (see Table 1). Figure 1 shows a circular DNA molecule from *Spinacia* chloroplasts.

Recent technical advances in the isolation of high molecular weight DNA have allowed recoveries of high yields of circular ptDNA molecules from several plant species (Herrmann et al. 1975; Tewari et al. 1977). This suggests that DNA molecules of plastids may all be circular. The ptDNA molecules of one plant species fall into a homogeneous size class if dimer forms are disregarded. They are

Table 1. Sizes of circular monomers and mean G+C contents of plastid DNA's[a]

Species	Organelle modification	G+C content (%)	Mean circumference (μm)	Reference
Angiospermae				
Dicotyledons				
Antirrhinum majus	Chloroplast	38	46	Herrmann et al. (1975)
Beta vulgaris	Chloroplast	38	46	Herrmann et al. (1975)
Lactuca sp.	chloroplast	39	41	Kolodner and Tewari (1975b)
Oenothera hookeri	Chloroplast	38	45	Herrmann et al. (1975)
Phaseolus sp.	Chloroplast	39	40	Kolodner and Tewari (1975b)
Pisum sp.	Chloroplast	32.7–39	39	Kolodner and Tewari (1972b, 1975b)
Spinacia oleracea	Chloroplast	37	44	Manning et al. (1972)
Spinacia oleracea	Chloroplast	38	46	Herrmann et al. (1975)
Spinacia oleracea	Chloroplast	37–39	39.5	Kolodner and Tewari (1975b)
Tropaeolum majus	Chromoplast	—	44	Sitte (1977)
Monocotyledons				
Avena sativa	Chloroplast	39	37	Kolodner and Tewari (1975b)
Narcissus pseudonarcissus	Chromoplast	—	44	Falk et al. (1974)
Narcissus pseudonarcissus	Chromoplast	37	—	Herrmann (1972)
Tulipa gesneriana	Chromoplast	—	43.6	Wuttke (1976)
Zea mays	Chloroplast	38	43	Manning et al. (1972)
Zea mays	Chloroplast	39	38	Kolodner and Tewari (1975b)
Spirodela oligorrhiza	Chloroplast	37	54	van Ee et al. (1979)
Archegoniatae				
Filicinae				
Asplenium nidus	Chloroplast	38	44	Herrmann et al. (1980)
(*Pteris vittata*)	Chloroplast	38	43	Herrmann et al. (1980)
Musci				
Sphaerocarpos donnellii	Chloroplast	31	37	Herrmann et al. (1980)
Algae				
Euglena gracilis	Chloroplast	36	62	Behn and Herrmann (1977)
Euglenophyceae				
Euglena gracilis	Chloroplast	28	40/44	Manning et al. (1971) Richards and Manning (1975)
Chlorophyceae				
Chlamydomonas reinhardii				
Xanthophyceae				
Vaucheria sessilis	Chloroplast	38	36	Kowallik and Hennig (unpubl.)

[a] The comparison of the contour length of spinach ptDNA which has been determined in three laboratories suggests that the size differences for ptDNA recorded from different laboratories are mainly due to calibration differences

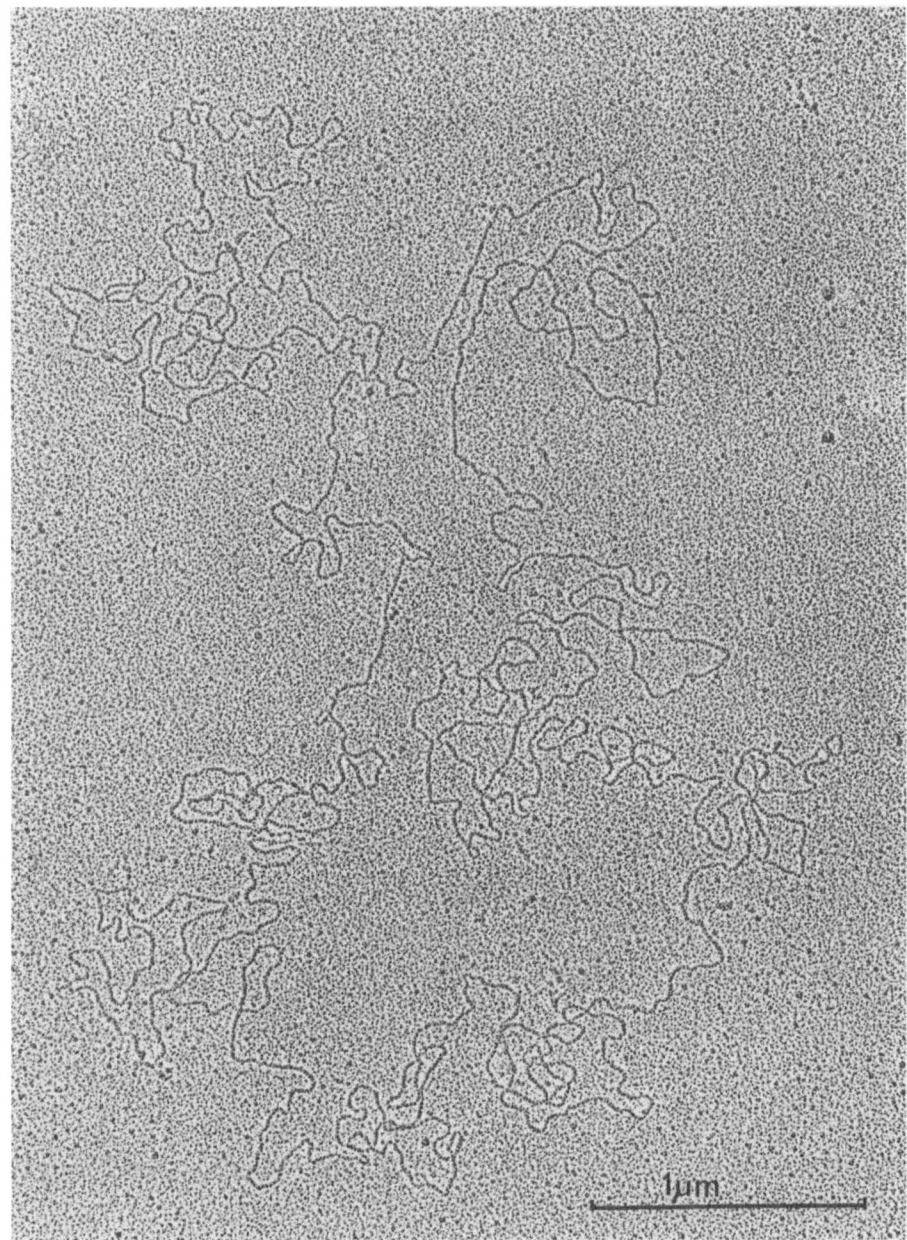

Fig. 1. Electron micrograph of an open circular DNA molecule (contour length 44.7μm) isolated from chloroplasts of *Spinacia oleracea*. (From Herrmann et al. 1975)

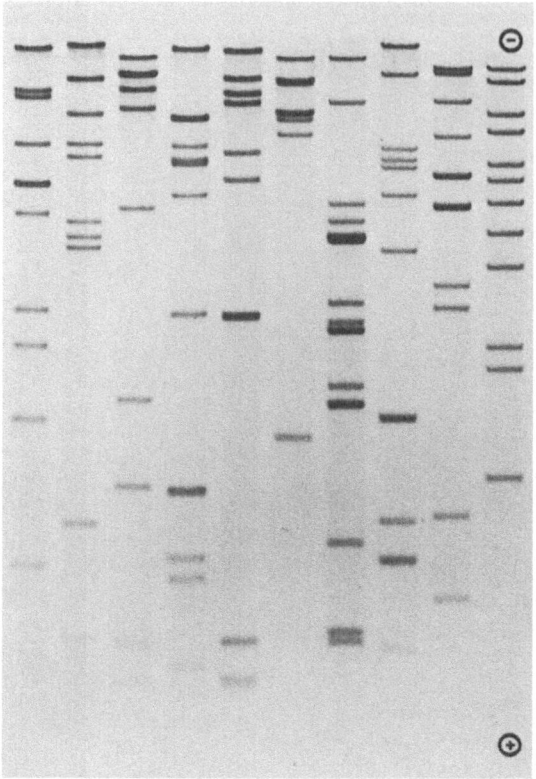

Fig. 2. Digestion of spinach ptDNA with ten different restriction endonucleases from left: Sal I, Pst I, Kpn I, Xho I, Pvu, I, Bgl I, Xma I, Pvu II, Hpa I and Sac I (Fragments separated on a 0.5% agarose gel). The molecular weights range from about 34 Md down to 0.8 Md. Smaller fragments have run off the gel. The sizes of the fragments obtained in the individual digestions add up to the molecular weight of the circle within the limits of errors of the method (Crouse et al. 1978). The stoichiometry of the fragments is not fully apparent since the lower parts of the gel have been overexposed to show the small fragments. Low backgrounds, especially between the largest fragments amounting to about 1/3 of the circle, indicate that DNA was almost undegraded

basically uniform in sequence and constitute monomers. This has now been established independently by means of restriction endonuclease analysis and denaturation mapping.

Endonuclease analysis has shown that (a) digestion of ptDNA with restriction endonucleases results in sets of unique fragments which occur in molar stoichiometries. (b) Addition of the molecular weights of the fragments of each set approximate within experimental error the size of the cyclic molecule derived from length measurements. (c) In all instances where the order of fragments has been determined a circular map of the monomer has been obtained (see Sect. IV). Figure 2 provides restriction patterns for *Spinacia* ptDNA and analogous data have been reported for ptDNA of *Zea mays* (Bedbrook and Bogard 1976), *Euglena* (Gray and Hallick 1976; Kopecka et al. 1977) and *Chlamydomonas* (Rochaix 1978). Similar conclusions can also be drawn from denaturation mapping as electron microscopic analysis of partially melted individual circular DNA molecules of a given species show similar patterns of early (A + T-rich) and late (G + C-rich) melting regions (Tewari et al. 1977). Taken together these observations establish that gross heterogeneities between individual ptDNA molecules are absent [minor heterogeneity has only been suggested for corn (Bedbrook and Bogard 1976) and *Chlamydomonas* ptDNA (Rochaix 1978)], and indicate that there is no other significant genetic component in the organelle. It is suggested that previous reports of linear DNA molecules from chloroplast fractions exceeding the monomer size up to fourfold (Tewari and Wildman 1970; Manning et al. 1972) can be traced back to contaminating nucDNA (Herrmann et al. 1975; see Sect.III.A). Thus, the coding potential of the plastome in the limited number of species so far investigated appears to be contained in a single circular DNA molecule. This has been called by some the chromosome, or taking account of its prokaryotic arrangement within the organelle, the genophor (Ris 1961).

Initially the genetic complexity of ptDNA's was determined using reassociation kinetics. This method was valuable in that it provided the first information on the genetic potential of ptDNA. However due to the inherent lack of precision of the method (e.g., Borst 1971) and because of pecularities of the organization of ptDNA in respect of sequence repetition and intramolecular heterogeneity in nucleotide composition (see Sect.IV.A), as well as possible contamination with nucDNA, it was not possible to prove that ptDNA consisted of a single sequence class equivalent to the circle (about 1×10^8 dalton for higher plants; Table 1). Apart from the fact that genetic complexities varied by a factor of about three [0.7–2.4×10^8 dalton, a tenfold higher complexity, 1.1–1.5×10^9 dalton, has been reported for DNA from chloroplast fractions of two *Acetabularia* species (Padmanabhan and Green 1978)], claims for both homophasic and biphasic kinetics based on reassociation data have been made for ptDNA even of the same organism (Wells and Birnstiel 1969; Stutz 1970; Tewari and Wildman 1970; Bastia et al. 1971; Wells and Sager 1971; Kolodner and Tewari 1972b, 1975; Herrmann et al. 1974; Rawson 1975; Siu et al. 1975; Slavik and Hershberger 1975). In some cases repetitive sequences escaped detection and later reassociation kinetics even favored their absence (Herrmann et al. 1974; Kolodner and Tewari 1975b). With ptDNA, standard reaction temperatures which allowed maximal reassociation rates have not always been optimal for stringent duplex formation, while accidental pairing between quasi-homologous segments and self-complementary single strands or unspecific self-folding contribute to "snap-back" and give high "zero-time" values. All impede genuine reassociation (Siu et al. 1975). Plastid DNA from several higher plants, *Chlamydomonas* and *Polytoma* are known to depart from true second-order kinetics after 30–50% reassociation, and many experiments have been interrupted in this range. More stringent temperatures which minimize pseudo-hybrid formation exert a sizeable differential effect in the relative reassociation velocities of G + C and A + T-rich sequences of ptDNA. Contaminating nucDNA, due to its high complexity and cross-reaction with ptDNA, also causes deviations from second-order kinetics, especially at later stages of reassociation. Incomplete strand separation by thermal denaturation can result in quick reannealing by zippering-up (Tewari and Wildman 1970). However, if sheared DNA and alkaline denaturation are used this is not likely to account for major deviations during reassociation. Unfortunately several of these variables are capable of partially compensating for each other. Thus by varying the experimental conditions it is possible to delineate several kinds of reassociation pattern. We believe that it is impossible to calculate precisely the complexity of ptDNA from reassociation analysis without supplementary data on the organelle DNA.

2. Conformation

DNA isolated from plastids consists of circular molecules of different conformations. As monomers they can be relaxed (open) circular duplexes (Fig. 1) or covalently closed (supercoiled) duplexes. They may occur as unicircular and perhaps interlocked dimers. Higher oligomers have not been observed. The individual conformations differ in their optical and physicochemical properties. Supercoiled DNA (Fig. 3a) may be converted to the open circular shape by single chain cuts which abolish the supercoiled structure by unwinding.

Although the majority of purified circular ptDNA is found in the open form, several observations indicate that the covalently closed duplex is the predominant species in situ. For example, supercoiled DNA is released almost exclusively from *Narcissus* chromoplasts after gentle osmotic shock (Falk et al. 1974), increased yields of supercoils have been obtained from *Pisum* chloroplasts after special precautions (Kolodner et al. 1976), and complexes of protein associated almost entirely with supercoiled DNA loops which are probably attached to membrane remnants have been obtained from partially lysed chloroplasts (Fig. 3a; Herrmann et al. 1974). This suggests that release from the membrane to which DNA molecules are attached several times is one of the critical steps in the isolation of supercoiled ptDNA.

Tewari and coworkers have shown that the empirical equations used for the physicochemical characterization and separation of DNA conformations developed by Vinograd and co-workers are applicable over a wider range of molecular weight and are suitable for molecules of the size of ptDNA. Supercoiled DNA differs from the relaxed form in having a different appearance in electron micrographs and faster sedimentation in neutral and alkaline solutions, by its resistance to thermal and alkaline denaturation, decreased intrinsic viscosity, elevated buoyant density in alkaline CsCl and by its restricted uptake of intercalating dyes such as ethidium bromide (Radloff et al. 1967) or propidium diiodide (Hudson et al. 1969). Since the dye lowers DNA densities, supercoiled molecules can be separated efficiently in CsCl equilibrium gradients from linear and open molecules.

Dimer molecules are usually present in ptDNA preparations in low quantities (1–3%; Kolodner and Tewari 1975b) but have been found in developing spinach leaves with a frequency of about 15% (Herrmann et al. 1975). These values generally should be considered as approximations since the isolation procedures tend to select against large molecules. In addition, the tendency of random overlayering during spreading causes technical problems which preclude the quantitative determination of relative amounts.

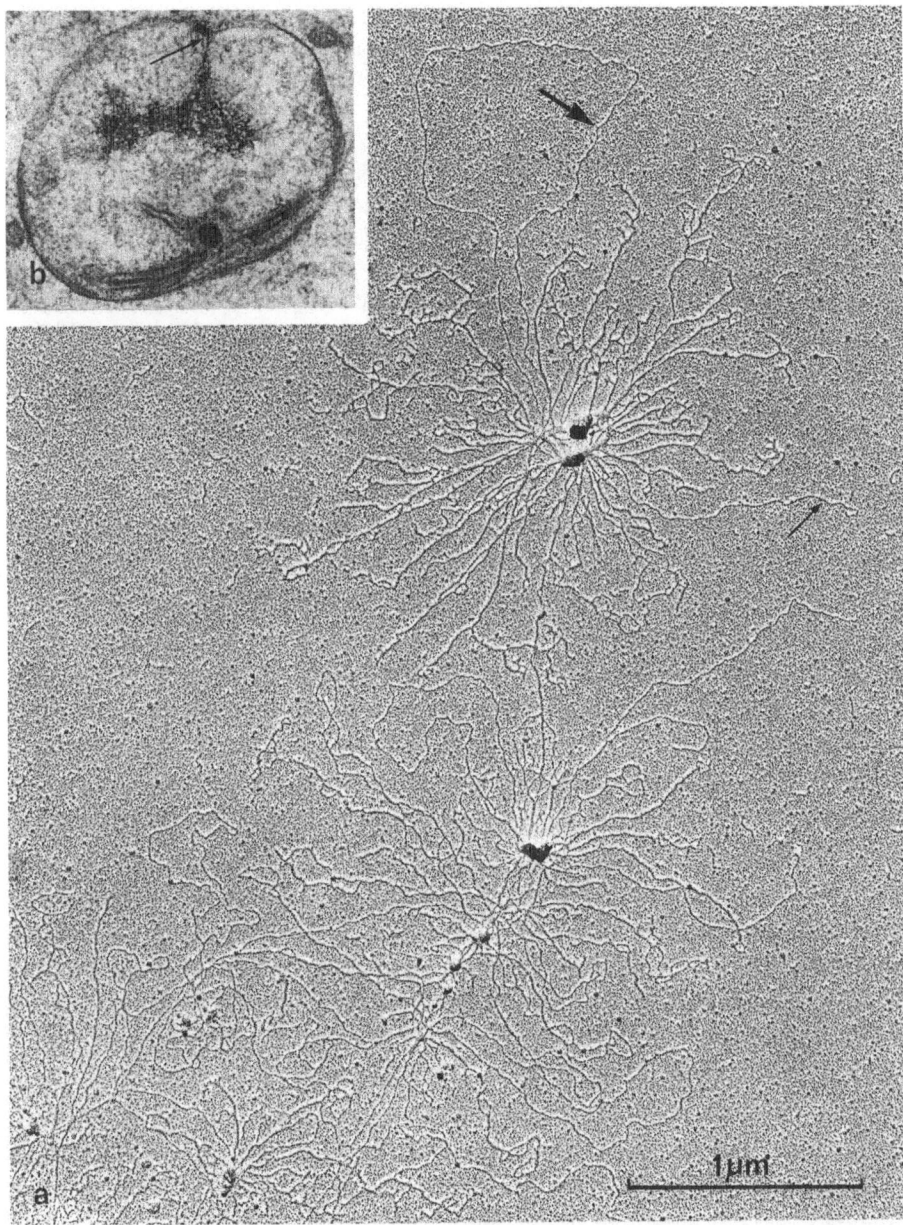

Fig. 3. a Visualization of putative membrane-associated DNA released by Triton X-100 lysis from *Spinacia oleracea* chloroplasts. The DNA protein complexes were isolated by centrifugation in an isopycnic CsCl equilibrium gradient and spread by the cytochrome monolayer technique. Note the coexistence of supertwisted and relaxed *(thick arrow)* loops in the same complex. Free ends are clearly discernible *(small arrow)* **b** Detail micrograph from a complete section series through a proplastid of *Beta vulgaris* showing a DNA-membrane association *(arrow)*. The matrix was partially removed by proteolytic digestion in order to expose DNA. The DNA is associated with protein remnants. Diameter of the plastid 1.1μm. (From Herrmann and Kowallik 1970)

Of the dimeric forms only the unicircular form is well established by its isodensity position in CsCl gradients with monomers, by length measurements (Herrmann et al. 1975; Kolodner and Tewari 1975b) and most conclusively by denaturation mapping. Mapping has established that both monomeric units are sequentially ligated head to tail (Tewari et al. 1977; Kolodner and Tewari 1979).

The situation is less clear with catenated dimers, consisting of two independent, interlocked monomers. Small proportions of this conformation have been reported to occur in ptDNA from several higher plants (Kolodner and Tewari 1975b). Their presence can only be established absolutely if the topological interlocking of the molecules can be visualized at high resolution by a combination of rotary and fixed-angle shadowing of the DNA on the grid (Hudson and Vinograd 1967). Entanglement of the large circles of ptDNA can readily occur or circular molecules may be linked by small amounts of protein (Bohnert and Herrmann 1974).

The origin of multimeric forms has not yet been established. Multimers may arise by crossing-over processes during replication or reciprocal recombination (Hudson and Vinograd 1967; Borst 1972). At present however there is no clear evidence for recombination in higher plant plastids (Kutzelnigg and Stubbe 1974). Since the amount of oligomers may alter with environmental and internal changes high levels have been interpreted as the result of structural or catalytic imbalance but this has recently been questioned (Borst 1972; Paoletti and Riou 1973).

3. Size

Constancy in the monomer sizes of ptDNA even from species that are phylogenetically widely separated was initially thought to occur as *Euglena* and all the representatives of mono- and dicotyledons examined had ptDNA molecules with circumferences in the range of 40–45μm (Manning et al. 1972). Ferns also have ptDNA of this size (Herrmann et al. 1979). However, considerable differences have now been found within Chlorophyta where there are ptDNA's both smaller and larger (Fig. 4), and in the xanthophycean species *Vaucheria sessilis*, which has ptDNA about 37μm in length (Kowallik and Hennig pers. commun.). A cyclic molecule with a circumference of about 62μm has been isolated from *Chlamydomonas reinhardii* ptDNA (Behn and Herrmann 1977). The complexity of this molecule has been confirmed by restriction endonuclease analysis (Rochaix 1978). Plastids of *Spirodela oligorrhiza* contain circular molecules of about 54μm (van Ee et al. 1979) and those of the liverwort *Sphaerocarpos donnellii* have been shown to have molecules of 37μm in length (Herrmann et al. 1980; Table 1). The size differences do not appear to follow simple phylogenetic relationships.

Circular molecules of a different order of magnitude have been reported for DNA from chloroplasts of *Euglena* which had a small number of covalently closed circles of 3μm circumference (Nass and Ben-Shaul 1972), *Acetabularia* has circles of 5μm (Green 1976) while relaxed molecules of variable sizes (0.5–4μm) have been found in *Nicotiana* (Wong and Wildman 1972). The origin of these miniature circles has not been fully established and their relationship to ptDNA needs to be substantiated by further work. Fibril measurements of the DNA of osmotically shocked *Acetabularia* chloroplasts (Green and Burton 1970; Woodcock and

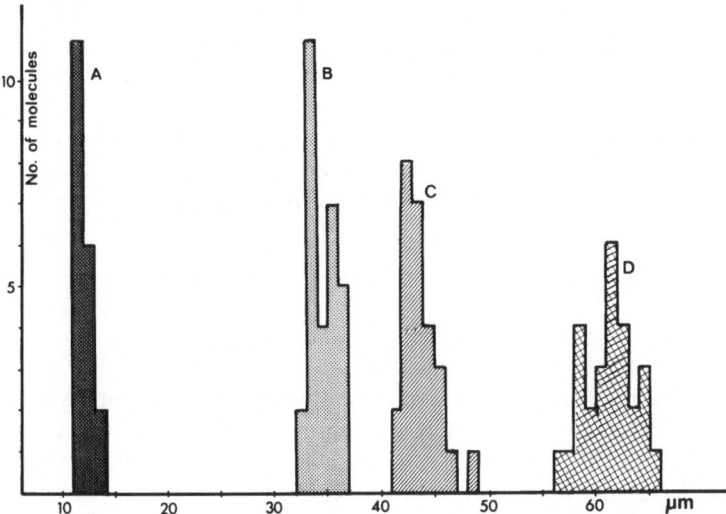

Fig. 4. Contour lengths of circular DNA monomers isolated from chloroplasts of different Chlorophyta species. *B: Sphaerocarpos donnellii* (36.9, S.D. 1.1 μm); *C: Spinacia oleracea* (43.9, S.D. 1.8 μm); *D: Chlamydomonas reinhardii* (61.8, S.D. 2.4 μm). Internal length standard: *A:* mitochondrial DNA of *Acanthamoeba castellanii* (12.5, S.D. 0.5 μm). The histogram was constructed from data obtained by mixing DNA species in pairs. The same preparation for each DNA species was used in this experiment. (From Behn and Herrmann 1977)

Bogorad 1970), the kinetic complexities of ptDNA of *Nicotiana* (Wells and Birnstiel 1969; Tewari and Wildman 1970) and of DNA from chloroplast fractions of *Acetabularia* (Padmanabhan and Green 1978), and the isolation of 45μm circles from *Euglena* chloroplasts (Manning et al. 1971; Richards and Manning 1975) and the restriction endonuclease analysis of this DNA (Gray and Hallick 1976; Kopecka et al. 1977) suggest strongly that the complexity of ptDNA of these species is different from that of the minicircles. Plastids are known to be difficult to isolate from *Acetabularia* (Bidwell 1972) so that ptDNA from this organism could be contaminated with other DNA's including those of plasmid origin as Green (1976) also suggests. Covalently closed 5 μm circles have recently been isolated from *Chlamydomonas reinhardii* mitochondria by Ryan et al. (1978). It is relevant here that the heterogeneity in nucleotide compositions of ptDNA (see Sect.IV.A) could allow for nonrandom excision at specific places as has been suggested for yeast mtDNA (Bernardi et al. 1976). Open circular molecules could also be of nuclear origin or parts of monomers with single-stranded ends, in which ring closure has been brought about by hydrogen bonding during isolation and spreading.

D. Reiteration

One of the outstanding features of the organization of plastomes is their high degree of reiteration. Evidence for this has come from two approaches, cytological studies of individual organelles and estimation of DNA levels in plastid populations. Plastome reiteration has quantitative, structural, functional, and developmental aspects. Although studies of sufficient detail on the physical organization of plastomes are confined to only a small number of species, notably *Beta* and *Spinacia,* the green algae *Acetabularia, Chlamydomonas,* the chrysophycean alga *Ochromonas,* and the dinoflagellate *Prorocentrum,* the similarity of the   data from this diverse group of organisms and supporting observations from representatives of almost all major plant phyla indicate common features of plastome organization.

1. DNA Quantity per Organelle and Cell

There are no absolute methods available to obtain accurate data on the quantities of DNA per organelle. One approach has been to extract whole-cell DNA, to subsequently separate pt- and nucDNA in CsCl equilibrium gradients, to determine the relative quantities by peak area integration in UV scans and to relate them to total cellular DNA quantity. This method can only be used where pt- and nucDNA of the cell differ sufficiently in $G + C$ content to separate in isopycnic salt gradients. In order to achieve meaningful data it is necessary to record density distributions at isoextinction wave lengths for AT- and GC base pairs, to monitor loss of DNA during purification by adding marker DNA and to check against preferential extraction of one of the DNA species. PtDNA, for example, may be attached to thylakoids (see Sect.III.D.2.b) and not easily released, an attribute which has been used to separate it from nucDNA (e.g., Brawerman and Eisenstadt 1964; Pascoe and Ingle 1978). A variation of this approach has been to determine the shift in second-order rate constant of the reassociation of radioactively labeled ptDNA in the presence of unlabeled total cell DNA (Rawson and Boerma 1976). In this way the relative proportion of ptDNA in total DNA extracts of *Euglena gracilis* cells grown under a range of different conditions has been determined.

Another approach is based on measuring DNA quantities in isolated plastids, usually measuring acid-insoluble deoxyribose with the diphenylamine reaction. There are several problems in evaluating data obtained in this way as no satisfactory tests have been devised to exclude the presence of nucDNA in chloroplast preparations. Even the application of the DNAase technique per se (Bennett and Radcliff 1975) does not guarantee its absence. "Multiorganelle complexes" (Larsson et al. 1971) which consist of aggregates of cytoplasm, organelles, and their fragments artificially entrained in membrane vesicles during homogenization can be inaccessible to the enzyme. These can represent a potential source of both nuclear and mitochondrial contamination (Blomquist et al. 1975; Schmitt and Herrmann 1977). These complexes arise most frequently in homogenates of developing tissue, which is a common source of material for the isolation of ptDNA. On the other hand, plastid envelopes when exposed to weakly hyper- or hypotonic solutions or injurious vacuolar contents such as tannins or soluble oxalates may no longer exclude DNAase (James and Jope 1979). Also stress during isolation and purification of organelles, especially during hypertonic gradient steps, may cause breakdown of envelopes and some of the organelle DNA may be lost during centrifugation and organelle resuspension. The quantitative extraction of DNA and the specificity of the colorimetric determination may also cause additional difficulties. Substances such as lipids are known to interfere in DNA determinations (Borst and Kroon 1969) and their levels in plastids may not be negligible in proportion to the level of DNA per organelle which is generally low (see Kirk and Tilney-Bassett 1978). Moreover, individual organelles within a population may possess widely differing amounts of DNA. It can be concluded that most current cell fractionation procedures are inadequate for obtaining pure fractions of structurally intact plastids for quantitative measurements of DNA and give data that are, at best, approximations.

Estimation of the DNA per organelle, determined with these methods for chloroplasts from various sources in different laboratories range from 1 to more than 200×10^{-16} g, which correspond to 1–200 chromosome copies. This extent of variation has also been reported for chloroplasts of the same organism, for example for *Euglena* (Edelman et al. 1964; Brawerman and Eisenstadt 1967; Rawson and Boerma 1976). Estimates for *Antirrhinum* chloroplasts, prepared nonaqueously and without significant contamination by nuclear material, are equivalent to 50–60 chromosomes (Ruppel 1967). *Acetabularia* where the possibility of nuclear contamination can be eliminated by enucleation has the lowest DNA content of any chloroplast so far investigated (about 40μm, Gibor and Izawa 1963). Woodcock and Bogorad (1970) and Coleman (1979) have suggested that 70% of the plastids of this organism may lack DNA. Also a low degree of reiteration has been found for *Narcissus* chromoplasts which on the average contain eight chromosomes (Liedvogel 1976). No estimates are available for the DNA content of the large chloroplasts of algal cells. Developmental changes in DNA quantities per organelle will be considered below.

2. Ultrastructure

The pioneering work by Ris (1961) on *Chlamydomonas, Zea, Elodea*, and *Anthoceros* chloroplasts established that the organization of the genetic material of plastids is similar to that of prokaryotes. Plastid nucleoplasms have therefore been called nucleoids or nuclear equivalents (Ris 1961).

In thin sections DNA is usually seen to be localized in one or several electron transparent areas of the organelle matrix. It is poorly differentiated and appears as networks of finely dispersed fibrils seemingly crossing at random. It displays considerable polymorphism with regard to outline and texture and can be more or less aggregated depending on fixation (Kislev et al. 1965). The diameter of individual fibrils is about 30Å and corresponds to that of a double-helix lacking histones. The fibrils tend to intermingle directly with the stroma. The extent to which they penetrate is difficult to establish in thin sections. The individual DNA regions are therefore rarely delineated clearly from the matrix. Occasionally, they may be surrounded by ribosomes which are possibly in contact with individual DNA fibrils (Gibbs et al. 1974a). In vivo nucleoplasms represent anionic gels. The forces which stabilize this dispersion state are largely unknown.

Some pecularities in the organization of plastid nucleoplasms have been reported from *Scripsiella* (Bibby and Dodge 1974). The nucleoids of this alga may be differentiated into a concentric "nucleolus-like" body embedded in an amorphous light matrix and surrounded by an envelope-like structure emanating from thylakoids. A transitory differentiation of a similar kind is also known from proplastid-like organelles of *Ochromonas* cells which may contain a single centrally located girdle thylakoid enclosing the DNA (Gibbs et al. 1974b).

Electron microscopic examination of plastid-containing tissues and autoradiographic observations of cells supplied with DNA precursor molecules have usually indicated that DNA is present in all plastids of cells (Wollgiehn and Mothes 1964; Yokomura 1967; Bisalputra and Bisalputra 1969; Herrmann 1970; Herrmann and Kowallik 1970; Gibbs et al. 1974b; Rose et al. 1974). A possible exception is *Acetabularia* as has been mentioned above.

a) The Nucleoid. Spatial reconstructions from electron microscopic serial sections and light microscopic autoradiographs have been used to obtain information on the three-dimensional organization of DNA in plastids. These studies suggest that the DNA of plastids may be concentrated within a single nucleoid which can be placed more or less centrally, it may be contained within a coherent, ring-shaped peripheral nucleoid or it is distributed in several nucleoids (Kowallik and Herrmann 1972a,b; Gibbs et al. 1974a). The first type of arrangement is observed in small organelle forms such as proplastids, chloroplasts of small-celled algae or chromoplasts (Fig. 3b, Jacobson 1968; Sprey 1968; Herrmann and Kowallik 1970; Woodcock and Bogorad 1970; Kowallik and Herrmann 1972; Bibby and Dodge 1974; Gibbs et al. 1974b; Liedvogel 1976). Ring-shaped nucleoids are present in five related classes of algae (Bacillariophyceae, Chrysophyceae, Phaeophyceae, Raphidophyceae, Xanthophyceae) in which plastids are characterized by the presence of a peripheral band of thylakoids located at the organelle rim, the girdle lamella (Gibbs et al. 1974a; Coleman 1979). The DNA encircles the organelle inside this girdle band. The third type represents a polyenergidic organelle and is most frequently found (see Sect.II). The number of nucleoids per plastid can be considerable and vary on a number of factors (see Sect.III.D.3). For example, Kowallik and Haberkorn (1971) estimate that there are 80–100 nucleoids in each of the large multilobate organelles of *Prorocentrum*. In polyenergidic forms, DNA regions are well separated in lamellate chloroplasts (Kowallik and Haberkorn 1971) but the strict individuality of each region is difficult to assess both in chloroplasts with grana and in organelle modifications which lack internal membrane structures. There is some evidence to be presented in the subsequent section that the DNA regions are in fact separate entities in grana-containing plastids, but it is not excluded that the individual areas may be joined by threads of DNA.

Nucleoids exhibit considerable polymorphism. They are usually spherical in plastid forms lacking an elaborate internal membrane structure. As the membrane system develops and the organelle grows nucleoplasms commonly become disc-like. They can also become elongated, branched or, in *Ochromonas* chloroplasts, they are ring-shaped. Because of the association of ring-shaped nucleoids with girdle lamellae, changes in nucleoid morphology tend to parallel those of girdle bands (Gibbs et al. 1974b). This and similar observations with chloroplasts of *Beta* (Herrmann and Kowallik 1970; Kowallik and Herrmann 1974) and *Prorocentrum* (Kowallik and Haberkorn 1971) suggest that membrane development and possibly

DNA/membrane attachment influence nucleoid shapes and arrangements. Variations in nucleoid shape are matched by similar variations in their size. In *Beta* plastids nucleoids may occupy volumes from 0.01–0.3 μm^3, in chloroplasts of *Prorocentrum* from 0.007–0.055 μm^3, in those of *Ochromonas* from 0.6–2.8 μm^3 and about 2 μm^3 in those of *Scrippsiella*.

Investigations on plastids of *Acetabularia, Ochromonas, Beta, Spinacia,* and *Prorocentrum* by different workers suggest that plastid nucleoids may be polyploid and that individual nucleoids of an organelle may differ by at least one order of magnitude in their ploidy level. Size estimates of DNA in surface-spread samples have indicated varying quantities from 40–1000μm in the small chloroplasts of *Acetabularia* containing only one nucleoid (Green and Burton 1970; Woodcock and Bogorad 1970). Gibbs et al. (1974b), on the basis of high-resolution autoradiography, have suggested that the chloroplasts of light-grown *Ochromonas* cells possess a minimum of 10–20 chromosomes. The differences in nucleoid volumes in *Beta* (Kowallik and Herrmann 1972a) and *Prorocentrum* (Kowallik and Haberkorn 1971) have been explained in a similar way. Evidence for variations in ploidy levels have also come from fibril measurements of individual nucleoids in electron micrographs of *Beta* chloroplasts cut tangentially to grana stacks, in which DNA has been selectively exposed after removal of the matrix with proteolytic enzymes (Kowallik and Herrmann 1972a). It is suggested that the chromosomes of one nucleoid may be attached to membranes in close proximity as complexes consisting of one to more than ten circular molecules have been isolated from chloroplasts of several higher plants (see Fig. 3a).

b) Structural Segregation. Ideas about segregation of DNA in plastids are largely based upon analogies with prokaryotic cells in which the membrane itself or closely contiguous structures are thought to play an active role in replication and orderly segregation of DNA (Ryter 1968; Pettijohn et al. 1973). The DNA is assumed to be in some way attached to membranes where replication occurs by passage through a membrane-located enzyme complex. Distribution of DNA could result from splitting of the attachment point possibly by intercalating membrane growth. Subsequently new matrix could then be built around the DNA. The obstinate retention of DNA in broken chloroplasts as well as electron microscope autoradiographic studies of cells labeled with tritiated thymidine suggest that such a DNA membrane association might exist (Rose and Possingham 1976a). Figure 3b demonstrates this association of DNA in a proplastid of *Beta*. A unique tongue-like thylakoid protruding into the nucleoid has been described as a possible attachment site for *Prorocentrum* chloroplasts (Kowallik and Haberkorn 1971). Sprey and Gietz (1973) illustrate it in osmotically shocked etioplasts. Further evidence has come from attempts to isolate and characterize membrane-associated DNA from chloroplasts. Under controlled lysis most of the DNA can be released into rapidly sedimenting structures containing a total filament length from a minimum of 40 to more than 500μm (Fig. 3a; Green and Burton 1970; Herrmann et al. 1974; Yoshida et al. 1978). These complexes consist of one to more than ten individual folded circular molecules. A major technical difficulty with this work is the lack of criteria to define a DNA membrane association but several lines of evidence suggest that these complexes are probably not artifacts. The structures are stable in a wide variety of conditions. Cyclic mtDNA added as probe before organelle lysis does not result in the generation of such structures. Fractions of complexes are capable of converting deoxyribonucleoside triphosphates into an insoluble, RNAase-insensitive and DNAase-sensitive form (Bohnert et al. 1974) and appear to be transcriptionally active (Hallick et al. 1976). In some cases these complexes may consist almost exclusively of supercoiled loops, one or more of which may separately unfold (Fig. 3a). This is indicative per se and suggests that rotational events may be unable to pass from one loop to the next and that multiple attachment of cyclic molecules may be involved in stabilizing the complete structure. Taken together these data suggest that the circular molecules within an organelle are organized into nucleoids and not spread randomly. Moreover, as the majority of structures have loops varying between an equivalent of 5–10 molecules, it is possible that the molecules of one nucleoid may be in close proximity or even attached to one another. If these results can be substantiated topological questions may be asked, about the internal interactions holding the loops at their core, the specificity in the constitution of loops, or transcription, replication, and segregation of chromosomes in condensed form.

3. Developmental Aspects

Consideration of changes in the amounts of ptDNA per organelle and per cell from a developmental point of view are complicated as events relating to the replication of the DNA of nucleoids, and the division of both plastids and cells are superimposed and almost certainly several control mechanisms operate at each level. The replication of ptDNA and the division of plastids may be partially independent of nucDNA replication and of cell division. However, these events cannot be completely uncoordinated, as both are influenced by genetic and environmental factors.

Changes in DNA quantities based on variations in the degree of polyenergidic organization can be accompanied by size changes of the organelle. In *Beta* nucleoids gradually increase in number during differentiation from proplastids through etioplast-like stages to chloroplasts. The reverse trend is seen in the de-differentiation of large chloroplasts to smaller organelles such as chromoplasts or leucoplasts. During the differentiation from proplastid-like organelles to chloroplasts in *Beta* there is a 25-fold increase in volume of the plastid. The data suggest that over this period there is a 10-fold increase, from 10 to more than 100 copies of chromosomes (Herrmann et al. 1974). Three-dimensional reconstructions from serial sections of this material indicate that chromosomes are concentrated in 1–2 regions in proplastids (Fig. 3b; Herrmann and Kowallik 1970) while they are found in 10–20 regions in mature chloroplasts (Kowallik and Herrmann 1972). From autoradiographic studies involving large numbers of plastids a correlation was obtained when the number of silver grains was plotted logarithmically against the corresponding organelle size (Herrmann 1970). In this study data was assembled from a range of different-sized *Beta vulgaris* leaves (from 0.5 to 11 cm in length) and from genetically different material (euploids and trisomic plants). Studies have been made of changes in the DNA contents of plastids during greening of bean (Gyldenholme 1968), corn (Mache et al. 1974), in the alga *Ochromonas danica* (Gibbs et al. 1974a,b) and in spinach (Possingham and Smith 1972; Possingham

1973b; Rose et al. 1974). During greening of bean no increase in plastid number occurred and no increase in ptDNA levels was recorded using methods that were unable to avoid nucDNA contamination of preparations. In corn both plastid numbers and ptDNA levels increased during the first four hours of greening. In greening *Ochromonas* the single peripheral nucleoids of the chloroplasts lengthen as the organelle perimeter increased. Concurrently plastids increased their DNA content since the DNA concentration per unit area as visualized electron microscopically appeared constant as nucleoid volume changed. Yokomura (1967) has noted that aging chloroplasts appear to have fewer DNA areas than do fully differentiated ones. Similar observations have been made on *Narcissus* chromoplasts (Kowallik and Herrmann 1972b; Liedvogel 1976). This suggests the possibility that small plastids of aged tissues might have less DNA.

Few studies have been made on the ptDNA of cryptogamic plants. However, the variety and changes in their organelles which occur in conjunction with metamorphosis from vegetative to sexual forms suggest that there might be similar changes in DNA content and DNA arrangement.

These results suggest that there is a general relationship between chloroplast size and their number of DNA areas. It is not known whether changes in size (or ploidy) of nucleoids influence plastid size and whether there is a relatively constant ratio of DNA quantity per nucleoid to matrix volume as there is between nucDNA amount and cytoplasmic volume (Butterfass, this vol.). The size of chloroplasts is, however, known to be influenced by age, genetic constitution, and by environmental factors such as light, temperature, and nutrition (Possingham 1970, 1973a, 1976; Possingham et al. 1972, 1975; Butterfass 1973; Cran and Possingham 1974; Marschner and Possingham 1975). It is not known whether any of these variables affect ptDNA levels either directly or indirectly via their effect on plastid size. However it is relevant here that autoradiographic studies on the large spinach plastids which form in green light indicate that they have a high level of thymidine incorporation (Possingham and Rose 1976). On the other hand the small plastids of spinach discs cultured for periods of up to ten days in darkness also have a high thymidine incorporation (Rose et al. 1975). The incorporation of thymidine into plastids of light-grown spinach discs during a period of active cell expansion can be correlated with the rate of chloroplast division. In this particular system all chloroplasts divide and a doubling of plastid number is associated with a doubling of total chloroplast silver grains per cell. Over this period chloroplast size remains approximately constant but chloroplast numbers change tenfold from 15 to 150 per cell (Rose et al. 1974; Possingham and Rose 1976). Thymidine incorporation studies indicate that both nucDNA and ptDNA are synthesized during greening. NucDNA synthesis usually precedes ptDNA synthesis and is probably necessary to support the increased population of chloroplasts. PtDNA synthesis is associated with chloroplast division but there is no obligatory coupling between the two processes (Kowallik and Herrmann 1974; Rose et al. 1975).

At least two mechanisms influence the degree of plastome reiteration at the cellular level. A correlation has been found between the volume of plastids on a cell basis and the level of nuclear ploidy. With doubling of the genetic material of the nucleus, plastids may respond in two ways. They may approximately double in number while retaining their size, or they may approximately double their volume without changing their number. Transition states between these two extremes may exist, especially in organisms of imbalanced genetic constitution (mutants, trisomic plants). A consequence of DNA content and organelle size being correlated is that an approximately constant ratio of genomes and plastomes may be maintained at least for a given differentiation level (Butterfass, this vol.).

The ratio of genomes and plastomes itself may however be subject to variation and fluctuate during ontogenesis and in response to environmental changes. For example, in many vascular plants parenchyma cell chloroplasts may number up to several hundred, while the number of proplastids in meristematic cells or in egg or generative cells is lower, being between 15 and 60 (Diers 1965, 1967; Possingham and Saurer 1969; Meyer and Stubbe 1974). In situations where the nucleus maintains its ploidy level during cell expansion plastid number and size and, implicitly, the ratio of pt-to nucDNA increases. For example, Lamppa and Bendich (1979) estimate 1.3% of ptDNA per total DNA in young shoots to 7.3% in fully green shoots in pea. Data on *Acetabularia* (Shepard 1965), *Euglena* and *Chlamydomonas* indicate that this mechanism may also apply to lower plants. Rawson and Boerma's (1976) data show that the number of ptDNA molecules per *Euglena* cell can vary from approximately 200 in heterotrophically dark-grown to about 500 in heterotrophically light-grown cells. Cells grown autotrophically in the light can contain approximately 1000 ptDNA molecules. Gametes and vegetative cells of *Chlamydomonas* contain approximately the same DNA quantity per cell (7.2×10^{10} dalton) but ptDNA represents 6% in the former and 14% in vegetative cells, i.e., about 25 or 55 chromosome copies, respectively, per chloroplast (Chiang and Sueoka 1967; Wells and Sager 1971). Similar data are available for *Sphaerocarpos* sperm and thallus tissue (Herrmann et al. 1980). The possibility that plastids can divide without concomitant DNA replication has been suggested by Boasson and Laetsch (1969). They found that the plastids of etiolated *Nicotiana* leaf discs can divide when cultured in the presence of the DNA synthesis inhibitor 5-fluorodeoxyuridine. Chloroplast division also takes place in the cotyledons of gamma plantlets of spinach (Rose and Possingham 1976). These were grown from seeds which had been exposed to 500 K.R. gamma irradiation which suppressed all cell division. Also relevant are observations such as those on regenerating moss where sudden increases in plastid number with concomitant decrease in plastid size take place after injury (Correns 1899; MacNutt and Maltzahn 1960). It seems certain that plastids may possess widely different amounts of DNA and still have a complete set of genes.

Problems of plastid continuity and reproduction during the sexual cycle are complex and involve both uni- and biparental inheritance. Information on the fate of ptDNA during this period is limited. Mainly morphological evidence suggests several mechanisms that may account for uniparental effects. A few examples may illustrate this. Male gametes may lack plastids because they are sorted out during pollen development or gametangiogenesis or may be eliminated from the sperm cell (*Catasetum, Phajus:* Chardard 1969; *Hordeum:* Cass and Karas 1975). Plastids may be lost or eliminated from the sperm (Clauhs and Grun 1977). In *Sphaerocarpos* neither sperm plastids nor mitochondria appear to enter the egg (Diers 1967). There is also the possibility that plastids of the female degenerate in the egg cytoplasm, as appears to occur in Gymnosperms (*Pinus* and *Larix:* Camefort 1969; *Biota:*

Chesnoy 1969). Even where plastid transfer of the male gametes is established (Meyer and Stubbe 1974) uniparental effects might still be expected if there are vastly unequal numbers of organelles in the zygote or if there are differences in the multiplication rate of organelles derived from the parents (Stubbe et al. 1978). In *Chlamydomonas* Sager and Lange (1972) have raised the possibility of temporal base modification in ptDNA as part of an organelle-located restriction/ modification system involved in uniparental inheritance. These authors observed that after fusion of isotope labeled gametes (including the fusion of the two chloroplasts, Cavalier-Smith 1970) DNA from the female (mt+) parent decreased in density in CsCl gradients, while that of the male (mt−) remained apparently unchanged and disappeared soon after mating. It should be noted that none of these mechanisms works absolutely. In fact, occasional biparental transmission in organisms showing uniparental inheritance of plastids has been repeatedly observed. It is also relevant in this context that in *Chlamydomonas* there is a discrepancy between a genetically diploid non-Mendelian linkage group affecting chloroplasts and a high number of circular ptDNA molecules per cell which requires still an explanation (Wells and Sager 1971).

In conclusion, most available evidence suggests that the developmental versatility of plastids is accompanied by changes in their genetic material. These may be qualitative in nature during the sexual phases. Almost certainly there are recurring patterns in DNA arrangement during the ontogenetic succession of the organelles and these can involve a progressive modification of DNA quantity. The large variations in organelle size and in DNA content per organelle and per cell during ontogenesis as well as the selective advantages of polyenergidic organization are poorly understood. The resolution of these questions will almost certainly contribute to our understanding the functional implications of the organelle fine structure. From a genetic point of view these changes are probably secondary phenomena. The common principle superseding morphological versatility of plastids is a high degree of plastome reiteration.

E. Replication

Little factual information is available about DNA synthesis in plastids. Structural details, localization, function and genetic origin of associated components are largely unknown. The replicative intermediates found in ptDNA suggest that DNA synthesis occurs within the organelle. Intermediates of the Cairns type have been reported from ptDNA of *Euglena* (Richards and Manning 1975) and those of the Cairns and rolling circle type from *Pisum* and *Zea* (Kolodner and Tewari 1975a). In the latter two ptDNA's replication appears to be initiated by the formation of two D-loops and to proceed bidirectionally. The replicative intermediates and $^{14}N/^{15}N$ transfer experiments in synchronized *Chlamydomonas* populations (Chiang and Sueoka 1967) and pulse-chase experiments with spinach (Rose et al. 1974) suggest that the replication mode is semi-conservative. In *Chlamydomonas* ptDNA is replicated at a different stage of the cell cycle to nucDNA. PtDNA is replicated during the light phase possibly utilizing photophosphorylation as energy source while nucDNA is replicated at the beginning of the dark phase (Chiang and Sueoka 1967). Autoradiographic studies indicate a similar timing in *Euglena* cells (Cook 1966). In spinach and wheat thymidine incorporation studies suggest that ptDNA is synthesized both in the light and in the dark, but particularly during the phase of cell expansion when plastid division occurs (Rose et al. 1974; Possingham and Rose 1976; Possingham 1976; Boffey et al. 1979). Thymidine kinase has been found in the soluble fraction of plastid proteins of *Chlamydomonas* (Swinton and Hanawalt 1972) and rye (Golazewsky et al. 1975). This enzyme has not been found in the nucleo-cytosolic compartment of the alga (Swinton and Hanawalt 1972).

Little information leading to an understanding of DNA synthesis in plastids has been provided by studies involving isolated organelles as the incorporation capacity of these organelles tends to be low. Although isolated plastids are known to incorporate externally added precursor nucleoside into DNA in a reaction that proceeds only in structurally intact organelles and which is strictly dependent on photophosphorylation (Bohnert et al. 1974), it is not known whether this incorporation reflects replication, repair activity or even terminal addition of nucleotides to DNA. Fractions of isolated chloroplasts of *Nicotiana* (Tewari and Wildman 1967), *Spinacia* (Spencer and Whitfeld 1969) and *Euglena* (Scott et al. 1968) have also been shown to incorporate radiolabeled deoxyribonucleoside triphosphates into insoluble material. The incorporation required the presence of all four nucleotides and magnesium ions and was inhibited by DNAase, but not by RNAase. A possible objection to studies on incorporation of labeled deoxyribonucleoside triphosphates which penetrate plastid envelopes with difficulty has been that soluble extraplastidic polymerases (see e. g., Stout and Arens 1970) of homogenates may become associated with DNA of broken organelles. There is thus a risk of their particle-bound sedimentation and subsequent reaction with organelle DNA (Bohnert et al. 1974). Moreover, in experiments with isolated *Nicotiana* chloroplast fractions a DNA component was synthesized which was probably mtDNA since it was separable from and heavier than nucDNA in CsCl equilibrium gradients. Subsequent investigations established that pt- and nucDNA of this plant are not sufficiently different in density to be separable in CsCl gradients (Whitfeld and Spencer 1968; Wells and Birnstiel 1969; Tewari and Wildman 1970). In spinach, in which the densities of both components also resemble each other closely, the analysis was based upon differences in reassociation velocity and accompanying band shifts in CsCl gradients (Spencer and Whitfeld 1969). However this approach suffers from the possibility of cross-reaction between both DNA's (see Sect.III.A).

IV. The Circular Chromosome

Plastid chromosomes with their species-specific 115–200 kbp are large enough to accomodate genes for 180–280 polypeptides with an average molecular weight of 20 kbp. This figure accounts for sequence repetition. It may be an overestimate if noncoding nucleotide sequences are present, it may be an underestimate when genes overlap as in virioms, both complementary strands bear overlapping information, or if splicing of sequences of different genes generates novel information.

Restriction endonuclease techniques have considerably widened our understanding of ptDNA. Because of the site-specific cleavage of type II restriction endonucleases the patterns produced are characteristic for a specific DNA molecule and small changes in nucleotide sequence may be reflected in changed fragment patterns. Each cut of such an enzyme represents therefore a physical marker that can be related to functions of the molecule or, on a comparative scale, to evolutionary events. Restriction endonucleases have therefore been used to estimate genetic complexities of ptDNA (see Sect.III.C.1), to construct physical maps of plastid chromosomes which may serve as basis for gene maps, to study mutant ptDNA, to obtain DNA fragments suitable for multiplying in plasmid vectors, to obtain information on the ontogenetic stability of ptDNA (see Sect.V.B) and on changes during species evolution, and for studies involving uni- or biparental transmission, sexual or somatic hybrids.

A. Anatomy, Evolution

Restriction endonuclease cleavage maps, constructed by serial ordering of DNA fragments, have been reported for the ptDNA of *Euglena* (Gray and Hallick 1978), *Chlamydomonas* (Rochaix 1978), *Zea* (Bedbrook and Bogorad 1976), *Spinacia* (Fig. 5, Hobom et al. 1977) and for the DNA's of the five *Euoenothera* wild plastomes (Gordon et al. 1980a,b; see Sect. V). Although the number of species studied in this respect is small, four interesting points with regard to anatomy and evolution of ptDNA emerge from the comparison of these maps and of additional physical parameters.

First, ptDNA contains large repetitive sequences. Sequence repetition first became apparent when Thomas and Tewari (1974) established that the gene dosage for rRNA was repeated. In all examined Chlorophyta DNA's this repetition consists of a duplication, in *Euglena* DNA of a triplication. The repeat regions make up about 20–30% of the monomer length. A second category of repetitions, i.e., relatively short and dispersed inverted sequences, has been reported for ptDNA of *Chlamydomonas* (Gelvin and Howell 1979). The significance of the repetitions is not understood.

Secondly, all ptDNAs are segmentally organized into repetitive and nonrepetitive regions. The Chlorophyta ptDNA's so far examined all share a common structure which is illustrated by Fig. 5. This is characterized by an inverse duplicated region, whose units are not contiguous but separated by two single-copy regions of varying lengths. This organization divides the molecule into four well-defined regions. The inverted orientation has been demonstrated by physical mapping (Bedbrook and Bogorad 1976; Hobom et al. 1977; Rochaix 1978), and visualized by electron microscopy (Bedbrook et al. 1977; Tewari et al. 1977). It is not known whether the orientation of both single-copy regions relative to each other is uniform or exists in two versions.

It appears that the repetitive and large single-copy regions of the various DNAs differ little in their size (13–15 Md and 48–52 Md, respectively). On the other hand the single-copy region which is situated close to the terminal of the rDNA regions (see Sect. IV.B) varies in length and measures about 52 Md in

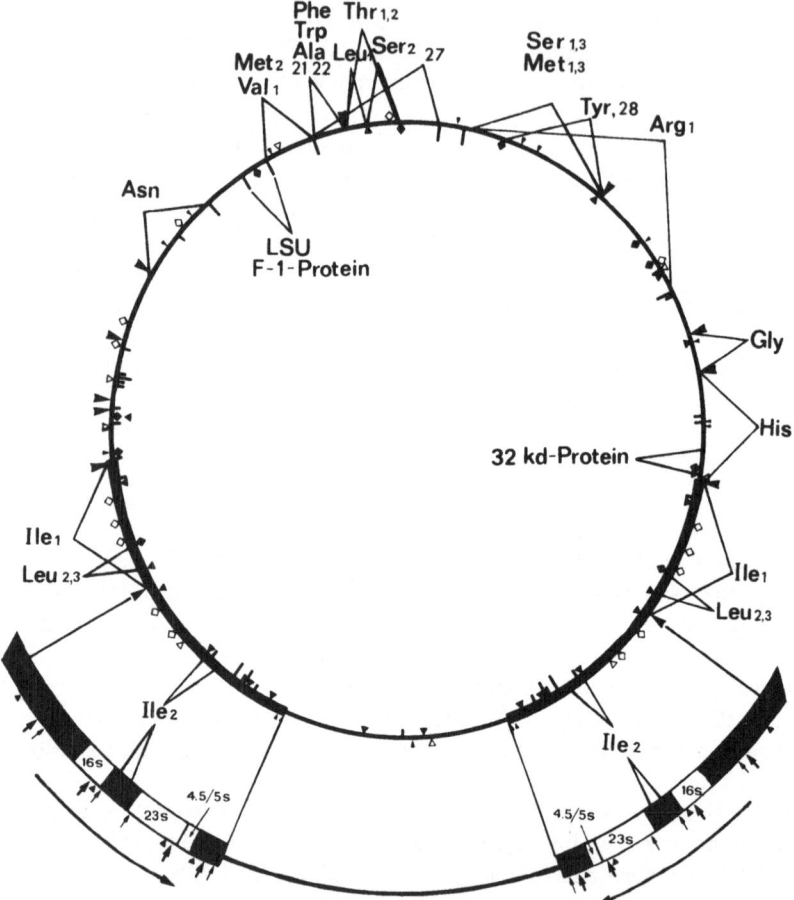

Fig. 5. The circular physical map of the *Spinacia oleracea* plastid chromosome. The chromosome accounts for about 95 MD or 140 kbp; the inverted repeat is demonstrated by thick arc regions. The map order of the various DNA fragments was determined by reciprocal digestion with several restriction endonucleases. The relative cut positions of endonucleases are drawn to scale and indicated by the following symbols: Sal I = ▼, Pst I = ♦ , Kpn I = Ⅰ, Xma I = ▾ , Xho I = ı, Bgl I = ℃ , Pvu I = ∘ , Sac I = ▾ . The location of the two sets of genes for ribosomal RNA within the inverted repeat is indicated in the expanded drawing. The symbols specify the cleavage sites of EcoRI = ↑, Bam HI = ↓, Sma I = ▾ . The polarity of transcription in the rDNA region preceeds from the 16S gene towards that of the 23S (*arrow*). For the mapping of tRNA and polypeptide genes the smallest DNA segment to which hybridization was observed is indicated for each single component

Chlamydomonas, 15 Md in *Spinacia* and 13 Md·in *Zea* ptDNA. It is therefore this region that is mainly responsible for the differing monomer contour lengths between species. It has not been established whether the size differences refer to a single and perhaps AT-rich sequence insertion as in mtDNA of various *Drosophila* species (Fauron and Wolstenholme 1976). Variations in the A+T contents of ptDNA from *Chlamydomonas* (65 mol%), *Zea* (61 mol%) and from *Spinacia* (63 mol%) would roughly relate to their respective contour length differences (Table 1).

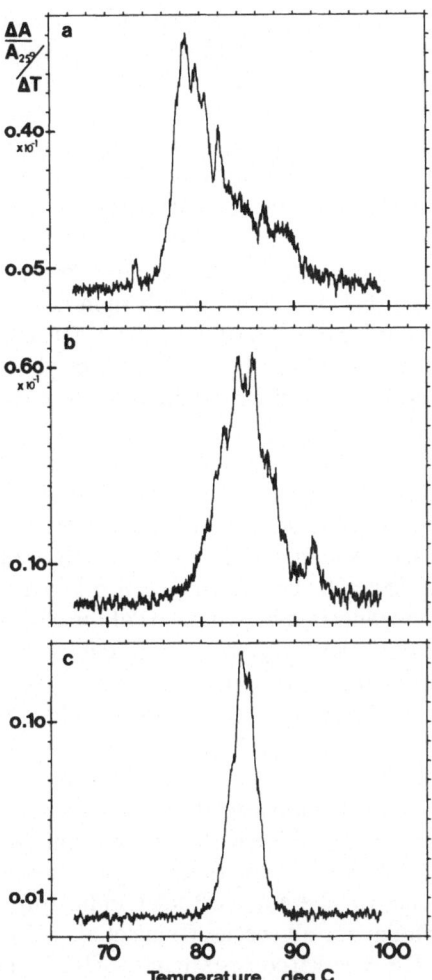

Fig. 6a–c. Derivative plot of high resolution thermal denaturation profiles of plastid DNA of a *Euglena gracilis,* **b** *Spinacia oleracea,* **c** T4 DNA. Note the broad and heterogenous helix-coil transition of the higher plant ptDNA. Solvent: standard saline citrate (0.15 M NaCl, 0.015 M Na-citrate, pH 7.2), heating rate $0.1°$ C min^{-1}, thermal increment $0.045°$ C, $A_{260, 25°C} = 0.85$ **a,** 0.91 **b,** 0.84 **c.** (From Crouse et al. 1978)

The plastid chromosome of *Euglena* shows fundamental differences in its anatomy from that of higher plants. It can be divided into a large single-copy region and a triplet repeat of 3.3 Md (5.6 kbp) arranged in tandem fashion. It is not known at present whether the repeat units are separated by short nonrepetitive sequences.

The third point of interest is that plastid chromosomes are characterized by prominent heterogeneities in their base composition. Compositional heterogeneity has been well documented from melting and pycnographic analysis for the ptDNA of *Chlamydomonas reinhardii* (Bastia et al. 1971; Wells and Sager 1971), *Euglena* (Fig. 6a; Slavik and Hershberger 1976; Stutz et al. 1976) and *Polytoma* (Siu et al.

1975). Corresponding data for higher plant ptDNA's are controversial. The temperature/absorbance curve of *Spinacia* ptDNA shown in Fig. 6b is broad, asymmetric, and clearly polymodal and the extent of its compositional heterogeneity, as deduced from profile width, is similar to that of *Euglena* ptDNA. Recent pycnographic analysis of DNA fragments of known map position confirms this heterogeneity and demonstrates that in both DNA's the heterogeneity occurs in blocks (each of which in turn exhibits heterogeneity) which coincide with repetitive and nonrepetitive regions (Crouse et al. 1978). Repetitive regions are characterized by a high $G+C$ content. By contrast, most of the reported thermal dissociation profiles of higher plant ptDNA's show smooth sigmoid transitions with spreads over varying degrees, even for ptDNA of the same organsism (Kung and Williams 1969; Kolodner and Tewari 1972b, 1975b; Knoth et al. 1974). It is likely that insufficient thermal resolution and contamination by nucDNA (see Sect.III.A) accounts for some of these differences.

There is good evidence to suggest that the topography of heterogeneity observed in *Spinacia* ptDNA is common to many Chlorophyta ptDNA's. Tewari et al. (1977) have mapped by electron microscopy the pattern of single- and double-stranded regions in partially denatured circular ptDNA of various mono- and dicotyledons including spinach. Except in *Pisum* ptDNA more rapidly looping (AT-rich) regions are concentrated in two thirds of the chromosome, between two distinct undenatured (GC-rich) regions which could constitute a duplicated segment. Also the thermal pattern of spinach ptDNA shown in Fig. 6b is similar to that of *Chlamydomonas* ptDNA in respect to profile width and mass fraction of the main subtransitions (Bastia et al. 1971; Wells and Sager 1971). Further evidence has been obtained from rRNA genes which are localized in the repetitive regions and which occupy stretches of the highest known $G+C$ content of the molecule (Crouse et al. 1978). Their number and position within the molecule appears to be conserved (see Sect.IV.B). Also ptDNA of a wide range of plant species will hybridize with apparently equal efficiency to different pt-rRNA's, indicating that the base sequence of rDNA is likewise similar for all species (Ingle et al. 1970; Thomas and Tewari 1974). Similarly, pt-4S RNA which consists mainly of tRNA hybridizes to relatively AT-rich areas of regions of average base composition in ptDNA of *Pisum* (Tewari et al. 1977) and *Spinacia* (Crouse et al. 1978). It is of interest to note that the association of rRNA and tRNA genes (Schwartzbach et al. 1976) with $G+C$ and $A+T$-rich stretches in *Euglena* follows a similar pattern.

The fourth point that can be made is that in spite of a certain evolutionary conservation in the anatomy of Chlorophyta ptDNA's the primary nucleotide sequences undergo fairly rapid changes. This implies that plastomes have evolved at significant rates. Comparisons of the cleavage patterns of various DNA's after digestion with restriction endonucleases have established this divergence. Even DNA's from closely related and still interbreedable species show identifiable variations in their fragment patterns. The five *Euoenothera* plastome DNA's, for example, still show only a few differences on the molecule (Fig. 7) most of which are located in two areas of the large single-copy region. These differences might have contributed to the specificity of the genome/plastome interaction and consequently to the speciation of this genus (Stubbe 1959; see Sect.V). The observation that the number but not the position of cuts of three restriction

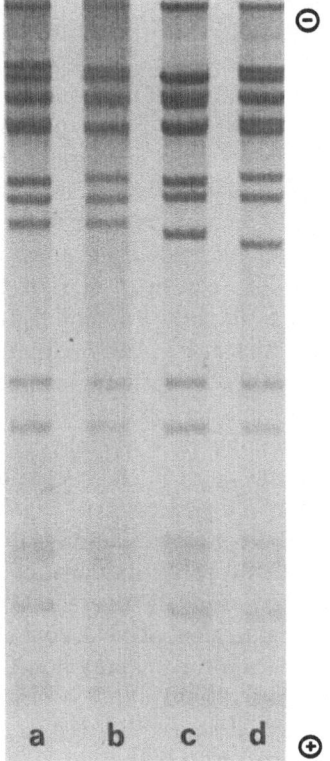

Fig. 7. Agarose gel (1.0%) of Sal I plus Pst I digests of *Euoenothera* ptDNAs. *a* Plastome I-, *b* II-, *c* IV-, *d* V-DNA. The fragment patterns are basically similar but specific differences between the different DNA's are apparent. Note that the phylogenetically closely related DNA's of plastome I and II show a high degree of similarity. (In collaboration with K. Gordon)

endonucleases remains constant in the five *Euoenothera* ptDNA's could indicate that recombination occurs in higher plant plastids though as a rare event. Comparisons of ptDNA of widely separated species such as *Chlamydomonas* and spinach by semi-quantitative co-reassociation and re-melting analysis indicate that these DNA's have diverged to an extent where only 20–30% of heteroduplex molecules can be formed (Herrmann, unpubl.).

Generalizations about the nature of all these differences cannot as yet be made although there is little doubt that they are of considerable functional and evolutionary importance. Atchison et al. (1976) suggest that the magnitude of these differences is correlated with cross-compatibilities between species. Furthermore, the general rate of plastome evolution has been claimed to be lower than that of the nucleus, a view which is compatible with the high degree of plastome reiteration (see Sect.III.D.3). Support for this view comes from the observation that several *Nicotiana* (Atchison et al. 1976; Vedel et al. 1977) and *Triticum* (Vedel and Quetier 1978) species possess apparently identical ptDNA's as indicated by restriction endonuclease analysis, or that *Euoenothera* plastome I is adapted to several A genotypes (Stubbe 1959; see Sect. V). However small differences between ptDNA's of closely related species may not be easily detectable biochemically or genetically. The observation that two related species of *Cucumis* and three of *Lupinus* possess different ptDNA's (Atchison et al. 1976) as well as the genome/plastome co-evolution model of Stubbe (see Sect. V.E) suggest that sequence divergence in ptDNA may not always follow a simple taxonomic pattern.

Collectively the present data suggest that ptDNA shares many properties with other genetic material. The differences in monomer sizes between species (Table 1; see Sect.III.C) demonstrate that evolutionary changes in ptDNA can involve loss and perhaps gain of nucleotide sequences. It is not known whether the size differences reflect differences in the coding potential and whether the functions of ptDNA's of different species are similar. Although only a few species have so far been examined, the fact that these include members of algae, mono- and dicotyledons suggest that the anatomy and gross intramolecular compositional heterogeneity of many Chlorophyta ptDNA's appear to have been conserved in an evolutionary sense. It would be of interest to know whether *Sphaerocarpos* ptDNA, which is known to be of reduced length or *Pisum* ptDNA, which appears to lack an inverted repeat (but as other higher plant ptDNA's contains two copies of genes for rRNA) also show this type of organization. The size reduction in the liverwort ptDNA of about 15% and low average $G + C$ content of this DNA compared with that of most higher plant ptDNA could have resulted from the loss of one of the repetitive units. Data for the only non-Chlorophyta studied in this respect, *Euglena*, suggests that plastid chromosomes from other phyla may have fundamentally different designs. PtDNA contains conservative and nonconservative regions both with respect to gene position and primary nucleotide sequence, and its genes can be differentially expressed as will be outlined subsequently. The comparison of positions of genes for 4S RNA's in the plastid chromosomes of spinach and *Chlamydomonas* indicates that the arrangement of functional regions is not completely conserved in the evolution of ptDNA (see Sect.IV.C) as one could infer from arrangements of rRNA genes and polypeptide genes (see Sects.IV.B and IV.D). In *Chlamydomonas* the genes for 4S RNA appear to be distributed throughout both single-copy segments while none has yet been detected in the small single-copy region of the spinach chromosome. Also, the primary sequence of the corn plastid 16S rRNA gene shows 76% homology to that of *E.coli* and is about 50 nucleotides shorter (Schwarz and Kössel 1979). A comparison of the 5'-terminal region of the 16S rRNA genes of corn and of *Euglena* ptDNA reveals, however, both considerable homology, including the possibility of hairpin formation as well as divergence in distinct parts of the gene (Schwarz et al. 1979; see Sect.IV.E).

B. The rRNA Operon

More information about the functions of plastid chromosomes is available for the RNA components of plastid ribosomes than for any other genes. In common with mitochondria, plastids possess their own complement of ribosomes which belong to the prokaryotic (70S) type and which are thus different in their functional and physicochemical properties to those of the cytosol which are 80S type. Unlike mitochondrial ribosomes (Leaver and Pope 1977) plastid ribosomes can provide about 50% of the total cellular ribosome complement and make a major contribution to cellular protein biosynthesis (Ellis 1976). By virtue of their accessibility genes for ribosomal RNA were the first plastid-specific gene products to be identified (Scott and Smillie 1967; Tewari and Wildman 1968). Their study gave the first indication that plastid genes are active (Hartley and Ellis 1973; Bohnert et al. 1974) and established the existence of gene repetition (Thomas and Tewari 1974), of polycistronic transcription units including transcribed spacers and RNA modification (Bohnert et al. 1974), and of sequence inserts within genes of ptDNA (Rochaix and Malnoe 1978).

The ribosomes of plastids of many higher plants commonly contain four RNA types. Apart from the high molecular weight 23S and 16S RNA there are two low molecular weight components, the 5S and a recently discovered 4.5S RNA (Bohnert et al. 1976; Dyer et al. 1977; Whitfeld et al. 1978b). The latter RNA species is not a tRNA nor related to 5S RNA (Dyer et al. 1977). The first three RNA species have also been described in plastid ribosomes from *Euglena* (for the 5S RNA see Phillips and Carr 1975) and *Chlamydomonas* and for the latter Rochaix and Malnoe (1978) have described additional 3S and 7S RNA's. All the different RNA types are complementary to ptDNA.

Hybridization plateaus (Thomas and Tewari 1974) and physical mapping of rRNA genes have established that there are two genes for 16S and 23S rRNA per Chlorophyta DNA molecule. Earlier data which indicated the presence of only one gene are incorrect or need to be re-evaluated. The gene dosage for the 4.5S and 5S rRNA's remains to be established. Current data vary between two and eight sites (Bohnert et al. 1979b; Hartley 1979; Hartley and Head 1979). There is now also agreement for three rRNA genes in *Euglena* ptDNA (Gray and Hallick 1978; Jenni and Stutz 1978; Rawson et al. 1978). Jenni and Stutz (1979) have recently raised the possibility of a fourth gene for 16S rRNA.

Genes for ribosomal RNA have been physically mapped on the plastid chromosome of *Zea* (Bedbrook and Bogorad 1976; Bedbrook et al. 1977), spinach (Crouse et al. 1978; Whitfeld et al. 1978), *Chlamydomonas* (Rochaix and Malnoe 1978) and *Euglena* (Gray and Hallick 1978; Jenni and Stutz 1978; Rawson et al. 1978) by electron microscopy or hybridization of radioiodinated RNA species to DNA fragments of known map position. In all cases the genes of the various rRNA's are closely associated with each other as they are combined into sets. This was previously suggested from transcription studies with isolated spinach chloroplasts (Hartley and Ellis 1973; Bohnert et al. 1974). The sets are positioned in the repetitive segments of the molecule (Fig. 5) and occupy about 35% of the repeat in Chlorophyta ptDNA's and about 80% in *Euglena* ptDNA. Their arrangement is conserved over a wide evolutionary range as in the higher plant DNA's studied they show a similar disposition in that they are oriented inversely towards the end of the duplicated segment, in the vicinity of the small single-copy region. [As mentioned *Pisum* ptDNA may be a possible exception (Kolodner and Tewari 1979)]. The rDNA sets are separated by about 26 kbp in spinach ptDNA, by 18 kbp in maize ptDNA and are located almost opposite in the *Chlamydomonas* chromosome due to the size similarity of its single-copy regions. The arrangement of the rDNA units in *Euglena* ptDNA is different, the gene sets occur as triplets in tandem fashion and are separated from each other by about 0.8 kbp (Gray and Hallick 1978; Jenni and Stutz 1978; Rawson et al. 1978). The fourth 16S rRNA gene suggested by Jenni and Stutz (1979) lies in the vicinity of the triple gene cluster close to another gene for 16S rRNA.

All plastid rRNA units hitherto examined exhibit a uniform and essentially bacterial type of organization. The sequence of genes is 16S, 23S and 5S, whereby the genes for the two large rRNA's are separated by a spacer sequence of largely unknown function (Fig. 5). An additional small spacer sequence of approximately 350 bp may separate the genes for 23S and 5S in spinach ptDNA (Bohnert et al. 1979b). The genes are encoded in only one strand (Bedbrook et al. 1977; Rochaix and Malnoe 1978) and are nonoverlapping (Thomas and Tewari 1974). In isolated spinach chloroplasts it has been shown that the rDNA region is transcribed polycistronically (Hartley and Ellis 1973; Bohnert et al. 1974, 1976) and the transcript which exceeds the mature rRNA's by about 30% in length is processed subsequently (Bohnert et al. 1974). The transcription polarity proceeds in the order given (Bohnert et al. 1976).

Demonstration of transcription and multistep processing in isolated chloroplasts requires both the preparation of unbroken organelles and that material from which chloroplasts are isolated is competent for rRNA synthesis. If transcript maturation is to be achieved it is essential to preserve not only the morphological but also the physiological integrity of organelles (Bohnert et al. 1974, 1976).

The existence and modification of polycistronic rRNA transcripts implies that the pt-rDNA region is limited by promotor and terminator sequences and that it contains signals for RNA processing (see Sect.IV.E). About half of the transcript excess sequences originates from the intercistronic region between the genes for 16S and 23S rRNA (thus constituting transcribed spacer, Bohnert et al. 1976) and a further part may represent leader sequence in front of the 16S region (Schwarz and Kössel 1979). The intercistronic spacer is about 200 bp in ptDNA of *Euglena* and, depending on the species, between 1750 and 2400 bp in Chlorophyta ptDNA's which is equivalent to about 0.1–1.5% of the chromosome length. The significance of the spacer variation which contrasts that of the rRNA genes is not understood. The spacer of spinach ptDNA carries centrally the gene for one of the two isoaccepting tRNA species for isoleucine (see following section). It is not known whether it is on the rRNA coding strand. The *Chlamydomonas* spacer also carries sites for unidentified 4S RNA (see following section).

Position, function and origin of the 4.5S rRNA which may represent a heterogeneous population of molecules is as yet unknown. It is conceivable that this RNA as well as 3S and 7S RNA species found in *Chlamydomonas* ribosomes are processing remnants or fragments of aging rRNA's. The nucleotide sequences of the latter two RNA's reside in the intercistronic spacer between the genes for 16S and 23S RNA (Rochaix and Malnoe 1978). In spinach the 4.5S RNA originates from positions either between the 23S and 5S RNA gene and/or from within the 23S RNA gene (Whitfeld et al. 1978; Bohnert et al. 1979b). It is not known whether maturation of the multimeric RNA transcript starts in vivo before transcription is completed, that is, still during the association of the transcript with DNA, as in prokaryotes(Perry 1976). Some processing events appear to occur on ribosome-like particles as p23 and p16, precursors to 23S and 16S rRNA, respectively, are found in populations of chloroplast ribosomes (Driesel 1975).

Posttranscriptional modification in *Chlamydomonas* chloroplasts may include splicing (ligation) in addition to nucleolytic activity. The 23S rRNA gene in this organism is interrupted by a 0.97 kbp intervening sequence lacking in the mature rRNA (Rochaix and Malnoe 1978; see Sect.IV.E). No intervening sequences have been found in the analogous region of the corn plastid chromosome (Bedbrook et al. 1977).

C. Transfer RNA Genes

The plastid matrix contains tRNA's and the cognate aminoacyl-tRNA syntheses which exert a central function in the information transfer from organelle DNA to protein. As plastid ribosomes they have largely prokaryotic properties and are therefore distinguishable from their cytosolic counterparts by their preferential acylation by homologous and prokaryotic synthetases, their chromatographic and pherographic properties, their nucleotide sequence where examined, and possibly also by the response of their genes to environmental variations such as light. Little is known about origin, site, and regulation of synthesis of the plastid-located aminoacyl-tRNA ligases. The available evidence indicates that at least some of them originate in the nucleo-cytosolic compartment. However a considerable number of individual tRNA's, including isoaccepting species and a formylated tRNA^{met}, from a variety of higher and lower plants, have been identified by amino acylation and chromatographic separation. Their complementarity to ptDNA has been shown by hybridization and remelting of hybrids. Methionine-accepting tRNA's, specifically capable of being enzymatically formylated, are involved in polypeptide chain initiation in prokaryotic systems. The cytosol lacks formylating activity. Little is known about the distinction of plastid tRNA's and their synthetases from their prokaryotic equivalents in plant mitochondria. In general, cross-hybridization of plastid tRNA's to nucDNA has not been observed. However even if hybridization conditions and purities are sufficiently controlled nuclear gene dosage may be too low to detect the putative hybrids. All these topics have been critically reviewed by Weil et al. (1977), Barnett et al. (1978) and Wollgiehn and Parthier (this vol.).

The central question of whether ptDNA codes for a full complement of tRNA genes, that is for tRNA's for all the 20 amino acids that constitute proteins plus one for chain initiation, remains to be established. Hybridization plateaus of total plastid 4S RNA from *Nicotiana* (Tewari and Wildman 1970), *Zea* (Haff and Bogorad 1976), *Phaseolus* (Steinmetz and Weil 1976), *Euglena* (Gruol and Haselkorn 1976; McCrea and Hershberger 1976; Schwartzbach et al. 1976) and *Pisum* (Tewari et al. 1977) range from 0.4–0.7% of the ptDNA equivalent to 20–30 cistrons. These values are approximations as not all low-molecular RNA's are necessarily tRNA and degradation products cannot be eliminated by RNA presaturation if the original RNA is unknown. An additional difficulty may arise from differences of about one order of magnitude in the specific activities of

Table 2. Identified chloroplast tRNA's and of tRNA genes mapped on the circular DNA of *Spinacia oleracea* chloroplasts. (From Driesel et al. 1979)

Amino acid	Number of isoaccepting tRNA's identified	Number of isoaccepting tRNA's whose genes have been localized on the physical map (Fig. 5)	Minimum number of genes mapped
Leu	3	3	3
Met	3	3	2
Ser	3	3	2
Arg	2	1	1
Gly	2	1	1
Ile	2	2	4
Thr	2	2	1
Val	2	1	1
Ala	1	1	1
Asn	1	1	1
His	1	1	1
Lys	1	—	—
Phe	1	1	1
Pro	1	—	—
Trp	1	1	1
Tyr	1	1	1
Asp	—	—	—
Cys	—	—	—
Glu	—	—	—
Gln	—	—	—
Total	27	22	21

individual tRNA's when radioiodinated RNA is used (Steinmetz, pers. commun.). There are no organisms in which plastid tRNA species have been identified for all amino acids, not all known 4S RNA species have been shown to be tRNA's, or complementarity to ptDNA has not been determined for all tRNA species. Thus, possible precursors or even the risk of tRNA import from the cytosol still remain to be excluded in individual cases.

The best insight into the complexity of plastid tRNA, synthetases, and tRNA genes has been gained from investigations of *Phaseolus, Zea,* and *Spinacia. Phaseolus* chloroplasts have facilities to accept at least 18 amino acids (Burkard et al. 1970). Transfer RNA genes for 16 amino acids have been established by amino acid analysis of ptDNA/aminoacyl tRNA hybrids in *Zea* (Haff and Bogorad 1976) and about 40 distinct low-molecular RNA species of unbroken *Spinacia* plastids have been resolved by electrophoresis in two-dimensional gels (Steinmetz et al. 1978; Driesel et al. 1979). Of these 27 species have been identified as tRNA's for 16 amino acids and of these at least 21 distinct loci have been assigned positions on the plastid chromosome by hybridization to DNA fragments (Table 2). Together with loci of unidentified 4S RNA's this number increases to about 30, which is close to the figure of 32 necessary to read all codons according to the wobble hypothesis (Crick 1966). The failure to identify tRNA's accepting for met, gln, asn and cys in corn chloroplasts, those for glu, gln, cys, and asp in spinach chloroplasts as well as synthetases for cys and trp in bean chloroplasts in all probability results from synthetase instability.

The genes for tRNA (spinach, Steinmetz et al. 1978; Driesel et al. 1979) and for unidentified 4S RNA (*Chlamydomonas*, Malnoe and Rochaix 1978; *Euglena*, Hallick et al. 1978) are distributed in the repetitive and nonrepetitive regions on the circular molecule and spaced nonrandomly. In spinach ptDNA the large single-copy region carries most of them, and they are largely clustered in two positions (Fig. 5). It would seem logical to transcribe these contiguous genes as multimeric precursors provided they are encoded in one strand. On the *Euglena* chromosome most 4S genes lie adjacent to the sequence triplication while in the *Chlamydomonas* DNA at least 12 coding regions seem to be scattered almost equally on both single-copy segments.

Several aspects have to be considered with regard to the gene dosage of individual tRNA species. Hybridization plateaus suggest that most tRNA's derive from only a single locus. In spinach ptDNA the duplicated sequence, however, carries at least the genes for both isoleucine-accepting tRNA species and genes of tRNA's for leucine which all are represented on both copies and thus present at least twice per circle (Fig. 5; Driesel et al. 1979). Mapping analysis has also established that isoaccepting tRNA species may be transcripts of different structural genes. In spinach ptDNA this is true for multiple species of the $tRNA^{leu}$, $tRNA^{ser}$, $tRNA^{met}$ as well as of $tRNA^{ile}$ (Table 2) where the assignment to distinct loci is based upon a certain differential stability of the DNA/RNA heteroduplexes against RNAase on Southern strips. Former competition experiments with isoaccepting *Phaseolus* pt-tRNA's had failed to decide whether they are transcripts from different though similar loci or arise by posttranscriptional modification (Steinmetz and Weil 1977). It would seem desirable to confirm the mapping results by nucleotide sequencing and to demonstrate whether the isoacceptors for a given amino acid recognize different codons. The currently available resolution does not answer whether isoaccepting species like $tRNA_1^{leu}$ and $tRNA_3^{leu}$, $tRNA_1^{ser}$ and $tRNA_3^{ser}$, $tRNA_1^{trp}$ and $tRNA_2^{trp}$, as well as $tRNA_1^{met}$ and $tRNA_3^{met}$ derive from different loci (Fig. 5). The minimum number of genes and the number of isoacceptors for individual tRNA species in spinach chloroplasts are summarized in Table 2.

The existence of at least one tRNA gene, one of the two iosoaccepting species for ile, in the intercistronic spacer between the genes for 16S and 23 S rRNA on the spinach chromosome is of particular interest (Bohnert et al. 1979a; Driesel et al. 1979). Since similar arrangements are known from other organisms this might reflect a common feature of rRNA operons. Best-documented are the rDNA regions of *E.coli* which carry either one (glu) or two (ile and ala) tRNA cistrons (Lund et al. 1976; Morgan et al. 1977) within the spacer as well as tRNA genes (e.g., asp) distal to the operon (Morgan et al. 1978). Genes for unidentified 4S RNA localized in the spacer are present in ptDNA of *Chlamydomonas* (Malnoe and Rochaix 1978), in mtDNA of *Neurospora* (Terpstra et al. 1977) and of animal cells (Wu et al. 1972; Angerer et al. 1976; Dawid et al. 1976; Saccone et al. 1977). Also they are localized in the rDNA region of *Euglena* ptDNA (Hallick et al. 1978), as flanking genes in HeLa cells (Wu et al. 1972), and in *Xenopus* mtDNA (Dawid et al. 1976).

The large number of tRNA species and genes, their small size and the relative ease with which these are now obtainable make them ideal material for probing into the phylogenetic stability of primary sequences and of gene arrangement on ptDNA as well as into functional aspects (see Sect.IV.A; Barnett et al. 1978).

D. Polypeptide Genes

Genes of RNA species of the organelle-located translation machinery can be allocated directly to the chromosome as they are end-products and complementary to DNA. This is not possible for polypeptide genes.

Four approaches are currently in use to characterize or map plastome-coded polypeptides:

1. ptDNA-programmed translation in coupled (Bottomley and Whitfeld 1979) or linked (Coen et al. 1977) heterologous cell-free systems, either with total ptDNA (Bottomley and Whitfeld 1979) or with DNA fragments of known map position (Coen et al. 1977; Rochaix and Malnoe 1978; Rochaix et al. 1979).

2. Translation of plastid mRNA in heterologous reconstituted cell-free systems and identification of their transcriptional origin by hybridization to DNA or by hybrid arrested translation (Paterson et al. 1977; Bedbrook et al. 1978; Driesel 1979). This approach may also be used to probe into temporary changes of messenger populations.

Both approaches pre-suppose proper recognition of signals on plastid nucleic acids in the heterologous in vitro systems. Additional difficulties may arise in DNA-directed assays through the presence of intervening sequences.

3. The synthesis of polypeptides in organelles. This method is based on the assumption that polypeptides synthesized on plastid ribosomes are coded for by ptDNA. There is in fact no convincing evidence for messenger import from the nucleo-cytosolic compartment into these organelles.

4. Identification of altered genes or gene products in plastome mutants.

Newly synthesized material is usually identified by means of its electrophoretic properties, immunological methods and/or partial proteolysis. The problem that plastomecoded polypeptides are usually integral parts of complex structures and cannot be isolated without impairing their functional properties is not relevant in gene mapping.

Only two polypeptide genes have so far been unequivocally mapped on ptDNA and only one of them has been identified that is the gene for the large subunit of Fraction-1-Protein (corn: Coen et al. 1977; *Chlamydomonas:* Rochaix and Malnoe 1978; spinach: Driesel 1979; see Bottomley, this vol.). The other is the gene of a fast-turning-over thylakoid polypeptide of yet unknown function with a molecular weight between 32 and 35 kd (corn: Bedbrook et al. 1978; spinach: Driesel 1979, *Chlamydomonas:* Rochaix et al. 1979; for the properties of this polypeptide see Edelman and Reisfeld 1978; Ellis and Barraclough 1978). The genes for both proteins are located in the large single-copy region (Fig. 5) and their positions appear to be phylogenetically conserved as are those of rRNA genes. There appears to be only one gene for each of these two proteins per circular molecule. Evidence for two additional genes for membrane polypeptides exists for

Chlamydomonas ptDNA (Rochaix et al. 1979). The template activities for antigenic determinants of the large subunit polypeptide of Fraction-1-Protein and that of the fast-turning-over polypeptide have been resolved from one another and from RNA's other than rRNA by sedimentation (Edelman and Reisfeld 1978; Driesel 1979). Both mRNA's seem to be 1.5 to 2 times larger than the minimum gene size required for the corresponding protein (Reisfeld et al. 1978). The thylakoid protein is synthesized as a precursor (33.5–35 kd; Edelman and Reisfeld 1978; Grebanier et al. 1978) and changed into the final product (about 32 kd) perhaps during its incorporation into the membrane. No such precursor has been observed with the large subunit polypeptide of Fraction-1-Protein.

Both polypeptide genes are not constitutively expressed but seem to be under at least transcriptional control (Edelman and Reisfeld 1978). The gene activities appear to be related in magnitude and time to functional and developmental events (Bedbrook et al. 1978). The synthesis of the thylakoid polypeptide appears to be induced by light. No extractable messenger for this protein exists in etioplasts measured by translation fidelity and hybridization intensity to the encoding DNA fragment (Bedbrook et al. 1978). Its pool size increases considerably during the transformation from etioplasts to chloroplasts. The differential expression of the gene for the large Fraction-1-Protein subunit polypeptide is of different nature. Knowledge here is based upon the known functional and morphological dimorphism between chloroplasts in bundle-sheath and mesophyll cells of plants with a C4 pathway for CO_2 fixation. Mesophyll cell chloroplasts lack both the Fraction-1-Protein (Huber et al. 1976) and the corresponding mRNA (Link et al. 1978) but both components are present in bundle-sheath plastids. As both chloroplast types appear to contain the same DNA the gene is apparently arrested in alternative functional states in these cells.

On the basis of existing methods identification and mapping of further plastome-coded polypeptides genes may be expected in the near future. Isolated intact chloroplasts synthesize about 90 soluble and insoluble polypeptides of which the former have been resolved into 80 components on two-dimensional gels (Ellis and Barraclough 1978). Most of them have not been identified. If unique and originating from ptDNA, they could altogether account for about half the coding potential of the cyclic chromosome. There may be more species since thylakoid polypeptides have not been separated two-dimensionally and organelle modifications other than chloroplasts and etioplasts have not been studied. Among the identified polypeptides which appear to be made by isolated chloroplasts, carry f-met as N-terminal amino acid or are uniparentally inherited, are three (Mendiola-Morgenthaler et al. 1976; Ellis 1977) or four (Nelson et al. 1980) of the five subunits of the membrane-bound ATP synthetase (GF_1) the small DCCD-binding proteolipid of CF_0 (Sebald et al. 1979), the cytochromes f (Doherty and Gray 1979) and b 559 (Zielinski and Price 1980), polypeptides of photosystem I and II (Chua and Gillham 1977) and, representing the soluble fraction, the elongation factors EF-Tu and EF-G (Tiboni et al. 1978). Genetic evidence exists that one or two polypeptides of plastid ribosomes are plastome-coded (Bogorad et al. 1977).

The fourth method, the study of plastome mutants, has been little used. It has the difficulty that systems of segregation are not readily available and those of recombination are restricted to *Chlamydomonas* (Sager 1977). Other difficulties involve the frequently pleiotropic nature of mutants, such as those with ribosome-deficient plastids, the problem of fractionating mutated labile organelles (Schmitt and Herrmann 1977), and, in higher plants also the production of sufficient and uniform tissue for biochemical analysis. Pleiotropic expression of a mutation may be so complex that the primary lesion is obscured. Changes in nucleotide sequence of ptDNA have not yet been correlated with changes in amino acid sequence as no mutants with altered DNA have been observed. In spite of these present limitations of the genetic approach it is to be expected that the interdisciplinary study of suitable mutants and of interspecific sexual or somatic hybrids will become a most promising way not only for analyzing genetic functions of ptDNA including its regulatory role in the integration of activities of both genome and plastome (see Sect. V) but also for obtaining information about the nature of membrane translocation and about trimming and assembly processes of polypeptides.

E. Regulatory Genes

The isolation of replicative intermediates (Kolodner and Tewari 1975a; Richards and Manning 1975), transcriptional, and posttranscriptional events indicate regulatory processes on organelle DNA. Position, structure and function of replication origins are unknown. The temporal gene activities, the identification of the multimeric rDNA transcript in spinach chloroplasts and of mRNA's (for references see foregoing sections) suggest that promotor and terminator sequences seem to exist for the rRNA operon, the large Fraction-1-Protein subunit polypeptide and the fast-turning-over thylakoid polypeptide genes. Signals for nucleolytic post-trancriptional modification occur at least in the rDNA region (Bohnert et al. 1976). In *Chlamydomonas* the intervening sequence in the gene for 23S RNA is flanked by two identical triplets, 5'-CGT-3', and there is evidence for an inverted repeated sequence with structural properties similar to those of bacterial IS elements (Rochaix and Allet 1979). Attempts to identify the promotor site for the rRNA operon on corn ptDNA by binding with *E.coli* RNA polymerase and DNA sequencing remained inconclusive (Schwarz and Kössel 1979). However the sequences near the 3'-terminus of the 16S rRNA gene of maize ptDNA show structural properties for initiation regions of prokaryotic mRNA's (Schwarz and Kössel 1979). The possibility of hairpin formation close to the 5'-end of the 16S rRNA gene has been reported for corn and *Euglena* ptDNA and the sequences compared with those of *E.coli* rRNA (Schwarz et al. 1979), but structural and functional implications are still unknown. Nothing is known about eukaryote-specific signals involved in intercompartmental communication.

V. The Plastome

The genetic compartments of eukaryotic cells can influence each other through their gene products. A common type of co-operation is the formation of complex organelle-located structures which have a dual origin with regard to their coding site. An example is Fraction-1-Protein, the small subunit of which is nuclear-coded, cytoplasmically-translated and transported into plastids, while the large subunit is both encoded and translated within plastids (see Bottomley, this vol.). Other examples are the organelle-specific translation machinery, including ribosomes (Bogorad et al. 1977), tRNA and their cognate aminoacyl tRNA ligases (see Sect.IV.C), possibly elongation factors (Tiboni et al. 1978), as well as several polypeptides which form part of enzyme complexes associated with thylakoids (see foregoing section).

Many aspects of plastome function and especially their interrelationships with the nucleus remain obscure. For example no information is available for nearly 90% of the potential coding capacity of ptDNA. It is also not known how the products from these compartments are synthesized in matching amounts and brought together at the correct time and place within the cell to form complex structures, nor has it yet been possible to estimate the influence of ptDNA on the development of eukaryotic organisms by using the approaches of molecular biology. Some indications, however, have come from genetic experiments involving the genus *Oenothera*.

A. The Oenothera System

Since the pioneering work of Renner, Schwemmle, Cleland, Oehlkers, and their students the genus *Oenothera* has been used for studies of interspecific hybrids.

In *Oenothera* the transfer of plastids from one species into an environment of an alien nucleus and vice versa is facilitated by a combination of unique features in the genetic structure of this genus (see Stubbe 1966; Cleland 1972; Kutzelnigg and Stubbe 1974). These features include:

1. Biparental transmission of plastids which in crossing experiments have been shown to occur between a wide range of species and even between different subsections. Biparental transmission has recently been shown in an investigation involving electron-microscopic observation of serial sections of *Oenothera erythrosepala* zygotes (Meyer and Stubbe 1974).

2. Complex heterozygosity which largely eliminates the effects of nuclear recombination and prevents multifactorial segregation which occurs during meiosis interfering in the analysis of plastid genetics. Using the advantage of complex heterozygous combination of hybrid genotypes it is easy to produce any desired genome/plastome combination.

3. A variety of defective plastome mutants which can be used to mark the plastid pattern of one of the parents in crosses.

The following brief outline will be restricted to *Euoenothera*, genetically the most intensely studied subsection of this genus. The subsection *Euoenothera* can be separated into three well-characterized genomes (Cleland 1950, 1972, denoted A, B, C in the terminology of Stubbe 1959). These can be combined into six different genotypes, homozygous (AA, BB, CC) or complex heterozygous (AB, BC, AC) all of which are morphologically distinct. There are as well five genetically distinct plastid types denoted I–V by Stubbe (1959). The different plastid types give normal pigmentation in naturally occurring combinations with one or more nuclear

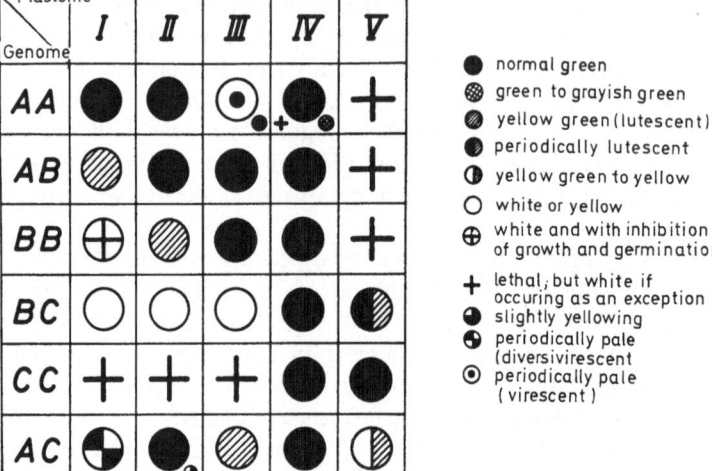

Fig. 8. Compatibility relations between different genotypes and plastomes of the subsection *Euoenothera* (Modified with permission from Kutzelnigg and Stubbe 1974). The combinations in natural species are: AA I: *Oe.elata, Oe.hookeri, Oe.jamesii, Oe.longissima, Oe.strigosa;* AB II: *Oe.biennis* ssp. *caecarium, Oe.biennis* ssp. *centralis* (part); AB III: *Oe.biennis* ssp. *centralis* (part), *Oe.biennis* ssp. *erythrosepala;* BB III: *Oe.biennis* ssp. *austromontana, Oe.grandiflora;* BC IV: *Oe.parviflora* ssp. *angustissima;* AC IV: *Oe.parviflora* ssp. *parviflora;* CC V: *Oe.argillicola*

genomes but reveal their inheritable individuality in the presence of other nuclear genomes. Hybrids, resulting from interspecific crosses, are frequently either bleached or variegated, bleached and green (Renner 1924). "Hybrid bleaching" reflects incompatibility between an alien genome and a plastome combined in one cell. "Hybrid variegation" takes place when one of the two parental wild-type plastids contributed to the zygote is compatible with the nucleus while the other is not. As a consequence of somatic segregation ("sorting-out") of both types of plastids, more or less stable chimeras develop. The relationship between the *Euoenothera* plastomes and genomes is presented in Fig. 8.

Even when present in the same cytosol, genetically different plastid types commonly remain morphologically distinct and give rise to variegated tissue and "mixed cells". The co-existence of differing plastid types illustrates that differences reside in the organelles themselves and that impaired structural and functional development of one plastid type cannot be simply overcome by an exchange of metabolites.

Hybrid bleaching is reversible and a temporary feature of incompatibility between plastid and nucleus. Impaired plastids from disharmonic combinations restore their original morphology and functionality when recombined with their natural or another compatible genome even after many generations under the influence of an incompatible nucleus. However this is impossible if bleaching is caused by defective mutations which can also occur spontaneously in plastomes. In such cases none of the *Euoenothera* genomes is capable of compensating for the resulting plastid alterations. These observations indicate that the development of autotrophic eukaryotes depends on at least two distinct though interdependent genetic compartments. In addition the five *Euoenothera* plastome DNA's show differences when digested with restriction endonucleases (Fig. 7). Although it is not yet proven, it is reasonable to assume that the plastome differences which have been located genetically reside in ptDNA.

B. Transmission of Plastid DNA

Restriction analysis of ptDNA from organelles that are associated with different *Oenothera* genotypes (isonuclear, plastome different strains; Fig. 8) showed that a given plastome DNA retains its characteristics (Fig. 7) under different genotypes, implying that plastids apparently transmit their genetic material independent of the nucleus. The inference that ptDNA is ontogenetically stable is in line with investigations demonstrating that physiological diversity of plastids is not accompanied by qualitative changes in their gene material. For example, the circular molecules isolated from chromoplasts of several monocotyledons (Falk et al. 1974; Wuttke 1976; Sitte 1977) do not differ in their circumference from those of higher plant chloroplasts (Table 1). DNA from different organelle modifications of one species, such as bundle-sheath and mesophyll chloroplasts in *Panicum* (Callis and Walbot 1977), corn (Link et al. 1978), from plastids of auto- or mixotrophically grown *Euglena* cells (Kopecka et al. 1977), chloroplasts and chromoplasts of *Narcissus* and *Oenothera* or etioplasts and chloroplasts from spinach (Herrmann 1977) have the same restriction fragment patterns, respectively.

Attempts to test for the presence or absence of ptDNA copies in nucDNA by molecular hybridization have generally not been conclusive due to problems in cell fractionation and possible cross-reaction between pt- and nucDNA (see Sect.III.A). However, work with heat-bleached mutants of *Euglena* which lack plastids suggests that they have less than 0.3% of a circular plastid chromosome per nucleus (Rawson and Boerma 1976). *Euglena* is unusual in that it appears to be almost the only known autotrophic eukaryote that is available in an aplastidic form (Gibbs 1978). For other aspects of transmission of ptDNA see Sect.III.D.3.

C. Specificity of Genome/Plastome Interaction

The fact that in *Oenothera* plastids even of related species may differ genetically and may not be arbitrarily exchangeable means that the interplay between genomes and plastomes is specific. This important fact is frequently ignored. An exchange of one of the components obviously disturbs an intracellular genetic balance which in extremely dysfunctional combinations such as the genome AA and the plastome V may even culminate in lethality (Fig. 8; Stubbe 1963). Hybrid bleaching and hybrid variegation are widespread (Kirk and Tilney-Bassett 1978), but the extent of specialization may differ between phylogenetic groups. No disharmonies are known, for example, from nuclear implantation experiments with *Acetabularia* species. It would be of considerable interest to know whether genome/plastome specificity operates at the level of structural and/or regulatory genes.

D. Extraplastidic Effects of the Plastome

The most obvious plastid abnormalities in interspecific hybrids and plastome mutants concern pigmentation and the lack of differentiation into fully functional chloroplasts. These observations support the current concept that organelle DNA co-operates with the nucleus in the biogenesis of plastids. The investigation of interspecific hybrids of *Oenothera* has established that the plastome influences a number of characters in plants which manifest themselves outside plastids. These include both the sporophytic (diploid) as well as gametophytic (haploid) phases of plant development. In *Oenothera* observations have been made of plastome-dependent misdevelopment of ovules, embryosacs, embryos, nonermination of viable pollen, the failure of seed germination, and the shape of starch grains and leaves (Kutzelnigg and Stubbe 1974).

E. Plastome/Genome Co-Evolution

The existence of plants with genetically distinct plastid types and the results of restriction endonuclease cleavage analysis showing that ptDNA of even related species may be different (Fig. 7; Atchison et al. 1976; Vedel et al. 1977) demonstrate that plastomes evolve. The phylogenetic relationship of the five *Euoenothera* plastomes proposed by Stubbe (1959) is based on compatibility relations and on differences in multiplication rates between the individual plastid types. These rates are largely independent of compatibility factors and the genetic system which controls them is localized in plastids. The plastome pedigree is an integral factor in the evolution of this subsection which is also influenced by other genetic compartments, particularly the nucleus. Genetic interaction between compartments probably places constraints with regard to progressive alteration of the system, because mutation in any one of the compartments has to satisfy the requirement that the entire system remains fully functional. It is therefore likely that plastome/genome co-evolution is a more general principle in the evolution of autotrophic eukaryotes.

From an evolutionary point of view in *Oenothera* the faster multiplying type will supplant the slower one, if continual outcrossing occurs. This is a general principle in nature. The more rapidly multiplying plastid types are found in all those instances where a single genome tolerates more than one plastome. Stabilization of species with slower multiplying plastid types (e.g., AB-II, BC-IV and AC-IV; Fig. 8) is achieved by incompatibility relations creating genetic barriers to gene flow. Since these species cannot be constructed nowadays from natural homozygous combinations AA, BB, and CC because of the incompatibility of their present plastomes with the heterozygous genotypes apparently evolution of the basic homozygous *Euoenothera* genomes preceded plastome divergence. The fastest multiplying plastid types are strictly adapted to only one genotype, e.g., I to AA or V to CC (Fig. 8), indicating that the phylogenetic trend of co-evolution in this subsection has apparently led to a closer adaptation between the separate genetic compartments. This probably represents a higher degree of functional specialization.

Acknowledgments. The authors acknowledge the help they have received from many colleagues with whom they have discussed specific aspects of this paper and particularly those colleagues who have read and made comments on sections of this paper. We thank Prof. Th. Butterfass, Prof. K.V. Kowallik, Prof. W. Stubbe, Prof. J.H. Weil, Dr. E. Crouse, Dr. U.J. Santore and Dr. P. Whitfeld. As well we thank Miss B. Schiller and Miss M. Streubel for their help with photographic work.

References

Akoyunoglou, G., Argyroudi-Akoyunoglou, J.H. (eds.): Chloroplast Development, 888 p. Amsterdam: Elsevier/North Holland 1978

Anet, R., Strayer, R.: Density gradient relaxation: A method for preparative buoyant density separations of DNA. Biochem. Biophys. Res. Commun. 34, 328-334 (1969)

Angerer, L., Davidson, N., Murphy, W., Lynch, D., Attardi, G.: An electron microscope study of the relative positions of the 4S and ribosomal RNA genes in HeLa cell mitochondrial DNA. Cell 9, 81-90 (1976)

Atchison, B.A., Whitfeld, P.R., Bottomley, W.: Comparison of chloroplast DNAs by specific fragmentation with EcoRI endonuclease. Mol. Gen. Genet. 148, 263-269 (1976)

Bard, St.A., Gordon, M.P.: Studies on spinach chloroplast and nuclear DNA using large-scale tissue preparations. Plant Physiol. 44, 377-384 (1969)

Barnett, W.E., Schwartzbach, S.D., Hecker, L.I.: The transfer RNAs of eukaryotic organelles. In: Progress in Nucleic Acid Research and Molecular Biology, Cohn, W.E. (ed.), pp. 143-179. London-New York: Academic Press 1978

Bastia, D., Chiang, K-S., Swift, H., Siersma, P.: Heterogeneity, complexity, and repetition of the chloroplast DNA of *Chlamydomonas reinhardtii*. Proc. Natl. Acad. Sci. USA 68, 1157-1161 (1971)

Baxter, R., Kirk, J.T.O.: Base composition of DNA from chloroplasts and nuclei of *Phaseolus vulgaris*. Nature (London) 222, 272-273 (1969)

Bedbrook, J.R., Bogorad, L.: Endonuclease recognition sites mapped on *Zea mays* chloroplast DNA. Proc. Natl. Acad. Sci. USA 73, 4309-4313 (1976)

Bedbrook, J.R., Kolodner, R., Bogorad, L.: *Zea mays* chloroplast ribosomal RNA genes are part of a 22.000 base pair inverted repeat. Cell 11, 739-749 (1977)

Bedbrook, J.R., Link, G., Coen, D.M., Bogorad, L., Rich, A.: Maize plastid gene expressed during photoregulated development. Proc. Natl. Acad. Sci. USA 75, 3060-3064 (1978)

Behn, W., Herrmann, R.G.: Circular molecules in the β-satellite DNA of *Chlamydomonas reinhardii*. Mol. Gen. Genet. 157, 25-30 (1977)

Bennett, J., Radcliff, C.: Plastid DNA replication and plastid division in the garden pea. FEBS Lett. 56, 222-225 (1975)

Bernardi, G., Prunell, A., Fonty, G., Kopecka, H., Strauss, F.: The mitochondrial genome of yeast: organization, evolution and the petite mutation. In: The Genetic Function of Mitochondrial DNA, Saccone, C., Kroon, A.M. (eds.), pp. 185-198. Amsterdam: Elsevier/North Holland 1976

Bibby, B.T., Dodge, J.D.: The fine structure of the chloroplast nucleoid in *Scrippsiella sweeneyae* (Dinophyceae). J. Ultrastruct. Res. 48, 153-161 (1974)

Bidwell, R.G.S.: Impurities in preparations of *Acetabularia* chloroplasts. Nature (London) 237, 169 (1972)

Bird, I.F., Cornelius, M.J., Dyer, T.A., Keys, A.J.: The purity of chloroplasts isolated. J. Exp. Bot. 24, 211-215 (1973)

Bisalputra, T., Bisalputra, A.A.: The ultrastructure of chloroplast of a brown alga *Sphacelaria* sp. I. Plastid DNA configuration—the chloroplast genophore. J. Ultrastruct. Res. 29, 151-170 (1969)

Blair, G.E., Ellis, R.J.: Protein synthesis in chloroplasts. I. Light-driven synthesis of the large subunit of Fraction I Protein by isolated pea chloroplasts. Biochim. Biophys. Acta 319, 223-234 (1973)

Blin, N., Gabain, A.v., Bujard, H.: Isolation of large molecular weight DNA from agarose gels for further digestion by restriction enzymes. FEBS Lett. 53, 84-86 (1975)

Blomquist, G., Larsson, Ch., Albertsson, P.Å.: A study of DNA from chloroplasts separated by countercurrent distribution. Acta Chem. Scand. B 29, 838-842 (1975)

Boasson, R., Laetsch, W.M.: Chloroplast replication and growth in tobacco. Science 166, 749-751 (1969)

Boffey, A.A., Ellis, J.R., Sellden, G., Leech, R.M.: Chloroplast division and DNA synthesis in light grown wheat leaves. Plant Physiol. 64, 502-505 (1979)

Bogorad, L., Davidson, J.N., Hanson, M.R.: The genetics of the chloroplast ribosome in *Chlamydomonas reinhardii*. In: Nucleic Acids and Protein Synthesis in Plants, Bogorad, L., Weil, J.H. (eds.), pp. 135-154. New York-London: Plenum Press 1977

Bohnert, H.J., Herrmann, R.G.: The genomic complexity of *Acanthamoeba castellanii* mitochondrial DNA. Eur. J. Biochem. 50, 83-90 (1974)

Bohnert, H.J., Schmitt, J.M., Herrmann, R.G.: Structural and functional aspects of the plastome. Port. Acta Biol. Ser. A 14, 71-90 (1974)

Bohnert, H.J., Driesel, A.J., Herrmann, R.G.: Characterization of the RNA compounds synthesized by isolated chloroplasts. In: Genetics and Biogenesis of Chloroplasts and Mitochondria, Bücher, Th., Neupert, W., Sebald, W., Werner, S. (eds.), pp. 629-636. Amsterdam: Elsevier/North Holland 1976

Bohnert, H.J., Driesel, A.J., Crouse, E.J., Gordon, K., Herrmann, R.G., Steinmetz, A., Mubumbila, A., Keller, M., Burkard, G., Weil, J.H.: Presence of a transfer RNA gene in the spacer sequence between the 16S and 23S rRNA genes of spinach chloroplast DNA. FEBS Lett 103, 52-56 (1979a)

Bohnert, H.J., Driesel, A., Crouse, E.J., Gordon, K., Mertens, I., Schmitt, J.M.: The chloroplast rRNA operon. Hoppe Seyler's Z. Physiol. Chem. 360, 237 (1979b)

Borst, P.: Size, structure and information content of mitochondrial DNA. In: Autonomy and Biogenesis of Mitochondria and Chloroplasts, Boardman, N.K., Linnane, A.W., Smillie, R.M. (eds.), pp. 260-266. Amsterdam: North Holland 1971

Borst, P.: Mitochondrial nucleic acids. Annu. Rev. Biochem. 41, 333-376 (1972)

Borst, P., Kroon, A.M.: Mitochondrial DNA: physico-chemical properties, replication and genetic function. Int. Rev. Cytol. 26, 107-190 (1969)

Borst, P., Kroon, A.M., Ruttenberg, G.J.C.M.: Mitochondrial DNA and other forms of cytoplasmic DNA. In: Genetic Elements, Properties and Function, Shugar, D. (ed.), pp. 81-116. London-Warsaw: Academic Press/Polish Scientific Publ. 1967

Bottomley, W., Whitfeld, P.R.: Cell-free transcription and translation of total spinach chloroplast DNA. Eur. J. Biochem. 93, 31-39 (1979)

Bottomley, W., Spencer, D., Whitfeld, P.R.: Protein synthesis in isolated spinach chloroplasts: comparison of light-driven and ATP-driven synthesis. Arch. Biochem. Biophys. 164, 106-117 (1974)

Brawerman, G., Eisenstadt, J.M.: Deoxyribonucleic acid from the chloroplasts of *Euglena gracilis*. Biochim. Biophys. Acta 91, 477-485 (1964)

Brawerman, G., Eisenstadt, J.M.: The nucleic acids associated with chloroplasts of *Euglena gracilis* and their role in protein synthesis. In: Organisational Biosynthesis. Vogel, J., Lampen, J.O., Bryson, V. (eds.), pp. 419-437. New York-London: Academic Press 1967

Burkard, G., Guillemaut, P., Weil, J.H.: Comparative studies on the tRNAs and the aminoacyl-tRNA synthetases from the cytoplasm and the chloroplasts of *Phaseolus vulgaris*. Biochim. Biophys. Acta 224, 184-198 (1970)

Butterfass, Th.: Control of plastid division by means of nuclear DNA amount. Protoplasma 76, 167-195 (1973)

Callis, J., Walbot, V.: Determination of the number of ribosomal cistrons in chloroplasts of C_3 and C_4 plants. In: Colloq. Int. C.N.R.S., Acides Nucléiques et Synthèse des Protéines chez les végétaux, Bogorad, L., Weil, J.H. (eds), pp. 137-141. Paris: Ed. CNRS 1977

Camefort, H.: Fecondation et proembryogénèse chez les Abiétacées. Rev. Cytol. Biol. Veg. 32, 253-271 (1969)

Cass, D.D., Karas, I.: Development of sperm cells in barley. Can. J. Bot. 53, 1051-1062 (1975)

Cavalier-Smith, T.: Electron microscopic evidence for chloroplast fusion in zygotes of *Chlamydomonas reinhardii*. Nature (London) 228, 333-335 (1970)

Chardard, R.: Aspects infrastructuraux de la maturation des grains de pollen de quelques Orchidacées. Rev. Cytol. Biol. Veg. 32, 67-100 (1969)

Chesnoy, L.: Sur la participation du gamè te mâle a la constitution du cytoplasma de l'embryon chez le *Biota orientalis*. Rev. Cytol. Biol. Veg. 32, 273-294 (1969)

Chiang, K.-S., Sueoka, N.: Replication of chloroplast DNA in *Chlamydomonas reinhardii* during vegetative cell cycle, its mode and regulation. Proc. Natl. Acad. Sci. USA 57, 1506-1513 (1967)

Chua, N.-H., Gillham, N.W.: The sites of synthesis of the principal thylakoid membrane polypeptides in *Chlamydomonas reinhardii*. J. Cell Biol. 74, 441-452 (1977)

Chua, N.-H., Schmidt, G.W.: Transport of proteins into mitochondria and chloroplasts. J. Cell Biol. 81, 461-483 (1979)

Clauhs, R.P., Grun, P.: Changes in plastid and mitochondria content during maturation of generative cells of *Solanum* (Solanaceae). Am. J. Bot. 64, 377-383 (1977)

Cleland, R.E.: Studies in *Oenothera* cytogenetics and phylogeny. Introduction and general summary. Indiana Univ. Publ. Sci. Ser. 16, 5-9 (1950)

Cleland, R.E.: *Oenothera*, Cytogenetics and Evolution. London-New York: Academic Press 1972

Coen, D.M., Bedbrook, R., Bogorad, L., Rich, A.: Maize chloroplast DNA fragment encoding the large subunit of ribulosebisphosphate carboxylase. Proc. Natl. Acad. Sci. USA 74, 5487-5491 (1977)

Coleman, A.W.: Use of the fluorochrome 4'6-diamino-2-phenylindole in genetic and developmental studies of chloroplast DNA. J. Cell Biol. 82, 299-305 (1979)

Cook, J.R.: The synthesis of cytoplasmic DNA in synchronized *Euglena*. J. Cell Biol. 29, 369-372 (1966)

Correns, C.: Untersuchungen über die Vermehrung der Laubmoose durch Brutorgane und Stecklinge. Jena: Fischer 1899

Cran, D.G., Possingham, J.V.: The effect of cell age on chloroplast structure and chlorophyll in cultured spinach leaf discs. Protoplasma 79, 197-213 (1974)

Crick, F.H.C.: Codon-anticodon pairing: The wobble hypothesis. J. Mol. Biol. 19, 548-555 (1966)

Crouse, E.J., Schmitt, J.M., Bohnert H.J., Gordon, K., Driesel, A.J., Herrmann, R.G.: Intramolecular compositional heterogeneity of *Spinacia* and *Euglena* chloroplast DNAs. In: Chloroplast Development. Akoyunoglou, G., Argyroudi-Akoyunoglou, J.H. (eds.), pp. 565-572. Amsterdam: Elsevier/North Holland 1978

Dann, O., Bergen, G., Demant, E., Votz, G.: Trypanocide Diamidine des 2-Phenyl-benzofurans, 2-Phenyl-indens und 2-Phenyl-indols. Justus Liebigs Ann. Chem. 749, 68-89 (1971)

Dawid, I.B., Klukas, C.K., Ohi, S., Ramirez, J.L., Upholt, W.B.: Structure and evolution of animal mitochondrial DNA. In: The Genetic Function of Mitochondrial DNA, Saccone, C., Kroon, A.M. (eds.), pp. 3-13. Amsterdam: Elsevier/North Holland 1976

Diener, T.O.: Viroids: The smallest known agents of infectious disease. Annu. Rev. Microbiol. 28, 23-39 (1974)

Diers, L.: Elektronenmikroskopische Beobachtungen zur Archegoniumentwicklung des Lebermooses *Sphaerocarpus donnellii* Aust. — Die Entwicklung des jungen Archegons bis zum Stadium der fertig ausgebildeten sekundären Zentralzelle. Planta 66, 165-190 (1965)

Diers, L.: Der Feinbau des Spermatozoids von *Sphaerocarpos donnellii*. Aust. (Hepaticae). Planta 72, 119-145 (1967)

Dobberstein, B., Blobel, G., Chua, N.H.: In vitro synthesis and processing of a putative precursor for the small subunit of ribulose-1,5-bisphosphate carboxylase of *Chlamydomonas reinhardii*. Proc. Natl. Acad. Sci. USA 74, 1082-1085 (1977)

Doherty, A., Gray, J.C.: Synthesis of cytochrome f by isolated pea chloroplasts. Eur. J. Biochem. 98, 87-92 (1979)

Driesel, A.J.: Ribosomen und ribosomale RNA aus Chloroplasten: Reindarstellung und Charakterisierung. Diplomarbeit, Univ. Köln (1975)

Driesel, A.J.: Identifizierung und physikalische Kartierung von Genen auf dem zirkulären Chromosom aus *Spinacia oleracea* Plastiden. Diss. Univ. Düsseldorf (1979)

Driesel, A.J., Crouse, E.J., Gordon, K., Bohnert, H.J., Herrmann, R.G., Steinmetz, A., Mubumbila, M., Keller, M., Burkard, M., Weil, J.H.: Fraction and identification of the individual spinach chloroplast transfer RNAs and mapping of their genes on the restriction endonuclease cleavage site map of chloroplast DNA. Gene 6, 285-306 (1979)

Dyer, T.A., Bowman, C.M., Payne, P.I.: The low-molecular-weight RNAs of plant ribosomes: their structure, function and evolution. In: Nucleic Acids and Protein Synthesis in Plants, Bogorad, L., Weil, J.H. (eds.), pp. 121-133. New York-London: Plenum Press 1977

Edelman, M., Reisfeld, A.: Characterization, translation and control of the 32000 dalton chloroplast membrane protein in *Spirodela*. In: Chloroplast Development, Akoyunoglou, G., Argyroudi-Akoyunoglou, J.H. (eds.), pp. 641-652. Amsterdam: Elsevier/North Holland 1978

Edelman, M, Cowan, Ch.A., Epstein, H.T., Schiff, J.A.: Studies of chloroplast development in *Euglena*. VIII. Chloroplast-associated DNA. Proc. Natl. Acad. Sci. USA 52, 1214-1219 (1964)

Ee van, Man in't Veld, W.A., Planta, R.J.: Chloroplast DNA from *Spirodela oligorrhiza*. In: Genome Organisation and Expression in Plants. Leaver, Ch. (ed.), p. 145. Edinburgh: NATO Advanced Studies Institute; FEBS Advanced Course 1979

Ellis, R.: The synthesis of chloroplast membranes in *Pisum sativum*. In: Membrane Biogenesis, Tzagoloff, A. (ed.), pp. 247-278. New York-London: Plenum Press 1975

Ellis, R.: Protein and nucleic acid synthesis by chloroplasts. In: The Intact Chloroplast, Barber, J. (ed.), pp. 335-364. Amsterdam: Elsevier/North Holland 1976

Ellis, R.: Protein synthesis by isolated chloroplasts. Biochim. Biophys. Acta 463, 185-215 (1977)

Ellis, R.J., Barraclough, R.: Synthesis and transport of chloroplast proteins inside and outside the cell. In: Chloroplast Development. Akoyunoglou, G., Argyroudi-Akoyunoglou, J.H. (eds.), pp. 185-194. Amsterdam: Elsevier/North Holland 1978

Falk, H., Liedvogel, B., Sitte, P.: Circular DNA in isolated chromoplasts. Z. Naturforsch. 29c, 541-544 (1974)

Fauron, C.M.-R., Wolstenholme, D.R.: Structural heterogeneity of mitochondrial DNA molecules within the genus *Drosophila*. Proc. Natl. Acad. Sci USA 73, 3623-3627 (1976)

Gelvin, S.R., Howell, S.H.: Small repeated sequences in the chloroplast genome of *Chlamydomonas reinhardii*. Mol. Gen. Genet. 173, 315-322 (1979)

Gibbs, S.P.: The chloroplasts of *Euglena* may have evolved from symbiotic green algae. Can. J. Bot. 56, 2883-2889 (1978)

Gibbs, S.P., Cheng, D., Slankis, T.: The chloroplast nucleoid in *Ochromonas danica*. I. Three-dimensional morphology in light- and dark-grown cells. J. Cell Sci. 16, 557-577 (1974a)

Gibbs, S.P., Mak, R., Ng, R., Slankis, T.: The chloroplast nucleoid in *Ochromonas danica*. II. Evidence for an increase in plastid DNA during greening. J. Cell Sci. 16, 579-591 (1974b)

Gibor, A., Izawa, M.: The DNA content of the chloroplasts of *Acetabularia*. Proc. Natl. Acad. Sci. USA 50, 1164-1169 (1963)

Golaszewski, T., Rytel, M., Rogzinski, J., Szarkowski, J.W.: Thymidine kinase activity in rye chloroplasts. FEBS Lett. 58, 370-379 (1975)

Gordon, K.H.J., Crouse, E.J., Bohnert, H.J., Herrmann, R.G.: Physical mapping of differences in chloroplast DNA of the five wild-type plastomes in *Oenothera* subsection *Euoenothera*. Theor. Appl. Genet. (submitted 1980a)

Gordon, K.H.J., Crouse, E.J., Bohnert H.J., Herrmann, R.G.: Restriction endonuclease cleavage site map of DNA from the *Oenothera parviflora* (plastome IV) Theor. Appl. Genet. (submitted 1980b)

Gray, P.W., Hallick, R.P.: Restriction endonuclease map of chloroplast DNA from *Euglena gracilis*. In: Genetics and Biogenesis of Chloroplasts and Mitochondria, Bücher, Th., Neupert, W., Sebald, W., Werner, S. (eds.), pp. 347-350. Amsterdam: Elsevier/North Holland 1976

Gray, P.W., Hallick, R.B.: Physical mapping of the *Euglena gracilis* chloroplast DNA and ribosomal RNA gene region. Biochemistry 17, 284-289 (1978)

Grebanier, A.E., Coen, D.M., Rich, A., Bogorad, L.: Membrane proteins synthesized but not proceeded by isolated maize chloroplasts. J. Cell Biol. 78, 734-746 (1978)

Green, B.R.: Covalently closed minicircular DNA associated with *Acetabularia* chloroplasts. Biochim. Biophys. Acta 447, 156-166 (1976)

Green, B.R., Burton, H.: *Acetabularia* chloroplast DNA: electron microscopic visualization. Science 168, 981-982 (1970)

Green, B., Heilporn, V., Limbosch, S., Boloukhere, M., Brachet, J.: The cytoplasmic DNA's of *Acetabularia mediterranea*. Proc. Natl. Acad. Sci. USA 58, 1351-1358 (1967)

Gruol, D.J., Haselkorn, R.: Counting the genes for stable RNA in the nucleus and chloroplasts of *Euglena*. Biochim. Biophys. Acta 447, 82-95 (1976)

Guerineau, M., Grandchamp, C., Slonimski, P.P.: Circular DNA of a yeast episome with two inverted repeats: Structural analysis by a restriction enzyme and electron microscopy. Proc. Natl. Acad. Sci. USA 73, 3030-3034 (1976)

Gyldenholme, A.O.: Macromolecular physiology of plastids. V. On the nucleic acid metabolism during chloroplast development. Hereditas (Lund) 59, 142-168 (1968)

Haff, L.A., Bogorad, L.: Hybridization of maize chloroplast DNA with transfer ribonucleic acids. Biochemistry 15, 4105-4109 (1976)

Hallick, R.B., Lipper, C., Richards, O.C., Rutter, W.J.: Isolation of a transcriptionally active chromosome from chloroplasts of *Euglena gracilis*. Biochemistry 15, 3039-3045 (1976)

Hallick, R.B., Gray, P.W., Chelm, B.K., Rushlow, K.E., Orozco, E.M.: *Euglena gracilis* chloroplast DNA structure, gene mapping, and RNA transcription. In: Chloroplast Development. Akoyunoglou, G., Argyroudi-Akoyunoglou, J.H. (eds.), pp. 619-622. Amsterdam: Elsevier/North Holland 1978

Hartley, M.R.: The synthesis and origin of chloroplast low-molecular-weight ribosomal ribonucleic acid in spinach. Eur. J. Biochem. 96, 311-320 (1979)

Hartley, M.R., Ellis, R.J.: Ribonucleic acid synthesis in chloroplasts. Biochem. J. 134, 249-262 (1973)

Hartley, M.R., Head, Ch.: The synthesis of chloroplast high-molecular-weight ribosomal ribonucleic acid in spinach. Eur. J. Biochem. 96, 301-309 (1979)

Herrmann, R.G.: Multiple amounts of DNA related to the size of chloroplasts. I. An autoradiographic study. Planta 90, 80-96 (1970)

Herrmann, R.G.: Do chromoplasts contain DNA? II. The isolation and characterization of DNA from chromoplasts, chloroplasts, mitochondria, and nuclei of *Narcissus*. Protoplasma 74, 7-17 (1972)

Herrmann, R.G.: Studies on *Oenothera* plastid DNAs. Progr. Regul. Dev. Proc. Plants, p. 48. Halle/Saale 1977

Herrmann, R.G., Kowallik, K.V.: Multiple amounts of DNA related to the size of chloroplasts. II. Comparison of electron-microscopic and autoradiographic data. Protoplasma 69, 365-372 (1970)

Herrmann, R.G., Kowallik, K.V., Bohnert, H.J.: Structural and functional aspects of the plastome. I. The organization of the plastome. Port. Acta Biol. Ser. A 14, 91-110 (1974)

Herrmann, R.G., Bohnert, H.-J., Kowallik, K.V., Schmitt, J.M.: Size, conformation and purity of chloroplast DNA of some higher plants. Biochim. Biophys. Acta 378, 305-317 (1975)

Herrmann, R.G., Palta, H.K., Kowallik, K.V.: Chloroplast DNA from three archegoniates. Planta 148, 319-327 (1980)

Hobom, G., Bohnert, H.J., Driesel, A., Herrmann, R.G.: Restriction fragment map of the circular plastid DNA from *Spinacia oleracea*. In: Acides Nucléiques et Synthèse de Proteines chez les Végétaux, Bogorad, L., Weil, J.H. (eds.), pp. 63-69. Paris: Ed. CNRS 1977

Hollenberg, C.P.: Recombination Recombinant DNA research, techniques and results. Prog. Bot. 40, 211-235 (1978)

Hollenberg, C.P., Decelmann, A., Kustermann-Kuhn, B., Royer, H.D.: Characterization of 2-μm DNA of *Saccharomyces cerevisiae* by restriction fragment analysis and integration in an *Escherichia coli* plasmid. Proc. Natl. Acad. Sci. USA 73, 2072-2076 (1976)

Hourcade, D., Dressler, D., Wolfson, J.: The nucleolus and the rolling circle. Cold Spring Harbor Symp. Quant. Biol. 23, 537-550 (1974)

Huber, C., Hall, T.C., Edwards, E.: Differential localization of Fraction I Protein between chloroplast types. Plant Physiol. 57, 730-733 (1976)

Hudson, B., Vinograd, J.: Catenated circular DNA molecules in HeLa cell mitochondria. Nature (London) 216, 647-652 (1967)

Hudson, B., Upholt, W.B., Devinny, J., Vinograd, J.: The use of an ethidium analogue in the dye-buoyant density procedure for the isolation of closed circular DNA: the variation of the superhelix density of mitochondrial DNA. Biochemistry 62, 813-820 (1969)

Ingle, J., Possingham, J.V., Wells, R., Leaver, C.J., Loening, U.E.: The properties of chloroplast ribosomal-RNA. In: Control of Organelle Development, Vol. 24, pp. 303-325. Cambridge: University Press 1970

Jacobson, A.B.: A procedure for isolation of proplastids from etiolated maize leaves. J. Cell Biol. 38, 238-244 (1968)

James, J.W., Jope, Ch.: Visualization by fluorescence of chloroplast DNA in higher plants by means of the DNA-specific probe 4'6 diamino-2-phenylindole. J. Cell Biol. 79, 623-630 (1979)

Jeffrey, S.W., Vesk, M.: Further evidence for a membrane-bound endosymbiont within the dinoflagellate *Peridinum foliaceum*. J. Phycol. 12, 450-455 (1976)

Jenni, B., Stutz, E.: Physical mapping of the ribosomal DNA region of *Euglena gracilis* chloroplast DNA. Eur. J. Biochem. 88, 127-134 (1978)

Jenni, B., Stutz, E.: *Euglena gracilis* chloroplast DNA contains a fourth 16S rDNA sequence. In: Genome Organisation and Expression in Plants. Leaver, Ch., (ed.), p. 142. Edinburgh: NATO Advanced Studies Institute; FEBS Advanced Course 1979

Kirk, J.T.O.: Effect of methylation of cytosine residues on the buoyant density of DNA in caesium chloride solution. J. Mol. Biol. 28, 171-172 (1967)

Kirk, J.T.O.: Will the real chloroplast DNA please stand up. In: Autonomy and Biogenesis of Mitochondria and Chloroplasts. Boardman, N.K., Linnane, A.W., Smillie, R.M. (eds.), pp. 267-276. Amsterdam: North Holland 1971

Kirk, J.T.O., Tilney-Bassett, R.A.E.: The plastids — their chemistry, structure, growth and inheritance. 2nd ed. 960 p. Amsterdam: North Holland 1978

Kislev, N., Swift, H., Bogorad, L.: Nucleic acids of chloroplasts and mitochondria in swiss chard. J. Cell Biol. 25, 327-344 (1965)

Knoth, R., Herrmann, F.H., Böttger, M., Börner, Th.: Struktur und Funktion der genetischen Information in den Plastiden. XI. DNA in normalen und mutierten Plastiden der Sorte "Mrs. Parker" von *Pelargonium zonale*. Biochem. Physiol. Pflanz. 166, 129-148 (1974)

Kolodner, R., Tewari, K.K.: Physicochemical characterization of mitochondrial DNA from pea leaves. Proc. Natl. Acad. Sci. USA 69, 1830-1834 (1972a)

Kolodner, R., Tewari, K.K.: Molecular size and conformation of chloroplast deoxyribonucleic acid from pea leaves. J. Biol. Chem. 247, 6355-6364 (1972b)

Kolodner, R.D., Tewari, K.K.: Chloroplast DNA from higher plants replicates by both the Cairns and the rolling circle mechanism. Nature (London) 256, 708-711 (1975a)

Kolodner, R., Tewari, K.K.: The molecular size and conformation of the chloroplast DNA from higher plants. Biochim. Biophys. Acta 402, 372-390 (1975b)

Kolodner, R., Tewari, K.K.: Inverted repeats in chloroplast DNA from higher plants. Proc. Natl. Acad. Sci. USA 76, 41-45 (1979)

Kolodner, R., Tewari, K.K., Warner, R.C.: Physical studies on the size and structure of the covalently closed circular chloroplast DNA from higher plants. Biochim. Biopyhys. Acta 447, 144-155 (1976)

Kopecka, H., Crouse, E.J., Stutz, E.: The *Euglena gracilis* chloroplast genome: analysis by restriction enzymes. Eur. J. Biochem. 72, 525-535 (1977)

Kowallik, K.V., Haberkorn, G.: The DNA-structures of the chloroplast of *Prorocentrum micans* (Dinophyceae). Arch. Mikrobiol. 80, 252-261 (1971)

Kowallik, K.V., Herrmann, R.G.: Variable amounts of DNA related to the size of chloroplasts. IV. Three-dimensional arrangement of DNA in fully differentiated chloroplasts of *Beta vulgaris* L. J. Cell Sci. 11, 357-377 (1972a)

Kowallik, K.V., Herrmann, R.G.: Do chromoplasts contain DNA? I. Electron-microscopic investigation of *Narcissus* chromoplasts. Protoplasma 74, 1-6 (1972b)

Kowallik, K.V., Herrmann, R.G.: Structural and functional aspects of the plastome. II. DNA regions during plastid development. Port. Acta Biol. Ser. A 14, 111-126 (1974)

Kung, S.D., Williams, J.P.: Chloroplast DNA from broad bean. Biochim. Biophys. Acta 195, 434-445 (1969)

Kung, S.D., Moscarello, M.A., Williams, J.P.: Studies with chloroplast and mitochondrial DNA. I. Evidence of sequence homology between chloroplast and nuclear DNA (broad bean) and between mitochondrial and nuclear DNA (rat liver). Biophys. J. 12, 474-483 (1972)

Kutzelnigg, H., Stubbe, W.: Investigations on plastome mutants in *Oenothera*. 1. General considerations. Sub. Cell. Biochem. 3, 73-89 (1974)

Lamppa, G.K., Bendich, A.J.: Changes in chloroplast DNA levels during development of pea *(Pisum sativum)*. Plant Physiol. 64, 126-130 (1969)

Larsson, Ch., Collin, C., Albertsson, R-A.: Characterization of three classes of chloroplasts obtained by counter-current distribution. Biochim. Biophys. Acta 245, 425-438 (1971)

Leaver, Ch. (ed.): Genome Organisation and Expression in Plants. Edinburgh: NATO Advanced Studies Institute; FEBS Advanced Course, 1979

Leaver, C.J., Pope, P.K.: Biosynthesis of plant mitochondrial proteins. In: Nucleic Acids and Protein Synthesis in Plants. Bogorad, L., Weil, J.H. (eds.), pp. 213-237. New York-London: Plenum Press 1977

Liedvogel, B.: DNA content and ploidy of chromoplasts. Naturwissenschaften, 63. Jg., 248 (1976)

Link, G., Coen, D.M., Bogorad, L.: Differential expression of the gene coding for the large subunit of ribulose-1,5-bisphosphate carboxylase in *Zea mays*. In: Chloroplast Development. Akoyunoglou, G., Argyroudi-Akoyunoglou, J.H. (eds.), pp. 559-564. Amsterdam: Elsevier/North Holland 1978

Lund, E., Dahlberg, J.E., Lindahl, L., Jaskunas, S.R., Dennis, P.P., Nomura, M.: Transfer RNA genes between 16S and 23S rRNA genes in rRNA transcription units of *E.coli*. Cell 7, 165-177 (1976)

Mache, R., Rozier, C., Loiseaux, S., Viala, M.: Synchronous division of plastids during the greening of cut leaves of maize. Nature (New Biol.) 242, 158-160 (1973)

MacNutt, M.M., Maltzahn, K.E.: Cellular dedifferentiation and redifferentiation in *Splachnum ampullaceum* (L). Hedw. Can. J. Bot. 38, 895-908 (1960)

Malnoe, P., Rochaix, J.D.: Localization of 4S RNA genes on chloroplast genome of *Chlamydomonas reinhardii*. Mol. Gen. Genet. 166, 269-275 (1978)

Manning, J.E., Wolstenholme, D.R., Ryan, R.S., Hunter, J.A., Richards, O.C.: Circular chloroplast DNA from *Euglena gracilis*. Proc. Natl. Acad. Sci. USA 68, 1169-1173 (1971)

Manning, J.E., Wolstenholme, D.R., Richards, O.C.: Circular DNA molecules associated with chloroplasts of spinach, *Spinacia oleracea*. J. Cell. Biol. 53, 594-601 (1972)

Marcu, K., Dudock, B.: Characterization of a highly efficient protein synthesizing system derived from commercial wheat germ. Nucl. Acids Res. 1, 1385-1397 (1974)

Marschner, H., Possingham, J.V.: Effect of K^+ and Na^+ on growth of leaf discs of sugar beet and spinach. Z. Pflanzenphysiol. 75, 6-16 (1975)

McCrea, J.M., Hershberger, C.L.: Chloroplast DNA codes for transfer RNA. Nucl. Acids Res. 3, 2005-2018 (1976)

Mendiola-Morgenthaler, L.R., Morgenthaler, J.J., Price, C.A.: Synthesis of coupling factor CF-1 protein by isolated spinach chloroplasts. FEBS Lett. 62, 96-100 (1976)

Meselson, M., Stahl, F.W., Vinograd, J.: Equilibrium sedimentation of macromolecules in density gradients. Proc. Natl. Acad. Sci. USA 43, 581-588 (1957)

Meyer, B., Stubbe, W.: Das Zahlenverhältnis von mütterlichen und väterlichen Plastiden in den Zygoten von *Oenothera erythrosepala* Borbas (syn. *Oe.lamarckiana*). Ber. Dtsch. Bot. Ges. 87, 29-38 (1974)

Morgan, E.A., Ikemura, T., Nomura, M.: Identification of spacer tRNA genes in individual ribosomal RNA transcription units of *Escherichia coli*. Proc. Natl. Acad. Sci. USA 74, 2710-2714 (1977)

Morgan, E.A., Ikemura, T., Lindahl, L., Fallon, A.M., Nomura, M.: Some rRNA operons in *E.coli* have tRNA genes at their distal ends. Cell 13, 335-344 (1978)

Morgenthaler, J.-J., Marsden, M.P.F., Price, C.A.: Factors affecting the separation of photosynthetically competent chloroplasts in gradients of silica sols. Arch. Biochem. Biophys. 168, 289-301 (1975)

Nass, M.M.K., Ben-Shaul, Y.: A novel closed circular duplex DNA in bleached mutant and green strains of *Euglena gracilis*. Biochem. Biophys. Acta 272, 130-136 (1972)

Nelson, N., Nelson, H., Schatz, G.: Biosynthesis and assembly of the proton-translocating adenosine triphosphate complex from chloroplasts. Proc. Natl. Acad. Sci. USA 77, 1361-1364 (1980)

Padmanabhan, U., Green, B.R.: The kinetic complexity of *Acetabularia* chloroplast DNA. Biochim. Biophys. Acta 521, 67-73 (1978)

Paoletti, C.A., Riou, G.: The mitochondrial DNA of malignant cells. Prog. Mol. Subcell Biol. 3, 203-248 (1973)

Pascoe, M.J., Ingle, J.: Distinction between nuclear satellite DNAs and chloroplast DNA in higher plants. Plant Physiol. 62, 975-977 (1978)

Paterson, B.M., Roberts, B.E., Kuff, E.L.: Structural gene identification and mapping by DNA-mRNA hybrid-arrested cell-free translation. Proc. Natl. Acad. Sci. USA 74, 4370-4374 (1977)

Pelham, H.R.B., Jackson, R.J.: An efficient mRNA-dependent translation system from reticulocyte lysates. Eur. J. Biochem. 67, 247-256 (1976)

Perry, R.P.: Processing of RNA. Annu. Rev. Biochem. 45, 605-629 (1976)

Pettijohn, D.E., Hecht, R.M., Stonington, O.G., Stamato, T.D.: Factors stabilizing DNA folding in bacterial chromosomes. In: DNA Synthesis *in vitro*. Wells, R.D., Inman, R.B. (eds.), pp. 145-162. Baltimore: Univ. Park Press 1973

Phillips, D.O., Carr, N.G.: Hybridization of prokaryotic and eukaryotic 5S rRNA to *Euglena gracilis* chloroplast DNA. FEBS Lett. 60, 94-97 (1975)

Possingham, J.V.: Some effects of mineral nutrient deficiencies on the chloroplasts of higher plants. 6th Int. Colloq. Plant Anal. Fertil. Probs. Tel Aviv 155-165 (1970)

Possingham, J.V.: Effect of light quality on chloroplast replication in spinach. J. Exp. Bot. 24, 1247-1260 (1973a)

Possingham, J.V.: Chloroplast growth and division during the greening of spinach leaf discs. Nature (New Biol.) 243, 93-94 (1973b)

Possingham, J.V.: Controls to chloroplast division in higher plants. J. Microsc. Biol. Cell 25, 283-288 (1976)

Possingham, J.V., Rose, R.J.: Chloroplast replication and chloroplast DNA synthesis in spinach leaves. Proc. R. Soc. London Ser. B 193, 295-305 (1976)

Possingham, J.V., Saurer, W.: Changes in chloroplast number per cell during leaf development in spinach. Planta 86, 186-194 (1969)

Possingham, J.V., Smith, J.W.: Factors affecting chloroplast replication in spinach. J. Exp. Bot. 23, 1050-1057 (1972)

Possingham, J.V., Cran, D.G., Rose, R.J., Loveys, B.R.: Effect of green light on the chloroplasts of spinach leaf discs. J. Exp. Bot. 26, 33-42 (1975)

Radloff, R., Bauer, W., Vinograd, J.: A dye-buoyant density method for the detection and isolation of closed circular duplex DNA: the closed circular DNA in HeLa cells. Proc. Natl. Acad. Sci. USA 57, 1514-1521 (1967)

Rawson, J.R.Y.: The characterization of *Euglena gracilis* DNA by its reassociation kinetics. Biochem. Biophys. Acta 402, 171-178 (1975)

Rawson, J.R.Y., Boerma, C.: Influence of growth conditions upon the number of chloroplast DNA molecules in *Euglena gracilis*. Proc. Natl. Acad. Sci. USA 73, 2401-2404 (1976)

Rawson, J.R., Kushner, S.R., Vapnek, D., Alton, N.K., Boerma, C.L.: Chloroplast ribosomal RNA genes in *Euglena gracilis* exist as three clustered tandem repeats. Gene 3, 191-209 (1978)

Reisfeld, A., Jakob, K.M., Edelman, M.: Characterization of the 32000 dalton chloroplast membrane protein. II. The molecular weight of chloroplast messenger RNAs translating the precursor to P-32000 and full-size RuDP carboxylase large subunit. In: Chloroplast Development. Akoyunoglou, G., Argyroudi-Akoyunoglou, J.H. (eds.), pp. 669-674. Amsterdam: Elsevier/North Holland 1978

Renner, O.: Die Scheckung der Oenotherenbastarde. Biol. Zentralbl. 44, 309-336 (1924)
Renner, O.: Die pflanzlichen Plastiden als selbständige Elemente der genetischen
  Konstitution. Ber. Verh. Sächs. Akad. Wiss. Leipzig. Math. Phys. Kl. 86, 241-266 (1934)
Richards, O.C.: Hybridization of *Euglena gracilis* chloroplast and nuclear DNA. Proc. Natl.
  Acad. Sci. USA 57, 156-163 (1967)
Richards, O.C., Manning, J.E.: Replication of chloroplast DNA in *Euglena gracilis*. Le
  Cycle Cellulaires et leur Blocage. Colloq. Int. C.N.R.S. 240, 213-221 (1975)
Ris, H.: Ultrastructure and molecular organization of genetic systems. Can. J. Gen. Cytol.
  3, 95-120 (1961)
Rochaix, J.D.: Restriction endonuclease map of the chloroplast DNA of *Chlamydomonas
  reinhardii*. J. Mol. Biol. 126, 597-617 (1978)
Rochaix, J.D., Allet, B.: Structural properties of the intervening sequence in the chloroplast
  ribosomal 23S RNA gene of *Chlamydomonas reinhardii*. In: Genome Organisation and
  Expression in Plants. Leaver, Ch. (ed.), p. 139. Edinburgh: NATO Advanced Studies
  Institute; FEBS Advanced Course 1979
Rochaix, J.D., Malnoe, P.: Anatomy of the chloroplast ribosomal DNA of *Chlamydomonas
  reinhardii*. Cell 15, 661-670 (1978)
Rochaix, J.D., Malnoe, P., Chua, N.H., Spahr, P.F.: Gene localization and transcription on
  the chloroplast DNA of *Chlamydomonas reinhardii*. In: Genome Organisation and
  Expression in Plants. Leaver, Ch. (ed.), p. 153. Edinburgh: NATO Advanced Studies
  Institute; FEBS Advanced Course 1979
Rose, R.J., Possingham, J.V.: The localization of (^{3}H) thymidine incorporation in the DNA
  of replicating spinach chloroplasts by electron microscope autoradiography. J. Cell. Sci.
  20, 341-355 (1976a)
Rose, R.J., Possingham, J.V.: Chloroplast growth and replication in germinating spinach
  cotelydons following massive γ irradiation of the seed. Plant Physiol. 57, 41-46 (1976b)
Rose, R.J., Cran, D.G., Possingham, J.V.: Distribution of DNA in dividing spinach
  chloroplasts. Nature (London) 251, 641-642 (1974)
Rose, R.J., Cran, D.G., Possingham, J.V.: Changes in DNA synthesis during cell growth and
  chloroplast replication in greening spinach leaf discs. J. Cell Sci. 17, 27-41 (1975)
Ruppel, H.G.: Nucleinsäuren in Chloroplasten. I. Charakterisierung der DNS und RNS von
  Antirrhinum majus. Z. Naturforsch. 22b, 1068-1076 (1967)
Ryan, R., Grant, D., Chiang, K.-S., Swift, H.: Isolation and characterization of
  mitochondrial DNA from *Chlamydomonas reinhardii*. Proc. Natl. Acad. Sci. USA 75,
  3268-3272 (1978)
Ryter, A.: Association of the nucleus and the membrane of bacteria: a morphological study.
  Bacteriol. Rev. 34, 39-54 (1968)
Saccone, C., Pepe, G., Bakker, H., Kroon, A.: The genetic organization of rat liver
  mitochondrial DNA. In: Mitochondria 1977, Bandlow, W., Schweyen, R.J., Wolf, K.,
  Kaudewitz, F. (eds.), pp. 303-315. Berlin: de Gruyter 1977
Sager, R.: Genetic analysis of chloroplast DNA in *Chlamydomonas*. In: Advances in
  Genetics, Caspari, E.W. (ed.), vol. 19, pp. 287-340. London-New York: Academic Press
  1977
Sager, R., Lane, D.: Molecular basis of maternal inheritance. Proc. Natl. Acad. Sci. USA 69,
  2410-2413 (1972)
Schell, J., van Montagu, M., De Picker, A., De Waele, D., Engler, G., Genetello, C.,
  Hernalsteens, J.P., Holsters, M., Messens, E., Silva, B., van den Elsacker, S., van
  Larebeke, N., Zaenen, I.: *Agrobacterium tumefaciens*: What segment of the plasmid is
  responsible for the induction of crown gall tumors? In: Nucleic Acids and Protein
  Synthesis in Plants, Bogorad, L., Weil, J.H. (eds.), pp. 329-342. New York-London:
  Plenum Press 1977
Schmitt, J.M., Herrmann, R.G.: Fractionation of cell organelles in silica sol gradients. In:
  Methods in Cell Biology, Prescott, M. (ed.), Vol. 15, pp. 177-200. London-New York:
  Academic Press 1977
Schnepf, E.: Organellen-Reduplikation und Zellkompartimentierung. In: Probleme der
  biologischen Reduplikation, Sitte, P. (ed.), pp. 372-393. Berlin-Heidelberg-New York:
  Springer 1966

Schwartzbach, S.D., Hecker, L.I., Barnett, W.E.: Transcriptional origin of *Euglena* chloroplast tRNAs. Proc. Natl. Acad. Sci. USA 73, 1984-1988 (1976)

Schwarz, Zs., Kössel, H.: Sequenzing of the 3' terminal region of a 16S rRNA gene from *Zea mays* chloroplast reveals homology with *E.coli* 16S rRNA. Nature (London) 279, 520-522 (1979)

Schwarz, Zs., Kössel, H., Graf, L., Stutz, E.: Primary structure of the 5' terminal region of a 16S rRNA gene from *Euglena gracilis* chloroplasts. In: Genome Organisation and Expression in Plants. Leaver, Ch. (ed.), p. 141. Edinburgh: NATO Advanced Studies Institute; FEBS Advanced Course 1979

Scott, N.S., Smillie, R.M.: Evidence for the direction of chloroplast ribosomal RNA synthesis by chloroplast DNA. Biochim. Biophys. Res. Commun. 28, 598-603 (1967)

Scott, N.S., Shah, V.C., Smillie, R.M.: Synthesis of chloroplast DNA in isolated chloroplasts. J. Cell Biol. 38, 151-157 (1968)

Sebald, W., Hoppe, J., Wachter, E.: Amino acid sequence of the ATPase proteolipid from mitochondria, chloroplasts and bacteria (wild type and mutants). In: Function and Molecular Aspects of Biomembrane Transport. Quagliariello, E., Palmieri, F., Papa, S., Klingenberg, M. (eds.), pp. 63-74. Amsterdam: Elsevier/North Holland 1979

Shapiro, A.L., Viñuela, E., Maizel, J.V.: Molecular weight estimation of polypeptide chains by electrophoresis in SDS polyacrylamide gels. Biochem. Biophys. Res. Commun. 28, 815-820 (1967)

Shepard, D.C.: Chloroplast multiplication and growth in the unicellular alga *Acetabularia mediterranea*. Exp. Cell Res. 37, 93-110 (1965)

Shepherd, R.J.: DNA viruses of higher plants. Adv. Virus Res. 20, 305-339 (1976)

Sinclair, J., Wells, R., Deumling, B., Ingle, J.: The complexity of satellite deoxyribonucleic acid in a higher plant. Biochem. J. 149, 31-38 (1975)

Sitte, P.: Chromoplasten – bunte Objekte der modernen Zellbiologie Biol. unserer Zeit 7. Jg., 65-74 (1977)

Siu, C.-H., Chiang, K.-S., Swift, H.: Characterization of cytoplasmic and nuclear genomes in the colorless alga *Polytoma*. V. Molecular structure and heterogeneity of leucoplast DNA. J. Mol. Biol. 98, 369-391 (1975)

Skinner, D.M., Beattie, W.G.: Characterization of a pair of isopycnic twin crustacean satellite deoxyribonucleic acids, one of which lacks on base in each strand. Biochemistry 13, 3922-3929 (1974)

Slavik, N.S., Hershberger, Ch.L.: The kinetic complexity of *Euglena gracilis* chloroplast DNA. FEBS Lett. 52, 171-174 (1975)

Slavik, N.S., Hershberger, Ch.L.: Internal structural organization of chloroplast DNA from *Euglena gracilis* Z. J. Mol. Biol. 103, 503-581 (1976)

Southern, E.M.: Detection of specific sequences among DNA fragments separated by gel electrophoresis. J. Mol. Biol. 98, 503-517 (1975)

Spencer, D., Whitfeld, P.R.: The characteristics of spinach chloroplast DNA polymerase. Arch. Biochem. Biophys. 132, 477-488 (1969)

Sprey, B.: Zum Verhalten DNS-haltiger Areale des Plastidenstromas bei der Plastidenteilung. Planta 78, 115-133 (1968)

Sprey, B., Gietz, N.: Isolierung von Etioplasten und elektronenmikroskopische Abbildung membranassoziierter Etioplasten-DNA. Z. Pflanzenphysiol. 68, 397-414 (1973)

Steinmetz, A., Weil, J.H.: Hybridization of bean chloroplast transfer RNAs to chloroplast DNA. Biochim. Biophys. Acta 454, 429-435 (1976)

Steinmetz, A., Weil, J.H.: Origin of chloroplast-specific tRNA[leu] in *Phaseolus vulgaris*. In: Colloq. Int. C.N.R.S. Acides Nucleiques et Synthèse des Proteines chez les Végetaux, Bogorad, L., Weil, J.H. (eds.), pp. 259-264. Paris: Ed. CNRS 1977

Steinmetz, A., Mubumbila, M., Keller, M., Burkard, G., Weil, J.H., Driesel, A.J., Crouse, E.J., Gordon, K., Bohnert, H.J., Herrmann, R.G.: Mapping of tRNA genes on the circular DNA molecule of *Spinacia oleracea* chloroplasts. In: Chloroplast Development. Akoyunoglou, G., Argyroudi-Akoyunoglou, J.H. (eds.), pp. 573-580. Amsterdam: Elsevier/North Holland 1978

Stout, E.R., Arens, M.Q.: DNA polymerase from maize seedlings. Biochim. Biophys. Acta 213, 90-100 (1970)

Stubbe, W.: Genetische Analyse des Zusammenwirkens von Genom und Plastom bei *Oenothera*. Z. Indukt. Abstamm. Vererbungsl. 90, 288-298 (1959)

Stubbe, W.: Extrem disharmonische Genom-Plastom-Kombinationen und väterliche Plastidenvererbung bei *Oenothera*. Z. Vererbungsl. 94, 392-411 (1963)

Stubbe, W.: Die Plastiden als Erbträger. In: Probleme der biologischen Reduplikation, Sitte, P. (ed.), pp. 273-288. Berlin-Heidelberg-New York: Springer 1966

Stubbe, W., Pietsch, B., Kowallik, K.V.: Cytologische Untersuchungen über väterliche Plastidenvererbung und plastomabhängige Degeneration der Samenanlagen bei einem *Oenothera*-Bastard. Biol. Zentralbl. 97, 39-52 (1978)

Stutz, E.: The kinetic complexity of *Euglena gracilis* chloroplasts DNA. FEBS Lett. 8, 25-28 (1970)

Stutz, E., Crouse, E.J., Graf, L., Jenni, B., Kopecka, H.: Structural and functional analysis of *Euglena gracilis* chloroplast DNA. In: Genetics and Biogenesis of Chloroplasts and Mitochondria, Bücher, Th., Neupert, W., Sebald, W., Werner, S. (eds.), pp. 339-346. Amsterdam: Elsevier/North Holland 1976

Swinton, D., Hanawalt, Ph.C.: *In vivo* specific labelling of *Chlamydomonas* chloroplast DNA. J. Cell Biol. 54, 592-597 (1972)

Terpstra, P., De Vries, H., Kroon, A.M.: Properties and genetic localization of mitochondrial transfer RNAs of *Neurospora crassa*. In: Mitochondria 1977, Bandlow, W., Schweyen, R.J., Wolf, K,, Kaudewitz, F. (eds.), pp. 291-302. Berlin: De Gruyter 1977

Tewari, K.K., Wildman, S.G.: Chloroplast DNA from tobacco leaves. Science 153, 1269-1271 (1966)

Tewari, K.K., Wildman, S.G.: DNA polymerase in isolated tobacco chloroplasts and nature of the polymerized product. Proc. Natl. Acad. Sci. USA 57, 689-696 (1967)

Tewari, K.K., Wildman, S.G.: Information content in the chloroplast DNA. In: Control of Organelle Development. Symp. Soc. Exp. Biol. Vol. 24, pp. 147-179. Cambridge: University Press 1970

Tewari, K.K., Wildman, S.G.: Function of chloroplast DNA. I. Hybridization studies involving nuclear and chloroplast DNA with RNA from cytoplasmic (80S) and chloroplast (70S) ribosomes. Proc. Natl. Acad. Sci. USA 59, 569-576 (1968)

Tewari, K.K., Kolodner, R., Chu, N.M., Meeker, R. R.: Structure of chloroplast DNA. In: Nucleic Acids and Protein Synthesis in Plants, Bogorad, L., Weil, J.H. (eds.), pp. 15-36. New York-London: Plenum Press 1977

Tiboni, O., di Pasquale, G., Ciferri, O.: Purification, characterization and site of synthesis of chloroplast elongation factors. In: Chloroplast Development. Akoyunoglou, G., Argyroudi-Akoyunoglou, J.H. (eds.), pp. 675-678. Amsterdam: Elsevier/North Holland 1978

Thomas, J.R., Tewari, K.K.: Ribosomal-RNA genes in the chloroplast DNA of pea leaves. Biochim. Biophys. Acta 361, 73-83 (1974)

Tomas, R.N., Cox, E.R.: Observations on the symbiosis of *Peridinium balticum* and its intracellular alga. I. Ultrastructure. J. Phycol. 9, 304-323 (1973)

Tzagoloff, A., Meagher, P.: Assembly of the mitochondrial membrane system V. Properties of a dispersed preparation of the rutamycin-sensitive adenosine triphosphatase of yeast mitochondria. J. Biol. Chem. 246, 7328-7336 (1971)

Vedel, F., Quetier, F.: Study of wheat phylogeny by Eco RI analysis of chloroplastic and mitochondrial DNAs. Plant Sci. Lett. 13, 97-102 (1978)

Vedel, F., Quetier, F., Bayen, M.: Restriction endonuclease analysis of chloroplast and mitochondrial DNAs of higher plants. In: Colloq. Int. C.N.R.S. Acides Nucléiques et Synthèse des Protéines chez les Végétaux, Bogorad, L., Weil, J.H. (eds.), pp. 71-75. Paris: Ed. CNRS 1977

Walker, D.A.: Chloroplasts (and grana): aqueous (including high carbon fixation ability). In: Methods in Enzymology, San Pietro, A. (ed.), Vol. 23, pp. 211-220. London-New York: Academic Press 1971

Weber, K., Osborn, M.: The reliability of molecular weight determination by dodecyl-sulfate-polyacrylamide gel electrophoresis. J. Biol. Chem. 244, 4406-4412 (1969)

Weil, J.H., Burkard, G., Guillemaut, P., Jeannin, G., Martin, R., Steinmetz, A.: tRNAs and aminoacyl-tRNA synthetases in plant cytoplasm, chloroplasts and mitochondria. In: Nucleic Acids and Protein Synthesis in Plants, Bogorad, L., Weil, J.H. (eds.), pp. 97-120. New York-London: Plenum Press 1977

Wells, R., Birnstiel, M.: Kinetic complexity of chloroplastal deoxyribonucleic acid and mitochondrial deoxyribonucleic acid from higher plants. Biochem. J. 112, 777-786 (1969)

Wells, R., Ingle, J.: The constancy of the buoyant density of chloroplast and mitochondrial deoxyribonucleic acids in a range of higher plants. Plant Physiol. 46, 178-179 (1970)

Wells, R., Sager, R.: Denaturation and the renaturation kinetics of chloroplast DNA from *Chlamydomonas reinhardi*. J. Mol. Biol. 58, 611-622 (1971)

Whitfeld, P.R., Spencer, D.: Buoyant density of tobacco and spinach chloroplast DNA. Biochim. Biophys. Acta 157, 333-343 (1968)

Whitfeld, P.R., Herrmann, R.G., Bottomley, W.: Mapping of the ribosomal RNA genes on spinach chloroplast DNA. Nucleic Acid. Res. 5, 1741-1751 (1978a)

Whitfeld, P.R., Leaver, Ch.J., Bottomley, W., Atchison, B.A.: Low-molecular-weight (4.5S) ribonucleic acid in higher-plant chloroplast ribosomes. Biochem. J. 175, 1103-1112 (1978b)

Winkler, H.: Vererbung und Ursache der Parthenogenese im Pflanzen- und Tierreich. Jena: Fischer 1920

Wollgiehn, R., Mothes, K.: Über die Incorporation von ^{3}H-Thymidin in die Chloroplasten-DNA von *Nicotiana rustica*. Exp. Cell. Res. 35, 52-57 (1964)

Wong, F.Y., Wildman, S.G.: Simple procedure for isolation of satellite DNA's from tobacco leaves in high yield and demonstration of minicircles. Biochem. Biophys. Acta 259, 5-12 (1972)

Woodcock, C.L.F., Bogorad, L.: Evidence for variation in the quantity of DNA among plastids of *Acetabularia*. J. Cell Biol. 44, 361-375 (1970)

Woodcock, C.L.F., Fernández-Morán, H.: Electron microscopy of DNA conformations in spinach chloroplasts. J. Mol. Biol. 31, 627-631 (1968)

Wu, M., Davidson, N., Attardi, G., Aloni, Y.: Expression of the mitochondrial genome in HeLa cells XIV. The relative positions of the 4 s RNA genes and of the ribosomal RNA genes in mitochondrial DNA. J. Mol. Biol. 71, 81-93 (1972)

Wuttke, H.-G.: Circular DNA in chromoplasts of *Tulipa gesneriana*. Planta 132, 317-319 (1976)

Yokumura, E.: An electron microscopic study of DNA-like fibrils in chloroplasts. Cytologia 32, 361-377 (1967)

Yoshida, Y., Laulhère, P., Rozier, C., Mache, R.: Visualization of folded chloroplast DNA from spinach. Biol. Cell 32, 187-190 (1978)

Zielinski, R.E., Price, C.A.: Synthesis of thylakoid membrane proteins by chloroplasts isolated from spinach. Cytochrome b559 and P700-chlorophyll a-protein. J. Cell Biol. 85, 435-445 (1980)

Zubay, G., Chambers, D.A., Cheong, L.C.: Cell-free studies on the regulation of the *lac* operon. In: The Lactose Operon, Beckwith, J.R., Zipser, D. (eds.), pp. 375-391. New York: Cold Spring Harbor Laboratory 1970

RNA and Protein Synthesis in Plastid Differentiation

R. WOLLGIEHN and B. PARTHIER

Akademie der Wissenschaften der DDR, Institut für Biochemie der Pflanzen
Halle/Saale, GDR

I. Introduction

Twelve years ago, when we wrote a similar survey on nucleic acids and protein synthesis in plastids (Parthier and Wollgiehn 1966), the experimental evidence available at that time was suited to prove the flow of genetic information from DNA into protein inside the organelles. These data favored suggestions on the genetic autonomy of plastids which were also supported by earlier investigations demonstrating the non-Mendelian mode of chloroplast inheritance. During the last decade, an extensive literature has been accumulated dealing with occurrence, properties, coding, and synthesis of those macromolecular plastid constituents necessary to replicate, transcribe, and translate the genetic information for the construction of functional photosynthetic organelles.

At present, only a gross and superficial look at the synthesis of plastid macromolecules and their biochemical incorporation into the whole biogenetic process would maintain the previous idea of plastid autonomy. Plastid differentiation, development, or specialization is heavily dependent on nuclear gene expression in the surrounding cytoplasm, although plastid DNA has a size sufficient to code for more than 150 proteins of 50,000 daltons molecular weight. This amount is roughly equivalent to the number of proteins present in the green organelles including the enzymes necessary to perform the main chloroplast functions in photosynthesis, syntheses of carbohydrates, lipids, and proteins. Thus from a qualitative point of view, plastids seem to possess the *potential* for a high degree of autonomy, but we have to consider the *actual* contribution of the plastidic synthetic machinery for the assembly of functional organelles. Presently, a concept is generally accepted which supports the idea that differentiation as well as replication of plastids is partly controlled by the genome and partly by the plastome (recent reviews: Woodcock and Bogorad 1971; Boulter et al. 1972; Börner 1973; Ciferri 1975; Parthier et al. 1975; Parthier 1976; Ellis 1977; Kung 1977).

Abbreviations. *Aa-Rs:* aminoacyl-tRNA synthetase; *RuBPCase:* Ribulose-bisphosphate carboxylase; *Leu-RS:* leucyl-tRNA synthetase, and respectively, *Phe-RS, Ile-RS* etc.; $tRNA^{Leu}$: leucine-specific tRNA species, and, respectively, $tRNA^{Phe}$, $tRNA^{Ile}$ etc.; m^7G: 7-methylguanosine; acp^3U: 3-(3-amino-3-carboxypropyl)uridine; ms^2i^6A: 2-methylthio-$N^6(\triangle^2$-isopentenyl)adenosine.

The aim of this survey is to show the relationship in plastids between gene expression and the differentiation processes after they have been triggered by external effectors, e.g., light. In many plants only illumination induces the transformation of precursor organelles (proplastids, etioplasts) into photosynthetically active organelles (chloroplasts). During this process, changes in the pattern of differential gene activation at the DNA level cannot be demonstrated as yet. Effector-dependent appearance (or disappearance) or changes in activity and/or conformation of the components and products of gene expression, however, may be a suitable way to extrapolate the results to processes which take place at the genetic level.

The intention of our review is devoted to the alterations of RNA and protein syntheses during light-induced plastid differentiation. However, since knowledge of the properties of the transcription and translation constituents seems to us a prerequisite for the understanding of the process, they will be briefly described in the first part of the paper. More detailed information is available in recent reviews (Ellis and Hartley 1974; Parthier et al. 1975; Ellis 1977; Whitfeld 1977). In the second part we will discuss the macromolecular events on transcription and translation niveau, which are observed after illumination and which obviously enable etiolated plastids to be transformed into mature green organelles (see also Sundqvist et al., this vol.).

II. Plastid Transcription and Translation Elements

A. Ribosomal RNA and Ribosomes

Cells of green tissues of higher plants and of eukaryotic algae contain two classes of ribosome, as was first shown by Lyttleton (1962). Together with the ribosomes of bacteria, blue-green algae and mitochondria, chloroplast particles belong to the 70S class of ribosomes. They differ from the 80S cytoplasmic ribosomes in molecular weight of ribosomal RNA, in composition of the protein constituents, in some physical properties, and in some functional aspects of protein synthesis, for example, in their sensitivity to inhibitors of protein synthesis (see Table 1).

Like bacterial ribosomes, chloroplast ribosomes dissociate into 50S and 30S subunits, if the Mg^{2+} concentration is less than 1 mM. 80S cytoplasmic ribosomes, on the other hand, are relatively stable and require prolonged dialysis in the absence of Mg^{2+} to dissociate into 55S and 35S subunits (see Ellis and Hartley 1974). The large subunit (50S) harbors a rRNA molecule of 1.1×10^6 daltons (23S) and additionally 0.04×10^6 rRNA (5S) and at least in flowering plants also 4.5S RNA (Bohnert et al. 1976; Dyer et al. 1977; Whitfeld et al. 1977); the small subunit (30S) contains rRNA of 0.56×10^6 daltons (16S). In contrast, cytoplasmic ribosomes contain rRNA of 1.3×10^6 (25S), 0.7×10^6 (18S), 0.04×10^6 (5S) and 5.8S RNA (see Boulter et al. 1972; Ellis and Hartley 1974).

During the last few years, the nucleotide sequences of several chloroplast ribosomal RNA's have been partially determined. It was found that in the 5S RNA there are great similarities between different higher plants, and the 4.5S RNA's are

Table 1. Characteristics of the transcription and translation components in chloroplasts and in the nucleo-cytoplasm of eukaryotic plants

Component	Property	Chloroplast	Nucleo-cytoplasm
Ribosomes	S values, also of subunits	70 (50 and 30)	80 (60 and 40)
rRNA	M.W. in 10^6 daltons	1.1 (23 S)	1.3 (25 S)
		0.56 (16 S)	0.7 (18 S)
		0.04 (5 S)	0.05 (5.8 S)
		0.035 (4.5 S)	0.04 (5 S)
	M.W. of the primary transcription product (10^6 daltons)	max. 2.7	> 2.9
r-Proteins	Electrophoretic pattern, immunological properties	Different from each other	
mRNA	3'Poly(A)sequences	− (but see p. 103)	+
tRNA	Recognition by cognate chloroplast synthetases	+	−
Aa-tRNA	Acylation of chloroplast tRNAs	+	−
RNA polymerase	Sensitivity to rifampicin	+	−
Protein synthesis	Chain initiation by F-Met-tRNA$_F$	+	−
	Elongation factors	Different from each other	
	Inhibition by chloramphenicol, lincomycin, spectinomycin, streptomycin	+	−
	Inhibition by cycloheximide, anisomycin	−	+

also related to one another, but there are more differences than between the 5S plastid rRNA's. The low molecular weight RNA's of the cytoplasmic ribosomes are quite distinct from those of the chloroplasts (Dyer and Bowman 1976; Dyer et al. 1977). 16S rRNA of *Euglena gracilis* chloroplasts also has been characterized in terms of its two-dimensional electrophoretic "finger-print". A significant number of oligonucleotide sequences are shared by the chloroplast and at least certain prokaryotes, but not by the chloroplast and eukaryotes. Thus, the 16S RNA is clearly related structurally and presumably also functionally to prokaryotic 16S rRNA (Zablen et al. 1975; Buetow et al. 1976).

Mature rRNA isolated from chloroplasts can be separated from cytoplasmic rRNA by means of polyacrylamide gel electrophoresis. The 23S rRNA is fairly labile in many plant species. As a rule, the molecule can be obtained in its native form only after Mg^{2+} containing buffers of high ionic strength and low temperature has been used during the preparation procedure. The cleavage of the 1.1×10^6 rRNA proceeds, under denaturing conditions, into distinct pieces due to hidden breaks which in situ are held together by noncovalent bonds. The hidden breaks are obviously caused by specific endogenous ribonuclease active during maturation or aging of the ribosomes (Munsche and Wollgiehn 1974; Mache et al. 1978); in contrast, 16S rRNA is a much more stable molecule. The properties of chloroplast rRNA have been reviewed elsewhere in more detail (Ellis and Hartley 1974; Parthier et al. 1975; Whitfeld 1977).

Studies with isolated chloroplasts of different plant species and of enucleated *Acetabularia* have shown that the organelles are able to synthesize RNA (see Parthier et al. 1975; Whitfeld 1977). Several recent investigations not only demonstrate the synthesis of 1.1 and 0.56×10^6 ribosomal RNA in isolated organelles but provide more evidence on the mechanism of rRNA synthesis in chloroplasts. In tobacco leaves (Munsche and Wollgiehn 1973), spinach leaves (Hartley and Ellis 1973; Mache et al. 1978), cultured spinach leaf tissue (Detchon and Possingham 1973), *Euglena gracilis* (Heizmann 1974a), *Chlamydomonas reinhardii* (Miller and McMahon 1974) and *Spirodela* (Posner and Rossner 1975) relatively stable rRNA precursor molecules of about 1.2 and 0.65×10^6, respectively, could be detected; they were transferred into mature 1.1 and 0.56×10^6 rRNA's during a following chase period. Similar precursors were found also in isolated chloroplasts from tobacco, spinach, and *Euglena gracilis* under conditions of DNA-dependent RNA synthesis (Wollgiehn and Munsche 1972; Carritt and Eisenstadt 1973; Hartley and Ellis 1973; Bohnert and Schmitt 1974).

A common high-molecular ribosomal precursor RNA molecule was detected under usual preparation conditions only in few cases. This primary transcription product of the ribosomal gene with M.W. of 2.7×10^6 daltons was extracted from *Spinacia* and *Spirodela* chloroplasts after pulse labeling in vivo (Hartley and Ellis 1973; Posner and Rosner 1975). This is approximately the same size as the recently detected primary transcription product of 2.1×10^6 daltons rRNA in *E.coli* (Nikolaev et al. 1974).

Using labeled nucleosides as precursors and isolated spinach chloroplasts, Bohnert and Schmitt (1973), Hartley and Ellis (1973), and Bohnert et al. (1976) have synthesized RNA in a light-dependent reaction as a consequence of the light-driven production of the nucleotide triphosphates. As a result they isolated several high molecular weight (2.7×10^6 as the largest product) ribosomal precursor RNA's as confirmed by competitive hybridization. From chase experiments a more-step maturation pathway for chloroplast rRNA in *Spinacia* was proposed (Bohnert et al. 1976; Hartley et al. 1977):

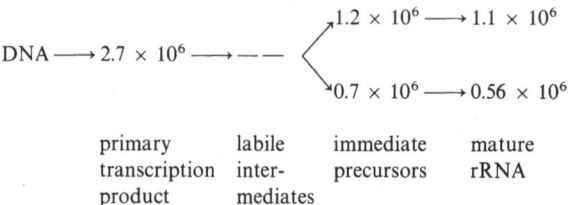

DNA $\longrightarrow 2.7 \times 10^6 \longrightarrow ---$
$\quad\quad 1.2 \times 10^6 \longrightarrow 1.1 \times 10^6$
$\quad\quad 0.7 \times 10^6 \longrightarrow 0.56 \times 10^6$

primary	labile	immediate	mature
transcription	inter-	precursors	rRNA
product	mediates		

Recently progress was made showing that in vitro the majority of transcription products of chloroplast DNA fragments is rRNA. It was further shown by these methods that chloroplast DNA of higher plants contains two (*Euglena* three) rRNA genes per circular DNA molecule (see Herrmann et al., this vol.).

It should be mentioned that the primary transcriptions product also contains the 5S and 4.5S RNA, which are both located in the large subunit of the ribosome (Bohnert et al. 1976).

The physical and biological homology in the rRNA constituents between plastid ribosomes and prokaryotic ribosomes does not appear to exist for the ribosomal proteins. Progress in two-dimensional gel electrophoresis demonstrated clear differences in protein content and number of r-proteins between chloroplast and cytoplasmic ribosomes of the same cell. On the other hand, chloroplast r-proteins also differ from prokaryotic 70S ribosome proteins (*E.coli*, blue-green algae). Approximately 75 proteins were found in wheat chloroplast ribosomes, and 85 spots were detected in cytoplasmic ribosome preparations (Jones et al. 1972). Pea chloroplast ribosomes contain 71 proteins (Oparin et al. 1975). *Euglena* 88S cytoplasmic ribosomes consist of 73–83 r-proteins and chloroplast 70S ribosomes of 56 to 60 proteins. Fourteen of the latter are acidic proteins. The large subunit of the chloroplast ribosome contains 30 to 34, the small subunit 22 proteins (Freyssinet 1975, 1977). Thus *Euglena* ribosomes show a higher number of r-protein constituents than *Chlamydomonas* (48–50 in chloroplast and 65 in cytoplasmic ribosomes (Freyssinet and Schiff 1974; Hanson et al. 1974; Spiess 1977). About 55 proteins can be separated from prokaryotic ribosomes (Wittmann 1972; Oparin et al. 1975). The available data point out minor coincidences in the electrophoretic and immunological properties of the ribosomal proteins from bacteria, blue-green algae, and chloroplasts and cytoplasm of different plant species (Gualerzi and Cammarano 1970; Vasconcelos and Bogorad 1971; Jones et al. 1972; Wittmann 1972; Oparin et al. 1975). However, in a few cases a small immunological cross reaction was found between proteins from the chloroplast and cytoplasmic ribosomes of the same organism (Gualerzi et al. 1974; Freyssinet et al. 1976; Freyessinet 1977).

Studies with translation inhibitors acting on 80S ribosomes (cycloheximide) or 70S ribosomes (chloramphenicol, lincomycin) showed that most of the chloroplast ribosomal proteins are synthesized at 80S cytoplasmic ribosomes and only few proteins at chloroplast ribosomes (Galling 1971; Ellis and Hartley 1971; Freyssinet 1977; Honeycutt and Margulies 1973; Kloppstech and Schweiger 1974). Genetic experiments with tobacco leaves and *Chlamydomonas* mutants, as well as nuclear transplantation experiments with different *Acetabularia* species, have clearly demonstrated that most of the chloroplast ribosomal proteins are coded for by nuclear DNA, but the genetic information for some proteins of the 70S particles are encoded in chloroplast DNA (Bourque and Wildman 1973; Kloppstech and Schweiger 1973; Bogorad et al. 1975). Genetic experiments with *Chlamydomonas* provide evidence that at least one structural gene for r-proteins is located in the organelle DNA (Bogorad et al. 1977).

B. Messenger RNA

The occurrence and synthesis of messenger RNA in plastids is now well established. Hybridization of chloroplast RNA fractions synthesized in vivo or in isolated chloroplasts from tobacco, spinach, corn, and pea to homologous chloroplast DNA indicated annealing of 20% to 50% of the chloroplast DNA (Tewari and Wildman 1970; Whitfeld et al. 1973; Haff and Bogorad 1976b; Tewari et al. 1977). *Euglena* DNA, in vitro labeled with ^{125}J, forms hybrids up to 32% with *Euglena* chloroplast total RNA (Rawson 1975). These results suggest that a much larger part of the chloroplast genome must be transcribed than the 10% equivalent to rRNA and tRNA genes.

Already in 1964, Brawermann and Eisenstadt isolated a RNA fraction from *Euglena* chloroplasts which stimulated protein synthesis in vitro using ribosomes and cytoplasm from *E.coli*. Much progress has been made in the last years in succeeding characterization of mRNA's for individual chloroplast proteins. Whitfeld et al. (1973) have developed a coupled transcription — translation system consisting of purified components from *E. coli* (ribosomes, high-speed supernatant and RNA polymerase) and chloroplast DNA from spinach. At least 90% of the resulting heterogeneous polypeptide population was dependent on the presence of chloroplast DNA and therefore on newly synthesized messenger RNA. The major polypeptide translated under these conditions was identified as the large subunit of ribulosebisphosphate carboxylase (RuBPCase) (Bottomley and Whitfeld 1978). In an elegant way Coen et al. (1977) have also shown that the large subunit from *Zea mays* is encoded in chloroplast DNA. In an in vitro transcription-translation system (RNA polymerase from *E.coli*, translation with a rabbit reticulocyte lysate) they were able to direct the synthesis of this protein using a BamHJ-generated chloroplast DNA sequence cloned in *E.coli*.

Hartley et al. (1975) and Bottomley et al. (1976) reported that the addition of spinach chloroplast total RNA to a cell-free system of *E.coli* results in the formation of several proteins with properties analogous to those synthesized in intact isolated chloroplasts. One of these proteins has been identified, by gel electrophoresis and finger-printing of the tryptic digest, as the large subunit of RuBPCase. In a further step, a 13 to 14S, poly(A)-free RNA from chloroplasts of *Spinacia*, *Spirodela*, and *Chlamydomonas* was translated in a cell-free protein-synthesizing system from *E.coli* (Rosner et al. 1975, 1977; Wheeler and Hartley 1975; Howell et al. 1976; Gelvin et al. 1977), and the analogous RNA isolated from *Euglena* chloroplasts in a wheat-germ system (Sagher et al. 1976). The product was identified by different methods (gel electrophoresis, M.W. estimation, isoelectric focusing, and precipitation with monospecific antibodies) as the large subunit of the RuBPCase (Fig. 1; Wheeler and Hartley 1975; Howell et al. 1976; Sagher et al. 1976; Gelvin et al. 1977). In *Chlamydomonas* translation of this messenger seems to occur on small polysomes (n = 2 to 5; Howell et al. 1976).

These results confirm the genetic claim that the large subunit of RuBPCase is encoded in chloroplast DNA (Chan and Wildman 1972), whereas the small subunit originates from and is genetically controlled by the nucleo-cytoplasmic compartment (Ellis 1977).

Spirodela chloroplast 13 to 14S poly(A)-free RNA contains the information also for a light-induced membrane protein of 32,000 daltons M.W. (Rosner et al. 1977). The size of the messenger for this protein has been determined as 0.5×10^6 daltons and for the large subunit of ribulosebisphosphate carboxylase as $0.6 - 0.7 \times 10^6$ daltons (Fig. 1; Edelman and Reisfeld 1978; Reisfeld et al. 1978). Chloroplasts contain also mRNA's for a number of other hydrophobic thylakoid polypeptides synthesized in vitro but not identified so far (Michaelis and Margulies 1975; Ellis 1975a).

A chloroplast RNA of M.W. 0.45 to 0.5×10^6 daltons has been demonstrated and characterized by several authors. There is reason to assume its identity with the mRNA for one of the two chloroplast polypeptides mentioned (Hartley and Ellis 1973; Grierson and Loening 1974; Posner and Rosner 1975; Wollgiehn et al. 1976,

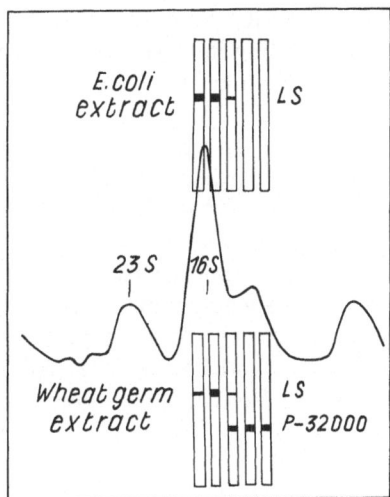

Fractionated ct-RNA

Fig. 1. Functional characterization of the two major mRNA activities in *Spirodela* chloroplasts. Chloroplast RNA extracts were separated by sucrose gradient centrifugation. Different RNA fractions were tested for template activity in either an *E.coli* or wheat germ cell free system. The distribution of the in vitro synthesized [³⁵S]-methionine labeled proteins were determined on polyacrylamide gels. *LS:* Large subunit of RuBPCase; *P-32000:* Membrane protein of M.W. 32,000. (After Rosner et al. 1977; Edelman and Reisfeld 1978)

1978). This RNA is located in the chloroplast fraction of *Phaseolus aureus* and *Nicotiana rustica* (Grierson and Loening 1974; Wollgiehn et al. 1978). In tobacco synthesis of 0.45×10^6 RNA is inhibited by rifamycin (Wollgiehn et al. 1978), showing that the RNA is transcribed on chloroplast DNA. The RNA is synthesized in chloroplasts of all green tobacco leaves independent of their age (Wollgiehn et al. 1976). The RNA does not contain a poly(A)-sequence (Rosner et al. 1975; Wollgiehn et al. 1978).

Although the messenger for the large subunit of the RuBPCase does not contain a poly(A)-sequence, it is not yet sufficiently documented whether or not other chloroplast mRNA's contain 3'poly(A)-sequences as eukaryotic mRNA species do. Few experimental data favor this possibility. Chloroplasts of at least one higher plant, wheat, contain poly(A)-polymerase (Burkard and Keller 1974) thus one can expect that some chloroplast mRNA's contain poly(A)-sequences. Using the iodine-labeling technique, Haff and Bogorad (1976b) have found that poly(A)-RNA comprises about 0.5% of the total RNA from purified corn chloroplasts. 65% of this poly(A)-RNA hybridizes to chloroplast DNA. The chloroplast poly(A) has an average length of about 45 nucleotides. The poly(A)-sequence is probably added to plastid RNA following their transcription, because poly(dT) was not detected in chloroplast DNA (Haff and Bogorad 1976b). According to Wheeler and Hartley (1975) ³²P-labeled total RNA isolated from spinach chloroplasts contains about 1.0% poly(A)-RNA which accounts for 0.003% poly(A)-sequences. The corresponding data for total leaf RNA are 10% and 0.08%, respectively. Tobacco chloroplasts also contain poly(A)-RNA, about 1% to 2% of the total chloroplast RNA radioactivity after ^{32}P-labeling in vivo. This RNA is heterogeneous in size after gel electrophoreses (Wollgiehn et al. 1978). Tewari et al. (1977) found that about 5% of the total chloroplast RNA contains poly(A), which hybridizes with 10%–20% of the chloroplast DNA. In a recent communication Bartolf and Price (1977) have shown that intact isolated spinach chloroplasts incorporate ^{32}P into poly(A)-RNA, the length of the poly(A)-sequence was 15 to 40 nucleotides.

It is not definitely known whether all mRNA's found in chloroplasts, especially those containing poly(A)-sequences, and translated in vivo on plastid ribosomes are transcribed at plastid DNA or whether messenger import from other parts of the cell exist. Some experimental data favor this hypothesis (Armstrong et al. 1971; Jennings and Ohad 1972; Sirevag and Levine 1972; Wildner 1976; Guignery and Duranton 1978; Heizmann and Verdier 1978), however at present there is no compelling evidence that it occurs.

C. Polysomes

The occurrence of polysomes in isolated chloroplasts was first described by Clark (1964) using density gradient centrifugation. This observation was confirmed in many other laboratories (see Parthier et al. 1975). The localization of the polysomes within the chloroplasts has been examined by electron microscopy, and further evidence came from analytical techniques. Both free and membrane-bound polysomes are present in chloroplasts (Chen and Wildman 1970; Chua et al. 1973a,b; Philippovich et al. 1973; Margulies and Michaels 1974, 1975; Michaels and Margulies 1975). The bound particles are attached to the membranes through nascent protein chains, and the connection can be abolished by treatment with inhibitors of protein synthesis (puromycin, chloramphenicol) as well as with salts of high ionic strength. The proportion between free and membrane-attached particles depends on the physiological state of the chloroplast or of the cell.

Electron micrographs of *Phaseolus vulgaris* leaf sections show polysomal whorls both at the outermost thylakoid membranes of the grana stacks and at single stroma thylakoids (Falk 1969). Helix-shaped polysomes have also been detected in free state in the matrix of wheat proplastids (Bartels and Weier 1967). In chloroplasts isolated from ten-day-old pea seedlings exposed to light for 19 h, circular polysome structures are localized in the thylakoids of the grana but are not bound to lamellae of the stroma (Philippovich et al. 1973). On the other hand, after illumination of dark-grown cells, in *Chlamydomonas reinhardii* chloroplast polysomes are attached to those parts of the thylakoid membranes which are not involved in the fusion to grana; they form closed, ring-like or polygonal penta- and hexamers. These bound polysomes can be released only by treatment with 500 mM KCl and 1 mM puromycin, suggesting that both ionic interactions and nascent poylpeptide chains are involved in the ribosome attachment to thylakoid membranes (Chua et al. 1973b; Margulies and Michaels 1974). Triton treatment of chloroplast membranes leaves a particulate component of residual membranes with attached polysomes which can be released by the nonionic detergent Nonidet P-40. This detergent removes some membrane polypeptides more completely than Triton, resulting in the release of polysomes (Margulies and Weistrop 1976).

Membrane-bound polysomes seem to be responsible for the synthesis of thylakoid proteins (Ellis 1975a). This was indicated by experimental data describing isolated membrane-bound polysomes from chloroplasts active in protein synthesis. The labeled products consist of membrane proteins inserted into the membranes apparently during the course of their synthesis (Micheals and Margulies 1975). On the other hand, the large subunit of RuBPCase is synthesized of free polysomes (Ellis 1975a; Howell et al. 1976; Dobberstein et al. 1977; see also Fig. 4).

D. Transfer RNA and Aminoacyl-tRNA Synthetases

The attachment of an amino acid to a specific tRNA by means of its cognate amino acid-activating enzyme (aminoacyl-tRNA-synthetase, Aa-RS) is a highly selective reaction which secures the genetically correct insertion of the amino acid into the nascent polypeptide chain at the polysome. The degeneration of the genetic code is reflected, at the translation level, by the existence of isoacceptor tRNA

Table 2. Chloroplast and cytoplasmic isoacceptor tRNA species of several plants

Isoacceptor	Number		Plant species	Reference
	Total	Chloroplast		
Glutamic acid	3	1	*Euglena gracilis*	Barnett et al. (1969)
Histidine	2	1	Cotton cotyledons	Merrick and Dure (1972)
Isoleucine	3–4	1	*Euglena gracilis*	Barnett et al. (1969), Goins et al. (1973)
Isoleucine	5	2 (1)	*Euglena gracilis*	Kislev et al. (1972)
Isoleucine	4	2	Cotton cotyledons	Merrick nd Dure (1972)
Leucine	5–6	2	*Euglena gracilis*	Parthier and Krauspe (1974)
Leucine	6 (9)	3	Bean leaves	Burkard et al. (1970); Guillemaut et al. (1975)
Leucine	6	2	Bean leaves	Williams et al. (1973)
Leucine	6	3	Cotton cotyledons	Merrick and Dure (1972)
Leucine	6	2	Tobacco leaves	Guderian et al. (1972)
Lysine	4	1	Cotton cotyledons	Merrick and Dure (1972)
Lysine	4	1	Bean leaves	Jeannin et al. (1976)
Methionine	3	1	*Euglena gracilis*	Goins et al. (1973)
Methionine	8	3	Bean leaves	Guillemaut and Weil (1975)
Methionine	5	2	Cotton cotyledons	Merrick and Dure (1972)
Methionine	5	2	Wheat leaves	Leis and Keller (1971)
Phenylalanine	2	1	*Euglena gracilis*	Barnett et al. (1969)
Phenylalanine	3	2	Bean leaves	Guillemaut et al. (1976)
Phenylalanine	4	2	Cotton cotyledons	Merrick and Dure (1972)
Proline	5	1	Bean leaves	Jeannin et al. (1976)
Threonine	2	1	*Euglena gracilis*	Parthier and Neumann (1977)
Tryptophan	3	2	Cotton cotyledons	Merrick and Dure (1972)
Tyrosine	4	2	Soybean cotyledons	Locy and Cherry (1978)
Valine	4	1	Bean leaves	Burkard et al. (1970)
Valine	5	2	Cotton cotyledons	Merrick and Dure (1972)

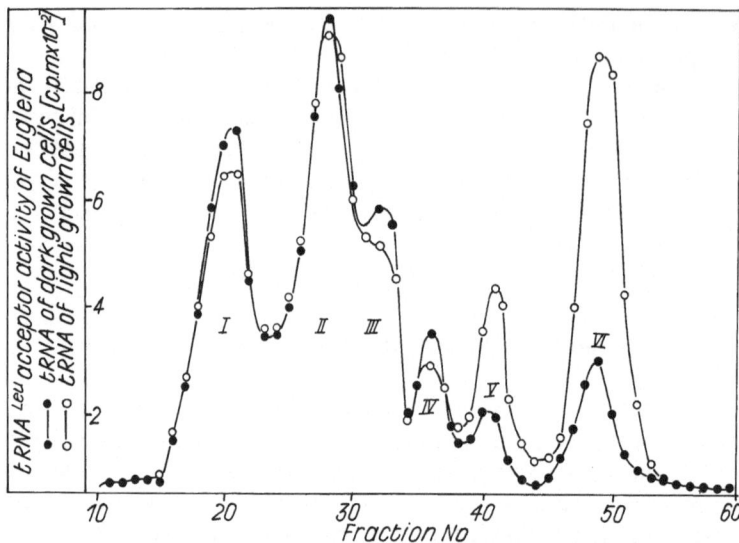

Fig. 2. Reversed-phase chromatography of the isoacceptor tRNA^Leu species of *Euglena gracilis*. Total tRNA fractions are prepared from bleached, dark-grown *(filled circles)* or green, illuminated cells *(open circles)* and charged with [^{14}C]-leucine by means of a crude leucyl-tRNA synthetase preparation from green cells (Parthier and Krauspe 1974). The [^{14}C]-leu-tRNA isoacceptors were separated on a RPC-5 system and the fractions counted for radioactivity. The tRNA of peaks *V* and *VI* is assumed to be of plastid origin

species. These isoaccepting tRNA's can recognize different code words for the same amino acid. Since the organelles of the eukaryotic cells contain their own sets of protein biosynthesis, one can predict at least 60 different tRNA species in a green plant cell, namely cytoplasmic, plastid, and mitochondrial species. This number might be still higher, because more than one isoaccepting tRNA occurs in a cell compartment, at least for amino acids with a high degeneration in their code words.

The presence of plastid-specific isoaccepting tRNA's has been demonstrated both in higher plants and algae (Table 2 and review by Weil 1978) using improved separation methods for tRNA fractions, e.g., chromatography on benzoylated DEAE-cellulose, or by the reversed phase systems (Fig. 2). Barnett and coworkers (1969) first demonstrated the presence of two isoacceptor species each for glutamic acid, phenylalanine, and isoleucine in green wild-type *Euglena gracilis,* but also the absence of each one isoacceptor tRNA species in plastid mutants. The plastid tRNA^Phe preparations from *Euglena* and bean lack the fluorescent base "Y", a hypermodified guanine derivative, adjacent to the anticodon; this tRNA species resembles tRNA^Phe of *E.coli* cells or of *Neurospora* mitochondria, but differs from the respective cytoplasmic tRNA^Phe species (Fairfield and Barnett 1971; Guillemaut et al. 1976; Hecker et al. 1976; Weil et al. 1977). At least five isoacceptor tRNA's for isoleucine (Kislev et al. 1972) or leucine (Parthier and Krauspe 1974) were shown to be present in the total tRNA fraction of *Euglena*. Two of them were found associated with chloroplasts when the tRNA preparations from green, dark-bleached and plastid mutant cells were compared. The chloroplast tRNA^Leu species but not the cytoplasmic tRNA^{Leu} species could be substituted by tRNA^{Leu} of the blue-green alga *Anacystis nidulans* in the aminoacylation reaction (Parthier and Krauspe 1974).

Compartment specificity of isoaccepting tRNA's in higher plants are likewise demonstrated in particular in J.H. Weil's laboratory. *Phaseolus* chloroplasts contain a specific tRNA for N-formyl-methionine, the initiator amino acid for polypeptide synthesis (Burkard et al. 1969). Only this plastid initiator tRNA^{Met} was formylated by endogenous transformylase or by the enzyme extracted from *E.coli*, but not from the cytoplasm of the higher plant. In certain other cases chloroplast isoacceptor tRNA's could be separated from their cytoplasmic and mitochondrial counterparts (Burkard et al. 1970, 1973; Guillemaut et al. 1973, 1975, 1976; Guillemaut and Weil 1975; Jeannin et al. 1976, 1978; see also review by Weil 1978). In general, one or two isoaccepting species were found in the green organelles (Table 2). In their chromatographic behavior, consequently in their primary structures, they seem to differ not only from the respective cytoplasmic species, but also from the mitochondrial species.

Chloroplast-specific tRNA's^{Phe}, tRNA's^{Lys} and tRNA's^{Leu} from *Phaseolus vulgaris* translate code words different from those recognized by their cytoplasmic counterparts (Ramiasa et al. 1977). At least in the case of tRNA^{Phe} it was shown by Guillemaut et al. (1975) that the chloroplast and cytoplasmic isoacceptor species differ in their primary structures apart from the anticodon triplet. Therefore, the organelle specificity of aminoacylation can be hardly interpreted as codon specificity. This idea is further supported by the existence of some isoacceptor tRNA^{Met} species which form base-pairs with the single known code triplet for methionine. On the other hand, the three cytoplasmic tRNA^{Leu} species of *Phaseolus* respond the three different codons; only one of them recognizes the same code word (UUG) which is recognized by chloroplast tRNA^{Leu} (Weil 1978). This example demonstrates the possibility of regulation of protein synthesis by the recognition specificity of tRNA species to cytoplasmic and organellar mRNA's.

The high specificity in aminoacylation reported for tRNA's^{Leu} (Parthier and Krauspe 1974) may not hold true for all 20 types of tRNA species. Studies on the charging specificity of plastid and cytoplasmic tRNA showed that the specificity found with Leu-RS is not observed for plastid Ph-RS, Ile-RS and few other synthetases of *Euglena* (Parthier and Krauspe 1973; Lesiewicz and Herson 1975; Parthier et al. 1978), cotton cotyledons (Brantner and Dure 1975), and bean plants (Jeannin et al. 1978).

Hecker et al. (1976) and Chang et al. (1976) using *Euglena* and Guillemaut et al. (1976) using *Phaseolus* purified chloroplast tRNA^{Phe} determined their base compositions. Close similarities exist to that of *E.coli*, as far as the minor hypermodified nucleosides (e.g., m^{7}G, acp^{3}U, or ms^{2}i^{6}A) are concerned; however, eight bases differ between the two chloroplast tRNA^{Phe} species (Guillemaut and Keith 1977). The differences are of subordinate significance because of purine — purine or pyrimidine — pyrimidine exchanges or methylations only. On the other hand, bean chloroplast tRNA^{Phe} differs from cytoplasmic wheat germ, lupine or pea tRNA^{Phe} in 25 base positions including the hypermodified G (Y base) in the anticodon loop. Purified spinach chloroplast tRNA do not contain ribosylzeatin which is present in total leaf tRNA (Vreman et al. 1978).

Progress has been made in the last years in order to prove the site of tRNA transcription on the plastid DNA. The most relevant information was contributed by tRNA-DNA hybridization studies using both plastid and nuclear DNA. Tewari and Wildman (1970) isolated a ^{32}P-labeled total tRNA mixture from tobacco chloroplasts, annealed it with chloroplast DNA of the same plant species and found hybridization values between 0.4% and 0.7%, corresponding with 4.4 to 7.9 × 10^5 daltons of the chloroplast DNA. This result was considered to prove tRNA synthesis inside the organelles.

On the other hand, in their studies with fractionated tRNA^{Leu} from bean leaves, Williams et al. (1973) did not obtain conclusive results which allow such a definite statement. Both chloroplast and cytoplasmic tRNA^{Leu} species were found to hybridize with chloroplast and nuclear DNA at about the same ratio (cDNA : nDNA) between 15 : 1 and 20 : 1 even under stringent hybridization conditions. Consequently, the authors interpreted their data as to the presence of complementary base sequences in both nuclear and chloroplast DNA responsible for the transcription of all tRNA^{Leu} isoacceptor species in bean leaves. The possibility that some of the plastidic tRNA species are coded for by nuclear DNA and transported into the organelles should not be excluded, however, this is not yet demonstrated experimentally. The number of plastid tRNA species transcribed at the plastid DNA (see below) is sufficient for translation at plastid ribosomes.

If tRNA transport through the plastid envelope is considered to take place, the plastid-specific tRNA's transcribed by nuclear genes might be specifically modified upon entering the organelle. Methylation of precursor tRNA's by plastid-specific tRNA methylases (Dubois et al. 1975) is one possibility of a tRNA modification which could sufficiently contribute to the chromatographic separation of isoacceptor species.

Most convincing are recent DNA-tRNA hybridization experiments from several laboratories demonstrating that the chloropalast genome does contain the cistrons for 20 to 26 tRNA species which were shown to be chloroplast-specific by several methods including ^{125}iodine labeling (Schwartzbach et al. 1976; McCrea and Hershberger 1976; both using *Euglena gracilis*; Haff and Bogorad 1976a, using corn; Weil et al. 1977, using bean leaves). All results are consistent with the hypothesis that the entire or nearly complete set of plastid tRNA's is transcribed from the organelle DNA (see also Herrmann et al., this vol.).

The interesting suggestion that certain plastid tRNA's (fraction-II-tRNA's) were exported into the cytoplasm was made by McCrea and Hershberger (1978). Their idea bases on the observations that these tRNA's isolated from *Euglena* cytoplasmic polyribosomes hybridize with chloroplast DNA and that these tRNA's are absent in isolated chloroplasts. The phenomenon might reflect a type of regulatory function of these tRNA's in the cytoplasmic synthesis for plastid polypeptides.

Plastid-specific aminoacyl-tRNA synthetases (Aa-RS), which specifically aminoacylate plastid tRNA, have been verified in green higher plants and algae (for references, see Parthier et al. 1975; Weil 1978). As soluble stroma enzymes they are easily leached out of the chloroplasts isolated in aqueous buffers, thus the unequivocal demonstration of the presence in these organelles is connected with some difficulties (Parthier et al. 1978). However, successful separation into

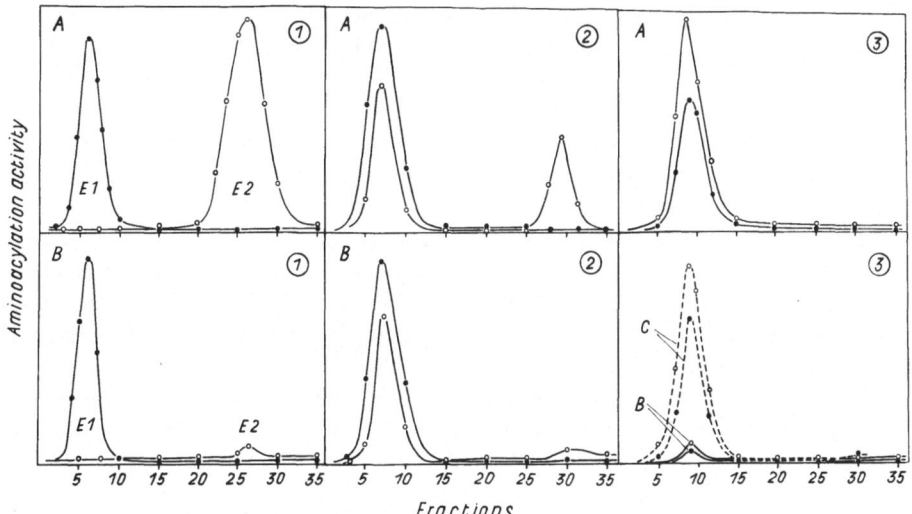

Fig. 3. Separation of chloroplast *(E1)* and cytoplasmic *(E2)* aminoacyl-tRNA synthetases of *Euglena gracilis* by hydroxyapatite chromatography. (From Parthier et al. 1978).*A* Enzyme activities of whole cells; *B* enzyme activities of isolated chloroplasts; *C* enzyme activities of aplastidic *Euglena* mutant W$_3$BUL; *1* enzyme group representing the synthetases for tRNA charging with Asp, Glu, Ile, Leu, Lys, Ser, Tyr, Val; *2* Phe-RS; *3* Ala-RS and Thr-RS. ● enzyme activities aminoacylating tRNA from the blue-green alga *Anacystis* (analogous to chloroplast tRNA); ○ enzyme activities aminoacylating *Euglena* cytoplasmic tRNA

organelle and cytoplasmic species using hydroxyapatite or DEAE-cellulose chromatography of the high-speed supernatant of the plant homogenate has been described for *Euglena gracilis* (Reger et al. 1970; Parthier et al. 1972; Krauspe and Parthier 1973; Hecker et al. 1974; Lesiewicz and Herson 1975), higher plants (Kanabus and Cherry 1971; Guderian et al. 1972; Brantner and Dure 1975; Guillemaut et al. 1975; Jeannin et al. 1976, 1978; Weil et al. 1977; Locy and Cherry 1978). With these methods it was possible to discriminate two or three peaks of activity charging the respective homologous tRNA fraction by exploiting the high recognition specificity between the cognate enzymes and tRNA species of prokaryotic and eukaryotic origin (Fig. 3).

Generally, the organellar enzymes elute at lower ionic strength than the respective cytoplasmic isotransferring species. Few exceptions were observed for *Euglena:* Thr-RS and Ala-RS activity was eluted as a single peak capable of charging both plastid and cytoplasmic tRNA (Parthier and Krauspe 1973; Parthier et al. 1978), and the absence of these two enzyme species in chloroplasts has been demonstrated (Fig. 3).

Little information is available about the properties of purified isotransferring enzyme species in plant cells, especially for comparisons between chloroplast and cytoplasmic Aa-RS. The high lability of most of these enzymes, particularly the cytoplasmic species, during the purification steps may be a reason for the lack of evidence. The two Leu-RS species of green *Euglena* cells show a number of differences in their physicochemical properties including a slight difference in the molecular weights around 100,000 daltons, but they are identical in the catalytic parameters (Krauspe and Parthier 1973, 1974). A similar situation is true for the two Val-RS species of *Euglena* (Sarantoglou et al. 1978). Locy and Cherry (1978) purified cytoplasmic and chloroplast Tyr-RS from soybean cotyledons and estimated M.W. of 98,000 daltons for the chloroplast enzyme and 126,000 daltons for the cytoplasmic synthetase.

Plastid-specific Aa-RS seem to be coded for in nuclear genes and are synthesized on cytoplasmic polyribosomes as demonstrated with plastid mutants of *Euglena* (Hecker et al. 1974; Parthier and Neumann 1977), by the use of inhibitors (Parthier 1973; Hecker et al. 1974) and by means of density labeling experiments (Nover 1976), see also p. 126.

The characterization of mitochondria-specific synthetases in green plant cells is more complicated. In some higher plants the activities of cytoplasmic and mitochondrial Aa-RS could not be distinguished from each other, as far as the recognition specificity of their cognate tRNA's was taken as a measure (Weil et al. 1977). However, by this method both cytoplasmic and mitochondrial synthetases were separated from the chloroplast Leu-RS (Guderian et al. 1972; Burkard et al. 1973; Brantner and Dure 1975) or Phe-RS (Jeannin et al. 1978). On the other hand, plastid-specific tRNA^{Pro} and tRNA^{Lys} could be charged by the cognate chloroplast, mitochondrial or *E.coli* snythetases, whereas the respective cytoplasm-specific enzymes charged cognate cytoplasmic tRNA species only (Jeannin et al. 1976). With *Euglena* preparations we were unable to discern Leu-RS of plastids and mitochondria by chromatographic or kinetic parameters (Krauspe, unpublished), while Phe-RS and Ile-RS have been separated into three activities (Hecker et al. 1974; Lesiewicz and Herson 1975).

In conclusion, the picture emerging from these data is that tRNA's and Aa-RS from chloroplasts and prokaryotes as bacteria or blue-green algae show a certain degree of similarity. The organelle-specific enzymes differ from their cytoplasmic counterparts in their intracellular localization, their chromatographic behavior and in their tRNA specificity. They are synthesized on cytoplasmic ribosomes and obviously are transported into the organelles. There is even the possibility that organelle and cytoplasmic enzymes are products of the same nuclear gene, and that the intraorganellar protein transport includes a modification of the enzyme such as limited proteolysis (Weil et al. 1977).

E. RNA Polymerase

Although the enzymology of transcription in plastids had made marked progress in the last years, we are far from being able to draw a unified picture of the chloroplast-specific DNA-dependent RNA polymerase. (We anticipate the occurrence on the organelles of only one enzyme species, but this is not yet completely clarified, see below.)

All communications about plastid RNA polymerase coincide with the observation that the enzyme is firmly attached to chloroplast membranes (e.g., Tewari and Wildman 1969; Bottomley 1970; Polya and Jagendorf 1971; Wollgiehn and Munsche 1972; Bogorad et al. 1973). The binding structures are probably thylakoid membranes, but an association with the membranes of plastid envelopes cannot be excluded. There is also evidence available on the binding to DNA and formation of a transcription complex (Hallick et al. 1976; Schiemann et al. 1977). On the other hand, little information is known whether the enzyme is directly attached to the membranes, or indirectly mediated by membrane-attached DNA. In spinach chloroplasts two-thirds of the DNA is accessible to externally applied *E.coli* RNA polymerase, and even more by DNase (Whitfeld et al. 1973).

The membrane-bound enzyme can be solubilized by addition of EDTA (Bottomley et al. 1971a), nonionic detergents as Triton X-100 (Joussaume 1973; Hallick et al. 1976; Schiemann et al. 1977), digitonin or N-lauroylsarcosine (Schiemann et al. 1978), high salt concentration (Polya and Jagendorf 1971), or even by simple osmotic shock in water and mercaptoethanol (Bennett and Ellis 1973).

The solubilization efficiency may depend on many factors including species specificity, however, removal of magnesium seems to be the most important one (Bottomley et al. 1971a). Recently, a transcription complex form *Euglena* chloroplasts could be solubilized by means of 1% Triton X-100 in the absence of magnesium ions (Hallick et al. 1976; Schiemann et al. 1977). This complex consists of DNA, RNA polymerase, and some unspecified proteins. It is highly active in RNA synthesis in vitro for 60 min or longer. More specified, Surzycki and Shellenbarger (1976) studied *Chlamydomonas* chloroplast RNA polymerase and isolated two sigma-like factors of which one (factor 2) was shown to be responsible for the initiation of transcription. The authors observed that this factor (M.W. 51,000 daltons) could replace the corresponding factor separated from *E.coli* RNA polymerase. Moreover, the heterologous *E.coli* RNA polymerase core enzyme was activated for RNA initiation by the *Chlamydomonas* sigma factor 2, though less than the homologous chloroplast core enzyme was. The nuclear RNA polymerase I and II from *Chlamydomonas* remained unaffected when the chloroplastic sigma factor was added.

After separation on SDS-polyacrylamide gels, purified RNA polymerase from corn chloroplasts appears to be multimeric. It consists of two large polypeptide subunits of 180,000 and 140,000 daltons besides some smaller polypeptides which still remain associated with the enzyme after several steps of enzyme purification (Smith and Bogorad 1974). Similar to the chloroplast enzyme the nuclear polymerase IIA also contains a subunit of 180,000 daltons. However, this similarity that the two enzymes contain a common polypeptide is hardly conclusive evidence. Nevertheless, RNA polymerase of different cell compartments possessing similar subunits might be a basis for interaction of organelle RNA synthesis and function within a eukaryotic cell during plastid differentiation (Smith and Bogorad 1974).

A second, stroma-localized RNA polymerase was observed in pea chloroplasts (Joussaume 1973). In contrast to the membrane-bound enzyme, the stroma enzyme was found strictly dependent on chloroplast DNA and showed an unusual optimum of activity at pH 9.1.

Plastid RNA polymerase is not inhibited by α-amanitin (Bottomley et al. 1971a,b; Munsche and Wollgiehn 1972; Brandt and Wiesner 1977). Rifamycins reduce the enzyme activity (RNA synthesis) in intact cells of green algae (Brown et al. 1970; Surzycki et al. 1970; Galling et al. 1973; Schiemann et al. 1978), of tobacco leaves (Munsche and Wollgiehn 1973), and of *Acetabularia* (Schweiger 1970). No effect in vivo was also reported (Bottomley et al. 1971b; Heizmann 1974a; Scott 1976). RNA synthesis in isolated chloroplasts was not inhibited by rifamycins even in high concentrations (Bottomley et al. 1971b; Wollgiehn and Munsche 1972; Carritt and Eisenstadt 1973; Schiemann et al. 1978). A partial inhibition of enzyme activity was observed with the solubilized crude plastid RNA polymerase (Bottomley et al. 1971a; Bogorad et al. 1973) or with a reconstituted purified enzyme system (Surzycki and Shellenbarger 1976), however, the drug-sensitivity was usually lost during enzyme purification (Polya and Jagendorf 1971; Bogorad et al. 1973; Hallick et al. 1976; Brandt and Wiessner 1977). In addition, the inhibition of RNA synthesis in vitro by rifamycin SV could be manipulated by the addition of exogenous DNA (Wollgiehn and Munsche 1972) or of NH_4Cl (Bogorad et al. 1973). Since rifamycins block initiation but not elongation steps catalyzed by prokaryotic RNA polymerases, any chloroplast or enzyme preparation lacking initiation factors must become insensitive toward the drug. Then precursor incorporation reflects merely an elongation of previously initiated RNA chains.

The controversy in the rifamycin action in vivo obtained in various laboratories seems to be due to different treatment conditions, e.g., the amount of drug per cell or per chloroplast varied largely. Species-specific uptake and transport peculiarities of the drug may be another reason for the different effects described in the literature. Finally, the light sensitivity of rifamycins has not always been taken into consideration.

Thus profound differences exist between plastid and nuclear RNA polymerases of the same plant material. In a comparison between isolated, enriched enzymes from maize seedlings, Bogorad and co-workers (1973) were able to demonstrate different properties, e.g., as shown by the influence of ions on enzyme activity. The chloroplast enzyme was distinguished by a temperature optimum at 48° C and recognized homologous DNA better than any other DNA tested. However, it was noted that denatured single-strand plastid DNA was used more effectively than native DNA (see also Bennett and Ellis 1973). On a unit DNA basis the plastid enzyme was found much more active than the corresponding nuclear polymerases.

The nuclear and chloroplast RNA polymerases from *Euglena gracilis* also differ in their temperature optima with 32°-33° C for the three nuclear polymerases and 28°-29° C for the plastidic enzyme (Brandt and Wiessner 1977).

F. Polypeptide Synthesizing System

In this chapter protein biosynthesis during plastid differentiation is focused on amino acid polymerization at the ribosomal level. Since the structural aspects of these nucleo-protein particles were already discussed above, the functional events will be considered in respect to plastid specificity of the components involved.

Amino acid activation was shown to be a highly selective system in plastids different to that of the cytoplasm (see p. 108f.). Are there further control factors in the course of plastid polypeptide formation? Both prokaryotic and eukaryotic protein-synthesizing machineries are modulated by protein factors necessary to accomplish initiation, elongation, and termination steps at the ribosomal subunits. The search for equivalent protein factors in chloroplast protein synthesis is at the early beginning. The reasons for this delay can be referred as to the high dilution of these factors within the cells and the necessity to have pure and intact chloroplasts as a source for the isolation of initiation or elongation factors. Otherwise these proteins are also found in the soluble post-ribosomal fractions of cell homogenates, from which they can be separated in a larger scale by chromatographic methods, as has been done successfully with the elongation factors of *Chlorella* (Ciferri and Tiboni 1973) and spinach chloroplasts (Tiboni et al. 1976). The authors were able to discriminate cytoplasmic and chloroplast elongation factors EF-G and EF-T_u using chromatography and poly(U)-dependent phenylalanine incorporation using cell-free ribosome systems. The site of synthesis of EF-G inside the chloroplasts was determined by means of inhibitors of protein synthesis (Ciferri and Tiboni 1976).

Analogous to protein synthesis initiation in prokaryotes, plastid polypeptide formation starts with N-formylmethionyl-tRNA$_F$, in contrast to methionyl-tRNA$_F$ specific for the cytoplasmic systems in eukaryotic cells. Detailed studies of initiating F-met-tRNA have been reported for bean chloroplasts (Burkard et al. 1969), wheat leaves (Leis and Keller 1971), cotton seedlings (Merrick and Dure 1971), and more recently, with pea chloroplasts (Highfield and Ellis 1978). These authors suggested from the formation of N-formyl ^{35}S-methionylpuromycin in isolated chloroplasts that the organelle ribosomes are able to initiate protein synthesis with plastid mRNA and likewise to carry out elongation and termination steps.

Of five tRNA^{Met} species detectable in the extracts, two are localized within chloroplasts, but only one of them charged with methionine is formylated by an endogenous enzyme or by *E.coli* formylase (Weil 1978).

Another aspect of plastid protein-synthesizing specificity is its sensitivity to selective inhibitors. Antibiotics such as D-threo-chloramphenicol, lincomycin, spectinomycin, and erythromycin are effective at 70S ribosomes both from plastids and from other sources, thus indicating the prokaryotic type of ribosome as a prerequisite for their modes of action (Ellis and Hartley 1971; Parthier et al. 1975). These antibiotics bind at the large subunit (50S) of bacterial ribosomes to ribosomal proteins in the vicinity of the peptidyl transferase or to this ribosomal enzyme itself. It may be suggested therefrom that a mode of protein assembly similar to that of bacterial ribosomes exists in plastid ribosomes. This is supported by the observation that the ribosomal subunits of chloroplasts and *E.coli* could be exchanged without a marked loss of polymerization activity (H.G. Wittmann, pers. commun.).

A third aspect is the energy source used in chloroplast protein synthesis. Since isolated organelles can generate ATP by both cyclic and noncyclic photophosphorylation, it was possible to demonstrate light-driven polypeptide formation (Blair and Ellis 1973; Bottomley et al. 1974). No additional ATP- or ATP-regenerating system is necessary under light conditions, and the addition of 3-(3,4-dichlorophenyl)-1,1-dimethylurea (DCMU), an inhibitor of photosynthesis, effectively reduced chloroplast protein synthesis. Exogenous ATP can only slightly penetrate intact chloroplast membranes, in contrast to certain amino acids (Heber 1974).

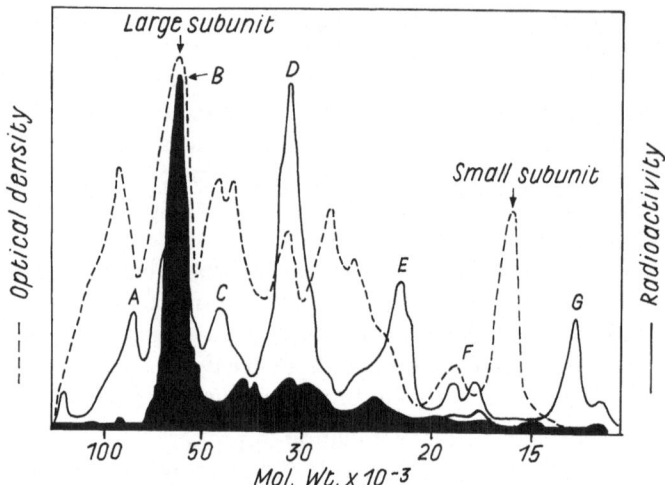

Fig. 4. Gel electrophoresis of pea chloroplast proteins synthesized in vitro. Whole chloroplasts or chloroplast 38,000 *g* supernatant fraction from lysed chloroplasts (free chloroplast ribosomes) were labeled in vitro with [^{14}C]-leucine in the light. The *solid line* shows the incorporation of leucine into the proteins by whole chloroplasts, the *black area* into soluble proteins by free chloroplast ribosomes. The letters *A* to *G* mark the discrete labeled peaks. *Large subunit* and *small subunit* refer to the subunits of RuBPCase. (After Ellis 1975a; Figs. 5 and 15)

In recent years experimental studies have succeeded in the identification of polypeptides synthesized in isolated etioplasts and chloroplasts (Blair and Ellis 1973; Harris et al. 1973; Whitfeld et al. 1973; Bottomley et al. 1974; Eaglesham and Ellis 1974; Siddell and Ellis 1975; Mendiola-Morgenthaler et al. 1976). Among the labeled polypeptides separated by polyacrylamide gel electrophoresis or tryptic finger printing the large subunit of Fraction I-protein (RuDPCase) was the abundant product of the soluble proteins (Fig. 4). A number of buffer-insoluble polypeptides were likewise synthesized in isolated chloroplasts (Apel and Schweiger 1973; Ellis 1975a; Vasconselos 1976; Vasconselos et al. 1976; Ellis 1977). Three of the five subunits of the membrane-bound coupling factor CF_1 (ATPase) are synthesized in isolated chloroplasts of spinach (Mendiola-Morgenthaler et al. 1976; Morgenthaler and Mendiola-Morgenthaler 1976) and of pea leaves (Ellis 1977). Thus there is no doubt that certain plastid polypeptides are coded for and synthesized inside the organelles. It was already shown on p. 102 that total RNA or mRNA extracted from chloroplasts are meaningfully translated in *E.coli* or reticulocyte lysate systems.

III. Sites of Coding and Synthesis of Plastid Proteins

The chloroplast consists of products of plastid and nuclear genomes. The question arises which of the macromolecular components are encoded in chloroplast DNA and are synthesized on plastid ribosomes and which of them are provided from the cytoplasmic surrounding.

Table 3. Identified chloroplast gene products

Ribosomal RNA's:	23 S, 16 S, 5 S, 4.5 S
Transfer RNA's:	25–30 species
Messenger RNA's:	for LS RuBPCase, for P-32,000
Proteins:	Ribosomal proteins: 2–5 species
	Soluble proteins (in vitro labeled ca. 90 bands):
	large subunit of RuBPCase
	cytochrome 552
	EFG; EFT$_u$ (?)
	fatty acid synthetase (?)
	Thylakoid-bound proteins (in vitro labeled >15 bands):
	protein P-32
	cytochrome f
	3 subunits of the CF$_1$ coupling factor (ATPase)
	certain subunits of PS I and PS II (reaction center, water splitting)

Evidence about the origin of chloroplast constituents was obtained by three methods:

1. Transcription and translation in vitro systems or within isolated plastids, and identification of the products.

2. Use of inhibitors of the gene expression specific for prokaryotic and eukaryotic transcription and translation processes.

3. Genetic and biochemical analysis of both genome and plastom mutants.

Of these methods the first one alone gives clear-cut evidence whether a macromolecule located inside the chloroplast is encoded in plastid DNA and synthesized within the organelle. This method has been successfully developed in the last years, but formerly the two other methods provided valuable knowledge on this problem, and they are still used in many instances. Objections to the use of inhibitors in vivo have occasionally been made (see Ellis et al. 1973; Parthier et al. 1975; Ellis 1977). Inhibitors may cause metabolic side-effects which act beyond the ribosome target, but nevertheless seriously influence protein synthesis secondarily. Even in the case where ribosome-specific inhibition of protein biosynthesis is definitely demonstrated, the suppression of a plastid protein can also be due to the inhibition of a regulatory protein or a factor necessary for the translation process itself.

The genetic approach has the disadvantage that while it can determine the coding site, it is unable to tell us where the protein is produced. Thus, incorrect localization of the mutation is possible. The pitfalls of the in vitro synthesis using isolated organelles are artifact formations and the leakage of substances necessary for transcription or translation.

Table 3 summarizes the present knowledge on identified chloroplast constituents which are unequivocally coded for by the chloroplast DNA and synthesized inside the organelle. The list will be extended in future, because a great number of in vitro chloroplast translation products have been demonstrated by gel electrophoresis but not further specified. On the other hand, results of several laboratories indicate that many of the stroma and thylakoid membrane proteins located within chloroplasts are encoded in the nucleus and formed on cytoplasmic ribosomes. This situation implies a transport of polypeptides or their immediate precursors through the plastid envelope membranes, e.g., the small subunit of RuBPCase (Highfield and Ellis 1978) and other enzymes of the Calvin cycle, nucleic acid polymerases, aminoacyl-tRNA synthetases (Parthier 1970; Börner 1973; Parthier et al. 1975).

There are several hypothesis on the mechanism of macromolecular transport into chloroplasts: transport via pinocytosis, signal hypothesis (Blobel and Dobberstein 1975) and as the most probable one, the envelope carrier hypothesis (Blair and Ellis 1973; Highfield and Ellis 1978; for more detailed discussion see Ellis 1977; Highfield and Ellis 1978).

IV. RNA and Protein Synthesis
During Light-Dependent Plastid Differentiation

A. General Aspects

In angiosperms and certain algae the biogenesis of fully developed chloroplasts is dependent on light. Nevertheless, plastid differentiation can take place also in darkness to chloroplasts in gymnosperms, ferns, mosses and the majority of algae. In angiosperm seedlings proplastids develop to etioplasts with a higher structural and functional organization, but the acquirement of the unique function of the chloroplast, the photosynthetic activity, requires the light-dependent formation of chlorophyll, grana membranes and the quantitative enhancement of photosynthetic enzymes.

Chloroplast formation in unicellular green algae is normally independent of light. However, certain species are distinguished by natural *(Euglena)* or acquired *(Chlamydomonas reinhardii,* y1-mutant) inability to green in darkness. Since in these organisms proplastids divide, alternating with cell division, the formation of etioplasts comparable with those in the cells of etiolated higher plants usually does not occur. In *Euglena gracilis* the light-induced transformation of proplastids to green and mature chloroplasts is associated with a 60-fold increase of the plastid volume.

In higher plants the etioplasts as immediate precursors of chloroplasts are usually much larger than proplastids and contain higher levels of plastid-specific constituents such as rRNA or RuBPCase and especially protochlorophyll.

The mechanism of the light-induced conversion of the etiolated plastids to chloroplasts requires not only de novo syntheses of lamellar proteins and photosynthetic enzymes, but also the synthesis of several RNA species. This suggestion is supported by studies on dark-grown plants exposed to inhibitors of the gene expression both in plastids and in the cytoplasmic compartment. In all cases the formation of functionally intact chloroplasts was prevented. Synthesis of RNA is one of the early events in chloroplast differentiation and therefore frequently investigated, at least to get information about the primary action of light in this process.

The necessity of RNA synthesis in light-dependent plastid differentiation does not mean that etiolated plastids are unable to synthesize RNA and protein species in darkness. It might be argued that these plastid-specific marcromolecules are coded for or are synthesized outside the organelles. For many protein components this seems to be the case; however, as will be demonstrated below, etiolated plastids do synthesize RNA and proteins in complete darkness, though at a lower rate than in light.

The question arises whether light induces the multiplication of those RNA and protein species already present in proplastids or etioplasts (and not sufficient for subsequent differentiation steps), or whether light triggers the de novo synthesis of a limited number of specific macromolecules which are inevitable for further differentiation to chloroplasts. The first possibility would express a quantitative, the second a qualitative aspect of the effector role of light. Thus we have to ask: are there new RNA and protein species induced by light, which are not present in etiolated plastids but necessary for the realization of chloroplast structure and function? What is the timing of light-dependent plastid-specific synthesis of RNA and protein species?

On the following pages we discuss the changes in transcription and translation in plastid differentiation upon illumination of etiolated cells.

From the data available, it seems difficult at the present time to make general statements, since due to the heterogeneity of the plant objects used the obtained results are rather conflicting. This is true not only between cotyledons of the dicotyledonous plants and primary leaves of the monocotyledonous plants, but becomes more severe if we compare plastid differentiation in higher plants and in algae.

B. Ribosomal RNA and Ribosomes

1. Higher Plants

From the results of electron microscopical and analytical investigations, it is known that not only chloroplasts from green leaves, but also plastids from etiolated leaves contain ribosomal particles (Jacobson et al. 1963; Clark 1964; Boardman 1966; Bartels and Weier 1967; Dyer et al. 1971). The course of plastid rRNA synthesis during germination in the dark has been studied in different plants. Very young, particularly embryonic, leaves contain abundant cytoplasmic rRNA but only very small amounts of plastid rRNA, sometimes not provable with the usual analytical methods (Ingle 1968b; Smith 1970; Vedel and D'Aoust 1970). During growth of cotyledons or leaves in the dark, the amount of plastid rRNA per cell and per etioplast increases gradually (Smith 1970), however, different in extent in dependency of plant species, tissue, and the varying experimental conditions. In some cases etioplast rRNA content can reach a level equivalent to that existing in light-grown plants. Cotyledons of cucumber seedlings (Vedel and D'Aoust 1970), corn leaves (Dyer et al. 1971), barley leaves (Poulson and Beevers 1970; Smith et al. 1970) and wheat leaves (Scott et al. 1971a) contain chloroplast and cytoplasmic rRNA in approximately the same proportion regardless of growth in light or in darkness. Dark-grown leaves of *Phaseolus vulgaris* also contain about the same amount of plastid ribosomes as green leaves (Boardman 1966). In other tissues, for example radish cotyledons (Ingle 1968b), *Vicia faba* leaves (Dyer et al. 1971),

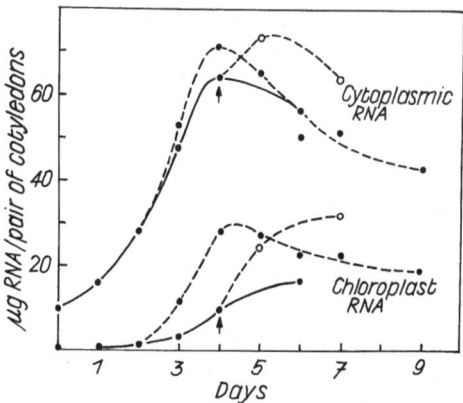

Fig. 5. Effect of light on the accumulation of cytoplasmic and chloroplast rRNA in radish cotyledons in the dark (●——●): in continuous light (●----●): and after transfer from dark to light (○----○) as indicated by *arrows* (Ingle 1968b)

mustard seedlings (Thien and Schopfer 1975), pea apices (Ellis and Hartley 1971) or isolated pumpkin cotyledons (Mikulovich 1978), the amount of plastid rRNA is much lower in dark-grown than in light-grown plants. Also the aquatic angiosperm *Spirodela oligorhiza*, one of the few higher plants which is able to grow and multiply when cultured in the dark in a mineral-sucrose medium, forms etioplasts in the dark, but their rRNA content and the number of plastid ribosomes are much lower than in light-grown plants (Rosner et al. 1974).

From these results one can conclude that light is not an inevitable requirement of plastid rRNA synthesis, although ribosomal RNA synthesis of plastids is strongly stimulated during illumination of etiolated leaves or during seed germination in the light. This is not only the case in tissues like radish cotyledons, *Vicia faba* leaves or pumpkin cotyledons which show a lower plastid rRNA content in dark-grown than in light-grown plants, but also in other material if illuminated at early stages of development, before etioplasts with a high rRNA content have been developed. A comparison of the development of radish cotyledons (Ingle 1968b) and barley leaves (Smith et al. 1970) demonstrates differences and coincidences of the two plant sources. Whereas in darkness radish cotyledons synthesize only 50% of the plastid rRNA of light-grown plants, they very quickly reach their normal rRNA content after illumination (Fig. 5). On the other hand, the plastids of barley are able to accumulate the same amount of chloroplast rRNA both in darkness and in light. Therefore light brings about a marked increase in plastid rRNA content only in young leaves, but the effect is diminished as the leaf grows (Fig. 6).

Not only rRNA accumulation but also the incorporation of labeled precursors into plastid rRNA is stimulated by light (Ingle 1968b; Poulson and Beevers 1970; Dyer et al. 1971; Ellis and Hartley 1971; Detchon and Possingham 1973; Harel and Bogorad 1973).

The present results indicate that the synthesis of new plastid ribosomes is no essential step in the light-stimulated plastid enzyme formation (Jacobson et al. 1963; Boardman 1966; Smith 1970; Pine and Klein 1972). The increased activity of

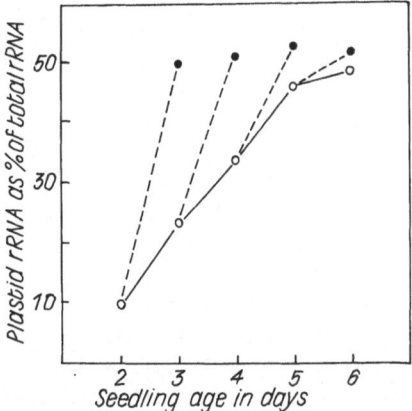

Fig. 6. Effect of 24-h light given at various times after soaking on the properties of total ribosomal RNA present as plastid RNA at various developmental stages of barley leaves. *Open circles* dark-grown controls; *closed circles* light treatment; *dotted lines* changes effected by light treatment (Smith et al. 1970)

several enzymes made on 70S plastid ribosomes of the older barley leaves is therefore not accompanied with an increase in plastid ribosomal RNA (Smith et al. 1970).

Light not only stimulates chloroplast RNA synthesis but also cytoplasmic rRNA synthesis in greening tissues (Ingle 1968b; Poulson and Beevers 1970; Ellis and Hartley 1971; Thien and Schopfer 1975; Mikulovich 1978). However, in most cases stimulation of cytoplasmic rRNA synthesis is lower compared with chloroplast rRNA, e.g., in radish cotyledons (Ingle 1968b). In other tissues such as corn (Harel and Bogorad 1973) or *Vicia faba* leaves (Dyer et al. 1971) light shows only a minimal influence on cytoplasmic rRNA synthesis.

The effect of light quality on rRNA synthesis is described in a number of papers. Scott et al. (1971b) compared pea seedlings irradiated with red light (662 nm) or far-red (730 nm) with those grown in darkness or under a white light regime. Red-light illumination caused a net increase of rRNA comparable with the increase after white-light illumination. In addition, there was a two- to threefold increase in etioplast rRNA relative to total ribosomal RNA. Since the red-light effect was reversed by far-red, the involvement of phytochrome in the regulation of plastid rRNA was suggested. More recently Thien and Schopfer (1975) confirmed these results but found a similar stimulation of the 1.3, 1.1, 0.7 and 0.56 × 10^6 rRNA's after continuous illumination of the seedlings with far-red light and short red-light pulses. It was concluded that the transcription of both plastid and cytoplasmic rRNA cistrons are under the photocontrol of the phytochrome system, and that the accumulation of plastid rRNA is independent of the formation of functional photosynthetic apparatus.

Plastid differentiation in higher plants is influenced by phytohormones such as cytokinins (Fig. 7). Several reports have shown that one of the cytokinin effects during plastid differentiation is the stimulation of rRNA synthesis (Rousseaux et al.

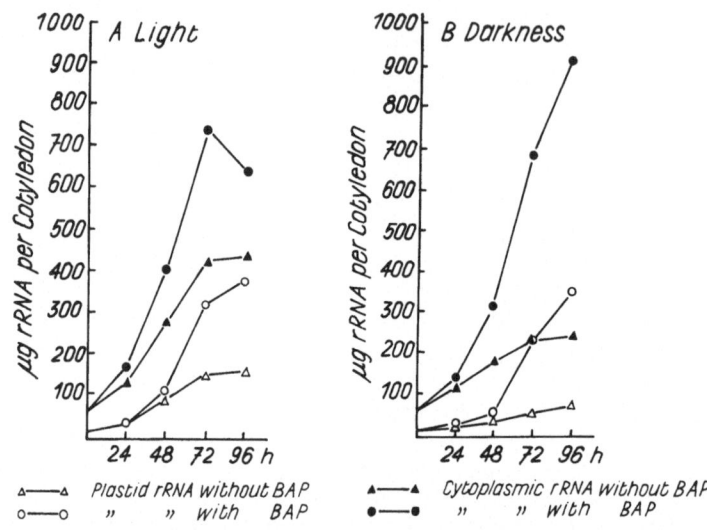

Fig. 7.A,B. Effect of 6-benzymaminopurine (BAP) and light on the accumulation of plastid and cytoplasmic rRNA in isolated *Cucurbita pepo* cotyledons (Mikulovich et al. 1978)

1976; Grierson et al. 1977; Mikulovich et al. 1978). However, it is difficult to decide whether this effect is a direct one. Since cytoplasmic ribosome content and, as a consequence, protein synthesis is also stimulated by cytokinin, any enhanced plastid rRNA formation during chloroplast development might be explained as a hormone-induced activation of the nucleo-cytoplasmic transcription-translation machinery, e.g., via an enhanced content of plastid RNA polymerase, which is produced on cytoplasmic ribosomes.

   Another aspect is the question whether rRNA synthesis stops after cessation of plastid differentiation, or whether chloroplasts in mature leaves maintain rRNA turnover and, consequently, continue in regeneration of the protein-synthesizing machinery. From results reported by Ingle, chloroplast ribosomal RNA synthesis in radish cotyledons and in the first leaf of wheat seedlings takes place at a very limited time during development of the leaf, namely during chloroplast formation from etioplasts (Ingle 1968a; Ingle et al. 1970, 1971). Chloroplast RNA synthesis seems to be a once-only phenomenon, since during growth of radish cotyledons ^{32}P incorporation into chloroplast rRNA was found exclusively during the first days of germination when new chloroplasts were formed. During the following periods when no net increase in either cytoplasmic or chloroplast RNA was observed, only a very slow turnover of chloroplast rRNA occurred as compared to the turnover of cytoplasmic rRNA. Investigations on the RNA metabolism in the first leaf of wheat seedlings confirm these results (Paterson and Smillie 1971). However, the results obtained with radish cotyledons and the first leaf of wheat seedlings should not be generalized. In mature or even aging leaves of *Nicotiana rustica*, chloroplasts do incorporate ^{32}P into rRNA (Wollgiehn et al. 1966, 1976). In mature tobacco leaves where chloroplast DNA is not synthesized and no chloroplast replication takes place, chloroplast rRNA synthesis proceeds, although ^{32}P incorporation into chloroplast rRNA as well as chloroplast rRNA content per leaf and RNA polymerase activity of the isolated chloroplasts are reduced in comparison to young leaves. It is unknown, however, whether the rRNA content per chloroplast changes during growth of the tobacco leaf (Wollgiehn et al. 1976). During a 14-day growth period of still-growing spinach leaves (the leaves increased in length from 2 to 7 cm), chloroplast rRNA content and chloroplast number per cell was found to increase several-fold, but the rRNA content per chloroplast remained constant (Detchon and Possingham 1972). Presumably the rRNA content of the chloroplasts reaches its highest level earlier in chloroplast differentiation, and the turnover of the chloroplast rRNA may differ in various plant materials. To our knowledge, only in one case plastid rRNA turnover was determined exactly. In *Lemna minor* the half-life time of chloroplast rRNA is 12 days, but 4 days for the corresponding cytoplasmic rRNA (Trewavas 1970).

If rRNA synthesis in leaves of different age is compared, the possibility should be kept in mind that precursor uptake or specific precursor pool sizes might change during chloroplast development and thus cause misinterpretations in the intensity of RNA metabolism. Such problems have not yet been reliably investigated in higher plants. In unicellular algae *(Chlorella, Scenedesmus)* it was shown that short-time labeling with uridine resulted in a specific labeling of chloroplast rRNA while cytoplasmic rRNA was left almost unlabeled. However, both chloroplast and cytoplasmic rRNA were labeled after guanosine application (Galling et al. 1975).

Plastid rRNA and therefore ribosomes are lacking or present only in extremely small quantities in the bleached cells of a *Pelargonium* plastom mutant (Börner et al. 1972), in chlorophyll-deficient barley mutants (Sprey 1972) and in heat-bleached rye seedlings (Feierabend and Schrader-Reichardt 1976). In all cases etioplasts appear surrounded by a normal double-layer envelope, but containing no thylakoid membranes. These observations were interpreted as participation of the nucleo-cytoplasmic compartment for the construction of the plastids, and for the formation of the plastid envelope in particular (Sprey 1972; Feierabend and Schrader-Reichardt 1976).

2. Algae

Wild-type cells of *Chlamydomonas reinhardii* are capable of synthesizing chlorophyll and thylakoid membranes even growing in complete darkness (Sager and Palade 1954). Nevertheless, during growth in continuous darkness a decrease in chloroplast rRNA content was observed (Matsuda 1974). After treatment with rifampicin *Chlamydomonas reinhardii* (Surzycki et al. 1970) and also *Chlorella pyrenoidosa* (Galling et al. 1973) do not contain detectable amounts of chloroplast ribosomes or 16S and 23S rRNA. Nevertheless, growth of the cells on an organic medium is unaffected, chlorophyll production is almost normal; only changes in the thylakoid arrangment of the chloroplast have been observed. However, photosynthetic activity is strongly reduced under such conditions. This shows that certain enzymes of photosynthesis are synthesized on chloroplast ribosomes or coded for by the plastid DNA, whereas the formation of the main structures of the chloroplast is independent of the plastid DNA transcription and of the presence of plastid ribosomes and protein synthesis.

Changes in the chloroplast and cytoplasmic rRNA content of the cells have been observed during the synchronous vegetative cell cycle of *Chlamydomonas* (Cattolico et al. 1973; Wilson and Chiang 1977). In a synchronously growing culture (12 : 12 h light : dark cycle) nearly 90% of the chloroplast and cytoplasmic rRNA's are transcribed in the nuclear G1 phase, which occurs during the light period, and only 10% of the rRNA's are transcribed in the dark period (Wilson and Chiang 1977). The transcriptional activity of both nuclear and chloroplast genomes closely approximate to each other as determined by rRNA content and ^{32}P incorporation. This is in contrast to the fact that chloroplast and nuclear DNA's were replicated out of phase, i.e., chloroplast DNA is synthesized during the 3rd and 7th h in the light period, and nuclear DNA between the 15th and 21st h during the dark period of the cell cycle. The transcription is not interrupted during chloroplast DNA synthesis. Thus the temporal programming of chloroplast DNA transcription appears to be prokaryotic. The bacterial DNA replication also does not change the RNA transcription rate in *E.coli* (Dennis 1972). On the other hand, transcription of cytoplasmic rRNA in *C. reinhardii* occurs preferentially in G1 and is repressed during the S-phase as in many other eukaryotes (Wilson and Chiang 1977). Interestingly, chloroplast 70S and cytoplasmic 80S ribosomes, as well as their RNA's, are very stable and conserved from generation to generation during generative growth of *Chlamydomonas* (Siersma and Chiang 1971).

Similarly, in *Chlorella*, plastid rRNA is synthesized in light as well as in darkness, though with different activity. Chloroplast ribosomes are synthesized throughout the whole cell cycle, but the ratio between chloroplastic and cytoplasmic rRNA changes within the cycle showing a maximum in the 8th and a minimum in the 20th of the 14 : 10 h light : dark cycle (Galling 1973, 1975).

In contrast to the wild-type cells the y-1 mutant of *Chlamydomonas* is unable to synthesize chlorophyll and thylakoid membranes, if growing in continuous darkness (Ohad et al. 1967a). After 8 to 10 h light, the normal membrane-filled chloroplast is restored. Chloroplast ribosomes are visible by electron microscopy in both green and yellow cells (Ohad et al. 1967b); however, the content of ribosomes per cell is higher in green cells after 6 h illumination than in yellow cells. The population of 68S chloroplast ribosomes increases faster and reaches a higher level during the greening process than does the 80S population (Hoober and Blobel 1969).

More detailed investigations (Matsuda 1974) on rRNA metabolism in a culture of nondividing cells have also shown that cells growing in darkness possess a lower amount of chloroplast rRNA (23/26S rRNA ratio about 35%) than cells growing in the light (47%). Therefore, during growth in darkness ^{14}C-uracil is incorporated to a higher degree into the cytoplasmic rRNA than into the chloroplast rRNA. Upon illumination the rRNA content of dark-grown cells begins to increase after a 1.5-h lag period, and soon the net synthesis of chloroplast rRNA (and ^{14}C-uracil incorporation into chloroplast rRNA) becomes higher than that of cytoplasmic rRNA. Finally, after 6 h light the 23S/26S rRNA ratio reaches the level of normally light-grown cells (48%).

Contrary to wild-type *Chlamydomonas*, *Euglena gracilis* forms chloroplasts only in light. Dark-grown *Euglena* cells actively synthesizing cytoplasmic rRNA contain only very small amounts of chloroplast ribosomes or ribosomal RNA,

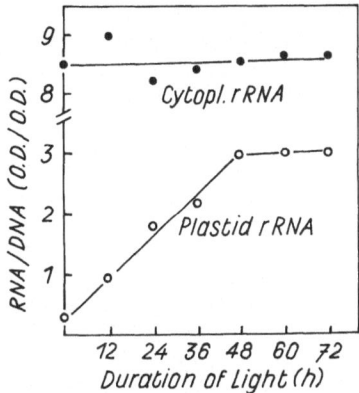

Fig. 8. A comparison between the changes in relative amounts of plastid rRNA and cytoplasmic rRNA during light-induced chloroplast development in resting *Euglena gracilis* cells. (After Cohen and Schiff 1976)

respectively (about 2% of total cellular rRNA). This rRNA is localized inside the proplastids (Brown and Haselkorn 1971; Hirvonen and Price 1971; Munns et al. 1972; Heizmann et al. 1975; Cohen and Schiff 1976).

Transfer of dark-grown cells into light results in a gradual increase in chloroplast rRNA content. After 24 h the concentration characteristic for light-grown cells is reached (up to 25% of the total cellular rRNA) (Brown and Haselkorn 1971). Incorporation of precursor radioactivity into plastid ribosomal RNA or other transcripts could not be detected after a 2-h pulse given to dark-growing cells. Incorporation into plastid rRNA was observed only after illumination (Brown and Haselkorn 1971).

A successful method to investigate chloroplast development in *Euglena* is the separation of chloroplast development from chloroplast replication in nondividing (resting) cells. If *Euglena* is grown in darkness on a heterotrophic medium to the stationary phase and then illuminated, chloroplast development proceeds in the absence of cell division. These resting cells show no appreciable synthesis of DNA during chloroplast development, synthesis of chlorophyll begins immediately after the cells were exposed to light. The content of cytoplasmic rRNA remains fairly constant (Cohen and Schiff 1976) or even decreases (Munns et al. 1972; Heizmann 1974b) during chloroplast development, but the 23S and 16S plastid rRNA's increase manyfold. After 48 h of development the maximum increase is obtained, representing 25% of the cellular rRNA (Fig. 8).

On the other hand, ^{32}P incorporation into plastid rRNA as well as into cytoplasmic rRNA is stimulated during chloroplast development. The labeling of chloroplast rRNA represents net synthesis, the labeling of cytoplasmic rRNA, however, indicates turnover, since there is no increase in the amount of cytoplasmic rRNA during this period. The extent of this turnover is about 60% after 36 h of illumination (Munns et al. 1972; Heizmann 1974b; Cohen and Schiff 1976). It can be assumed that the newly synthesized cytoplasmic rRNA enters the ribosomes of polysomes first, because the rRNA of cytoplasmic polysomes is higher labeled than

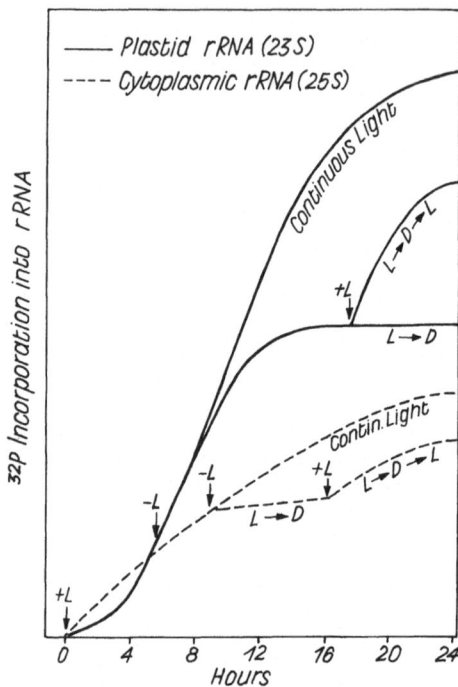

Fig. 9. Time course of ³²P-incorporation into chloroplast and cytoplasmic rRNA after illumination of dark-grown *Euglena gracilis* cells. *L* light; *D* darkness. (Schematic diagram based on experiments of Heizmann et al. 1975; Cohen and Schiff 1976)

that of the monosomes (Cohen and Schiff 1976). It may be possible that the cell is not able to activate preexisting monosomes, and the newly synthesized polysomes are responsible for protein synthesis associated with chloroplast development.

Plastid rRNA accumulation starts immediately after the cells were exposed to light, but with a reduced rate during the first few hours (Fig. 9). It might be referred to as an increased formation of the transcription complex in the organelles. On the other hand, a lag of 12 h was observed for other chloroplast constituents, e.g., ribulosebisphosphate carboxylase and cytochrome 552. These enzymes are synthesized, at least in part, on chloroplast ribosomes; therefore, synthesis of chloroplast ribosomes seems to be a prerequisite for the appearance of these proteins.

The control of formation of cytoplasmic rRNA and plastid rRNA is different (Fig. 9): labeling of both rRNA types is nearly linear during uninterrupted illumination, but the labeling of cytoplasmic rRNA is strictly light-dependent. When dark-grown resting cells are briefly illuminated followed by re-darkening, the formation of chloroplast rRNA continues in darkness for about 8 to 10 h, however, labeling of cytoplasmic rRNA ceases immediately. After a second illumination, both plastid and cytoplasmic rRNA syntheses start with similar rates, supporting the idea that an active transcription complex is now established in the chloroplasts (Heizmann et al. 1975; Cohen and Schiff 1976). The action spectrum indicated that a red- and blue-absorbing protochlorophyll(ide) system within the plastids controls the plastid-localized events including chlorophyll synthesis, but a blue-light-absorbing system controls the nuclear-cytoplasmic system of rRNA synthesis in *Euglena* cells (Cohen and Schiff 1976). A similar result was reported by Steup (1975) for the transcription of chloroplast and cytoplasmic rRNA in *Chlorella*.

C. Transfer RNA and Aminoacyl-tRNA Synthetases

It has been assumed that control of differentiation exists on the level of translation in addition to the control on the level of transcription. Possible candidates are isoaccepting tRNA species, aminoacyl-tRNA synthetases and the polypeptide synthesis on the ribosome, especially translation factors. Unfortunately for the process of plastogenesis causal relationship between the appearance of certain tRNA or Aa-RS species and chloroplast transformation from proplastids or etioplasts have not yet been demonstrated. The phenomena observed are limited to a description of parallel increase, after light induction, of plastid specific constituents and chloroplast structure or function. Differences in the sequence of events, however, may be valuable for a further analysis about their possible regulatory role at the translation level.

A several-fold increase of certain plastid-specific tRNA's was observed with dark-grown *Euglena* cells after illumination (Barnett et al. 1969; Reger et al. 1970; Goins et al. 1973; Krauspe and Parthier 1973; Parthier and Krauspe 1974). Similar to algal cells, a light-stimulated increase of the plastid-specific tRNA's was demonstrated by Burkard et al. (1972) when they compared the chromatographic profiles of tRNA^{Leu} from etioplasts and chloroplasts of *Phaseolus*. On the other hand, cotton cotyledon development was coupled with an enhancement of plastid tRNA species though this increase was not influenced by light (Merrick and Dure 1972). From the observations described, one should conclude that etioplast to chloroplast transformation in cotton cotyledons is hardly comparable with the proplastid to chloroplast differentiation in *Euglena* or etiolated leaf cells. In the flagellate, the light-induced greening process is coupled with an increase of plastid-specific tRNA^{Leu} (Fig. 10; see also Barnett et al. 1969). The kinetics of tRNA synthesis shows a lag period and resembles chlorophyll accumulation. A dark-grown, yellow plastid mutant (G$_1$BU) of *Euglena gracilis* contains the same level of plastid tRNA^{Met} and tRNA^{Ile} as light-grown, green wild-type cells (Goins et al. 1973). This finding is explained by light induction of plastid tRNA synthesis which is controlled by nuclear genes. The authors admitted, however: "Although the possibility does exist that the G$_1$BU mutant contains a nuclear chromosomal dominant mutation affecting control of organelle tRNA synthesis, it is not reasonable to assume that it is in fact a mutation in the plastid genome that gives rise to the constitutivity we have observed".

As can be seen from Figure 10, the activity of the plastid-specific leucyl-tRNA synthetase of dark-grown *Euglena* cells increases rapidly after illumination of the culture (see also Reger et al. 1970; Parthier et al. 1972; Parthier 1973; Krauspe and Parthier 1972, 1974; Hecker et al. 1974). The size of increase for the Leu-RS and several other Aa-RS ranges between 10- and 30-fold; however, other plastid Aa-RS may not exceed two to threefold increase in activity in the same greening cells (Parthier 1973).

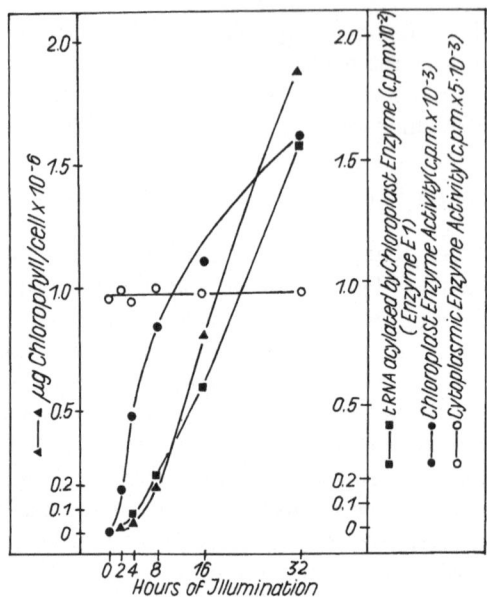

Fig. 10. Time course of plastid tRNA^Leu content, plastid and cytoplasmic Leu-tRNA synthetase activity, and chlorophyll content during the light-induced greening process of dark-grown *Euglena gracilis*. (For details see Parthier and Krauspe 1974)

The rapid rise in enzyme activity which clearly precedes chlorophyll accumulation and cognate tRNA formation seems to be a characteristic feature for most plastid stroma proteins that are synthesized in the cytoplasmic compartment (Schiff 1974). Indeed, the light-dependent stimulation of *Euglena gracilis* plastid Leu-RS is completely inhibited after the addition of 5 µg/ml cycloheximide to the culture. The inhibition can be observed at any stage of the greening process and appears within a few minutes. Since the drug does not affect enzyme activity itself, this experiment demonstrated the de novo biosynthesis at cytoplasmic ribosomes of Aa-RS during plastid differentiation in *Euglena* (Parthier 1973; Hecker et al. 1974). The assumption of nuclear coding of plastid Aa-RS is supported by the observation that bleached *Euglena* mutants show an increase of Aa-RS specific activity upon illumination (Parthier and Neumann 1977). This suggests a photocontrol, probably located in the nucleo-cytoplasmic compartment, of the plastid Aa-RS synthesis. Chloramphenicol, an inhibitor of protein biosynthesis at 70S plastid ribosomes, becomes the more inhibitory for plastid Leu-RS formation the longer the time of treatment proceeds (Parthier 1976). Even taking into account a much slower penetration of this antibiotic to the site of action than cycloheximide, the inhibition of plastid Leu-RS synthesis by chloramphenicol seems to reflect an indirect effect on the synthesis of the enzyme. In respect to antibiotic sensitivity, the growth phase of the *Euglena* culture seems to be more important than the differentiation stage of the plastids (Lesiewicz and Herson 1975). Finally, by means of D₂O labeling and CsCl density gradient analysis of the isotope-labeled enzymes we have shown that in *Euglena* the light-dependent increase of plastid Leu-RS activity is due to de novo synthesized enzyme and is not a result of proenzyme activation (Nover 1976).

One of the possibilities for the control of nuclear-coded and cytoplasm-synthesized plastid enzymes is the formation of a "derepressor" formed at plastid ribosomes but active in the nucleus in the transcription of mRNA for plastid proteins. In darkness, the "derepressor" becomes active after combination with a "co-repressor" that accumulates in darkness, but disappears rapidly upon illumination (Parthier 1973). Intermediates of the chlorophyll biosynthesis pathway might exert a function of "co-repressor" (Ohad 1975). On the basis of experiments with plastid transcription inhibitors, rifampicin and streptovaricin, Lesiewicz and Herson (1975) suggest the translation of plastid mRNA for plastid Phe-RS at cytoplasmic ribosomes in greening *Euglena* cells. The authors discussed likewise the existence of genes for this enzyme. The transcription of the nuclear synthetase gene might be controlled by a plastid transcription product. Translation of the nuclear-coded mRNA on cytoplasmic ribosomes could then account for a portion of the chloroplast Phe-RS synthesis, but the organelle, via RNA transcription, retains primary control over the synthesis of the plastid enzyme.

All the hypotheses mentioned presume light-dependent increased synthesis and transport of macromolecules through the double membranes of the plastid envelope. As shown for the import of the RuBPCase small subunit into chloroplasts (Highfield and Ellis 1978), it seems reasonable to suggest plastid Aa-RS penetration through the plastid envelope from their sites of synthesis on cytoplasmic polysomes. Interestingly, McCrea and Hershberger (1978) postulated the export of certain plastid-specific tRNA's to cytoplasmic polyribosomes. It was not shown whether or not they were aminoacylated. Nevertheless, other plastid isoacceptor tRNA species remain in the chloroplasts and need to be charged by the cognate synthetases. Transport of nuclear-coded tRNA's into mitochondria of yeasts and *Tetrahymena* was reported (Suyama and Hamada 1976), but RNA uptake into chloroplasts is generally regarded to be improbable. In the case where certain Aa-RS species seem not to be present in isolated *Euglena* chloroplasts, but the cognate tRNA species do, a "shuttling" of enzymes or Aa-tRNA between the cell compartments was discussed (Parthier et al. 1978).

In developing cotyledons the marked increase in activity of plastid Leu-, Ile- and Met-RS did not require an exposure to light (Brantner and Dure 1975). The increase in synthesis of enzyme but likewise rRNA and tRNA in this plant material can be regarded as a preprogrammed aspect of proplastid to etioplast development beginning after seed germination. Unlike enzyme synthesis during the light-dependent chloroplast formation in *Euglena*, in cotyledons it is difficult to determine whether the observed stimulation reflects an increase of plastids per cell or an increase in synthetase or tRNA molecules per plastid.

D. RNA Polymerase

The lag period observed for rRNA and tRNA synthesis in *Euglena* cells (Figs. 6, 9 and 10) suggests that plastid RNA polymerase might be delayed in de novo synthesis following illumination, and the very low level of RNA formation during this phase may be catalyzed by the enzyme molecules already present in the proplastids. Another explanation of the lag period is that the enzyme is present in plastids in sufficient amounts but not in the active form, e.g., nonassociated with thylakoid membranes. No clear decision can be made at present which favors one of the two possibilities. However, since plastid RNA polymerase formation is under the control of the nucleus, one should expect a rapid light-stimulated synthesis of the polymerase similar to other plastid enzymes coded for by nuclear genes.

An attempt to prove the alternatives was made by Apel and Bogorad (1976). The authors measured a four-fold increase of corn plastid polymerase activity after 16 h of illumination. This change was neither due to a quantitative enhancement of the enzyme amount nor it was the result of qualitative alterations of the purified enzyme. Other factors which are suggested to be light-inducible might be able to modify either DNA-enzyme interaction or RNA polymerase activity.

E. Messenger RNA, Polysomes, and Protein Biosynthesis

The light-stimulated synthesis of plastid proteins can be based on three processes:

1. Synthesis of new mRNA's and, consequently, new polysomes.

2. Enhancement of polysome content by light-induced formation of previously (in darkness) transcribed mRNA and ribosomal subunits (functionalization of present mRNA).

3. Activation of the translation machinery, i.e., of previously formed polysomes.

Since a great proportion of plastid proteins are synthesized on cytoplasmic polysomes both before and after illumination, cytoplasmic mRNA and polysome formation must be also taken into consideration, in addition to their plastidic counterparts.

An increase in the amount of polysomes after illumination has been described by several authors whithout distinction between plastid and cytoplasmic particles (Williams and Novelli 1968; Poulsen and Beevers 1970; Heizmann et al. 1972). First Clark (1964) and later Chua et al. (1973b) demonstrated the enhancement of the polysome fraction in chloroplasts after illumination. This process was found to be reversible (Clark 1964). Pine and Klein (1972), studying etiolated seedlings, clearly documented a light-stimulated increase of both plastid and cytoplasmic polysomes, the latter to a lower extent. Polysome accumulation was noted 2 h after onset of illumination; during this time no incorporation of radioactive precursors into the polysome fraction was observed. In this way, the authors were able to point out that the light-induced polysome formation was not related with a de novo synthesis of ribosomal RNA (see also Heizmann et al. 1972). In rye seedlings (Table 4) the amount of plastid rRNA found in the polysome fraction increases from 13% to 39% upon illumination, but cytoplasmic polysomes likewise increase from 29% to 49% (Feierabend and De Bore 1978). The elevated response of the plastid polysomes to light might be due to a higher intrinsic rate of RNA synthesis in the organelle than in the cytoplasm (Pine and Klein 1972).

The light-increased polyribosome formation is a phytochrome-mediated process (Smith 1976). Cytokinins also stimulate the polysome formation in chloroplasts. In rye seedlings growing in the dark the proportion of ribosomes present as polyribosomes increases after addition of kinetin, more in the chloroplasts than in the cytoplasm (Table 4).

Table 4. Influence of light and cytokinin on the proportion of chloroplast and cytoplasmic polyribosomes in the first leaves of 7-day-old derooted rye seedlings. (According to Feierabend and de Boer 1978)

	Percentage in polysomes of	
	Chloroplast rRNA	Cytoplasmic rRNA
Water (dark)	12.0	29.0
Water (light)	38.9	49.1
Kinetin (dark)	17.6	31.8

Rapidly formed "photo-activated polysomes" in the cytoplasm, containing poly(A)-mRNA's can be discriminated from "photo-induced polysomes" which appear delayed some hours after illumination of dark-grown *Euglena* cultures (Heizmann et al. 1972). It was, however, not clear whether the photo-induced polysomes originate only from the greening plastids or from both organelles and cytoplasm. More recently Nigon et al. (1978) demonstrated that the turnover of polysomes containing poly(A)-mRNA was much higher (shorter than 10 min half-life) than polysomes containing mRNA without longer poly(A) sequences. Only the former type responds rapidly to darkness with a stop of synthesis. The rapidity of the process point to a light-dependent aggregation or desaggregation, respectively, of cytoplasmic polysomes obviously involved in the chloroplast formation. Here the effect of light should be designated as "functionalization" of polysomes in contrast to the light-induced formation of plastid polysomes with a half-life of several hours.

Illumination of *Chlamydomonas* cells alters the relation between free and membrane-bound polysomes in the chloroplast. No attachment of ribosomes to thylakoid membranes was observed during the dark phase of the cell cycle where chloroplast membranes are not synthesized. However, in the light phase of the cycle, 20% to 30% of the chloroplast ribosome population became attached to the thylakoid membranes. Concomitant extensive growth of thylakoid membranes and the synthesis of chlorophyll and at least certain membrane proteins take place, e.g., cytochromes 552 and 559. This correlation suggests that the bound plastid ribosomes (polysomes) are involved in the synthesis of thylakoid membrane proteins (Chua et al. 1973b, 1976; Margulies and Michaels 1974; Michaels and Margulies 1975).

The last years provided progress in the elucidation of plastid mRNA's during light-induced chloroplast formation. Although in darkness only a small amount of precursor incorporation is observed into chloroplast RNA's, an appreciable amount of mRNA seems to be transcribed independent of light-induction in *Euglena* cells (Chelm and Hallick 1976; Rawson and Boerma 1976). Total cellular RNA isolated from cultures at various stages of chloroplast development was hybridized with ^{125}J-labeled chloroplast DNA. When the cells are grown in the dark 53% (7.2×10^4 nucleotide pairs) of the plastid DNA is transcribed. Five h after illumination of the cells, this value rises to 7.8×10^4 nucleotide pairs, and at completion of chloroplast development 47% of the chloroplast DNA (6.4×10^4

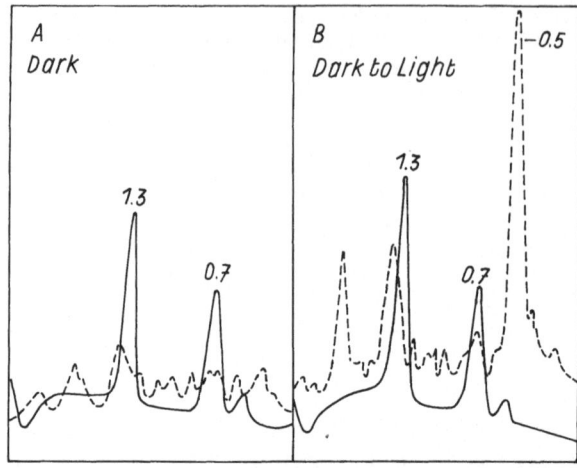

Electrophoretic mobility

Fig. 11A,B. Effect of light on the rate of synthesis of 0.5×10^6 RNA (mRNA for the P-32,000 chloroplast membrane protein) in *Spirodela*. **A** Dark-grown plants labeled with [³H]-uridine, 45 min. **B** 3 ½ h after transfer of dark-grown plants to light. Whole cell RNA was fractionated on polyacrylamide gels. 1.3 and 0.7 correspond with cytoplasmic rRNA's. See also Fig. 1. (After Rosner et al. 1977)

nucleotide pairs) are represented as RNA transcripts (Rawson and Boerma 1976). The corresponding data in Chelm and Hallick's (1976) experiments are 26.8% of DNA transcribed in the dark (this accounts for approximately 0.8% of the total cellular RNA), 21.3% after 4 h illumination, and 32.6% after 72 h light.

Hybridization experiments with total DNA and EcoRI restriction nuclease fragments have also shown that significant qualitative changes exist in chloroplast RNA synthesis during light-induced chloroplast development, i.e., DNA sequences not transcribed in dark-grown cells are transcribed in the light period of development. These new RNA sequences appear when morphological changes in the plastid structure become significant and when CO_2-fixation begins (Chelm and Hallick 1976; Rawson and Boerma 1976; Rawson et al. 1977).

Likewise, significantly more nuclear unique DNA is transcribed in light-grown than in dark-grown, etiolated *Euglena* cells (Curtis and Rawson 1977). In this connection the observations of Verdier et al. (1973) and Verdier (1975) are interesting. They found a rapid light-dependent increase of poly(A)-containing RNA in greening *Euglena* cells. They also confirmed that this mRNA was translated on cytoplasmic ribosomes, and that it probably originated from the nucleus. A similar dependency on light in poly(A)-rich RNA synthesis was noted in plastid mutants of *Euglena*; the observed stimulation was of the same degree as in the wild-type cells (Verdier 1975).

First direct evidence has been provided that a light-stimulated individual mRNA is transcribed on plastid DNA: Rosner et al. (1975) found an accumulation of the 0.5×10^6 daltons RNA after illumination of dark-grown *Spirodela* cells (Fig. 11). This RNA represents the messenger for a thylakoid membrane protein called P-32,000 (Rosner et al. 1977; Edelman and Reisfeld 1978).

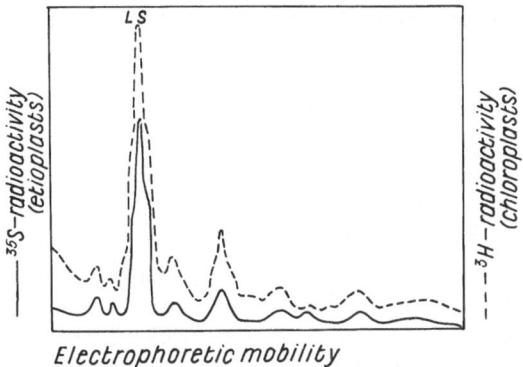

Fig. 12. Co-electrophoresis of pea etioplast and chloroplast proteins synthesized in vitro. Etioplast products were labeled in vitro with [³⁵S]-methionine and chloroplast products labeled in vitro with [³H]-leucine. Light was the energy source for isolated chloroplast protein synthesis. *LS* large subunit of RuBPCase. (After Siddell and Ellis 1975)

The photoregulated expression of a single plastid gene was studied by Bedbrook et al. (1978) using hybridization of mRNA from corn greening plastids and chloroplasts. This RNA which hybridized at restriction fragment Bam 8 of chloroplast DNA was absent in etioplasts. In vitro translation provided a 34,500 daltons thylakoid membrane polypeptide, thus indicating that chloroplasts but not etioplasts contain the mRNA for a specific thylakoid protein. Similarly, Howell (1978) demonstrated the presence of mRNA's for a membrane protein (M33) as well as for the large subunit of RuBPCase in *Chlamydomonas* chloroplasts. The synthesis of the latter polypeptide is markedly stimulated after illumination (see Bottomley, this vol.).

The enhancement by light of the protein-synthesizing activity in plastid differentiation might primarily reflect the induced synthesis of certain mRNA's which are transcribed at plastid DNA. Likewise, transcription and functionalization of mRNA of nuclear origin is involved. The products of these cytoplasmic light-induced mRNA's might be necessary for the inauguration of further steps in chloroplast formation. This is in accordance with the observations of Kaveh and Harel (1973) who showed that etiolated corn leaves 30 to 60 min after illumination did incorporate amino acids into a very small number of proteins, and this incorporation was stopped after the addition of inhibitors acting on cytoplasmic protein synthesis. Regarding the translation level, regulatory factors of the specific steps might be synthesized in connection with the light regime. Travis et al. (1972) claimed the light-stimulated ribosome activity to be related to the level of soluble proteins associated with (cytoplasmic) ribosomes. They could remove these protein factors from the particles by salt washings. Functional proof for initiation or elongation specificity, however, has not been made.

In this connection the study of protein biosynthesis in isolated plastids of various developmental states may offer valuable insight into the type of control of plastid proteins. Siddell and Ellis (1975) compared isolated plastids from etiolated and green pea plants in respect to their ability to incorporate precursors into plastid-specific polypeptides. The authors found remarkable quantitative but no qualitative differences in the gel pattern of the total protein fraction (Fig. 12). In both types of plastid only six polypeptides contained radioactivity. This result indicates that the pattern of synthesis at various stages of plastid differentiation is qualitatively not changed and gives rise to the idea that control of plastid protein synthesis can be exerted at the translation level. This is supported by the observation of Walden et al. (1978) who compared the translation of RuBPCase large subunit mRNA from etiolated and green cucumber in a cell-free *E.coli* system and found no difference, suggesting the lack of transcription control for this polypeptide in this plant material. On the other hand, the findings of Bedbrook et al. (1978) for the biosynthesis of a specific thylakoid protein in corn seedling chloroplasts and Alscher et al. (1977) for thylakoid proteins in pea chloroplasts provide evidence for mRNA transcription as the level of control. Apparently the synthesis of different plastid proteins can be regulated at different levels of gene expression.

V. Concluding Remarks

Chloroplast biogenesis is embedded in a network of regulatory processes of which two aspects seem to us worth discussing on the basis of the available evidence:

1. The genetic aspect: Cooperation of plastid and nuclear genomes for the ultrastructural and functional integration of different plastid constituents coded for by structure genes, however, likewise the control of the processes by the differential expression of regulatory genes.

2. The physiological (developmental) aspect: The fundamental role of light for the proplastid or etioplast to chloroplast transformation as well as the influence exerted by phytohormones or other endogenous factors.

In respect to the first aspect, Ellis (1977) proposed two principles which dominate at least for the control of the synthesis of organellar proteins, the multisubunit completion principle and the cytoplasmic control principle. The former is based on the fact that some subunits of a chloroplast protein are synthesized on plastid ribosomes whereas the other subunits are produced at cytoplasmic ribosomes with a subsequent import into the organelle for assembly.

Examples are RuBPCase, coupling factor CF_1 and thylakoid membrane proteins. No case is reported yet of a chloroplast protein completely made inside the organelle.

The cytoplasmic control principle states that cytoplasmic products control organellar protein synthesis, but that the converse does not occur, e.g., a possible control of the RuBPCase large subunit synthesis by the available amount of the small subunit inside the plastid (negative feedback control, cf. Ellis 1978). The control by cytoplasmic proteins as positive control factors on organellar transcription or translation is likewise discussed (Ellis 1975b).

Other observations, however, do not support Ellis' second principle: In heat-treated, plastid-ribosome-deficient rye seedlings the RuBPCase small subunit is accumulated in the absence of large subunit synthesis (Feierabend and Wildner 1978) indicating an independent regulation of the synthesis of the two subunits.

Furthermore, in *Chlamydomonas* plastid gene products seem to influence typical cytoplasmic processes such as cell division (Blamire et al. 1974), and recently McCrea and Hershberger (1978) claimed the export of certain tRNA species, hybridizing with plastid DNA, out of the organelles into the cytoplasm where they bind to polyribosomes. The authors speculated about their role in the biosynthesis of plastid proteins in the cytoplasmic compartment.

To the second the physiological aspect: The light effect on chloroplast formation appears different with respect to the plant material considered. In certain algae the proplastid-chloroplast conversion is fully light-dependent and is characterized by marked increase in all constituents of the organelle including the de novo synthesis of proteins enabling photosynthesis (e.g., *Euglena*, *Chlamydomonas* y-1). During the etioplast-chloroplast transformation in etiolated angiosperms illumination seems to cause little more than the photochemical reduction of protochlorophyll(ide), de novo synthesis of some thylakoid proteins, and the formation of key enzymes in porphyrine biosynthesis. Concerning stroma proteins such as RuBPCase, however, great quantitative differences do exist in various plant species. Finally, gymnosperm seedlings, ferns, and mosses need no light for chlorophyll accumulation and chloroplast formation, although certain reactions in the photosynthetic electron transport chain remain inactive during greening in darkness.

Before light induces the multiple process of plastid differentiation in terms of plastid and nuclear gene expression it must be absorbed. Whereas in *Euglena* the protochlorophyll(ide) itself seems to be receptor for both red and blue light, probably complemented by a blue-light receptor in the cytoplasm (Schiff 1974), in higher plants the phytochrome system mediates plastid development (Mohr 1977).

Our knowledge about the steps following light absorption is extremely limited. What substances play a role as transmitter between receptor and gene activation? Which are the primary metabolic reactions in a chain of events leading to the functional chloroplast?

Most of the available results are concomitant with the idea that one of the first light effects in dark-grown cells is the activation of a nucleo-cytoplasmic system of gene expression which provides the prerequisites for chloroplast formation, including energy, precursor supply for the macromolecular biosynthesis, and activation of enzymes to mobilize the energy.

The following sequence of early events after illumination, in the area of gene expression is discussed: (1) Appearance of mRNA transcribed at nuclear genes and formation of cytoplasmic polyribosomes. Some of their polypeptide products may have a function in the triggering of the plastid gene expression. (2) Synthesis of plastid mRNA which is qualitatively different from those synthesized in the dark period (Chelm and Hallick 1976; Rawson and Boerma 1976; Coen et al. 1977). (3) Mass synthesis of rRNA, ribosomal proteins, ribosomes, polysomes and the transcription and translation enzymes both in plastid and nucleo-cytoplasm and subsequent cooperated formation of the chloroplast constituents. There may exist large quantitative differences between plastid differentiation in algae and angiosperms, but the sequence of events after induction by light seems to be uniform.

The situation becomes still more complicated by the interference of light as a "substrate", especially during differentiation steps following the early phase. With the very beginning of photosynthesis light can contribute to an increased precursor supply. Therefore, in many studies the results have the drawback of overlapping of the two actions of light. Effector and "substrate" functions of light have been rarely separated by means of photosynthetic inhibitors (e.g., DCMU). The light-enhanced precursor supply for macromolecule synthesis could be a reason for the light-stimulated RNA synthesis. For example, the lag period of rRNA and tRNA synthesis observed in greening *Euglena* might reflect the deficiency in nucleoside triphosphates, as the energy source to form RNA is not sufficiently available in the early phase of illumination. However, according to Cohen and Schiff (1976) there are no precursor pool limitations for rRNA synthesis in resting cells of *Euglena* upon illumination. In vivo, it seems extremely difficult to decide from incorporation experiments and rate determination between precursor metabolism and synthesis or turnover of the macromolecules. Precursor pool formation and its availability ("channeling") multiply the difficulties emerging for the exact determination of the biosynthesis of plastid macromolecules.

Although remarkable progress has been made in the last years at the level of transcription and translation of chloroplast biogenesis, important questions cannot be answered at present: (1) Mechanisms of macromolecule transport through the plastid envelope and assembly of macromolecules to functional chloroplast constituents. (2) Primary events immediately after light absorption and the response in gene activation. (3) The exact sequence of events in plastid and cytoplasm of gene expressions and the control of cooperation of the two genetic systems. (4) In which way endogeneous factors (e.g., hormones) influence plastid differentiation.

References

Akoyunoglou, G., Argyroudi-Akoyunoglou, J.H. (eds.): Chloroplast Development, 888 p. Amsterdam: Elsevier/North-Holland 1978

Alscher, R., Jagendorf, A.T., Fonda, S.A.: The polyribosome-membrane association in chloroplasts. Plant Physiol. 59, Suppl. 9 (1977)

Apel, K., Bogorad, L.: Light-induced increase in the activity of maize plastid DNA-dependent RNA polymerase. Eur. J. Biochem. 67, 615-620 (1976)

Apel, K., Schweiger, H.-G.: Sites of synthesis of chloroplast-membrane proteins. Evidence for three types of ribosomes engaged in chloroplast protein synthesis. Eur. J. Biochem. 38, 373-383 (1973)

Armstrong, J.J., Surzycki, S.J., Moll, B., Levine, R.D.: Genetic transcription and translation specifying chloroplast components in *Chlamydomonas reinhardii*. Biochemistry 10, 692-701 (1971)

Barnett, W.E., Pennington, C.J., Fairfield, S.A.: Induction of *Euglena* transfer RNAs by light. Proc. Natl. Acad. Sci. USA 63, 1261-1268 (1969)

Bartels, P.G., Weier, T.E.: Particle arrangements in proplastids of *Triticum vulgare L.* seedlings. J. Cell Biol. 33, 243-253 (1967)

Bartolf, M., Price, C.A.: Synthesis of poly(A)-containing RNA by isolated spinach chloroplasts. Plant Physiol. 59, Suppl. 21 (1977)

Bedbrook, J.R., Link, G., Coen, D.M., Bogorad, L., Rich, A.: Maize plastid gene expressed during photoregulated development. Proc. Natl. Acad. Sci. USA 75, 3060-3064 (1978)

Bennet, J., Ellis, J.R.: Solubilization of the membrane-bound DNA-dependent RNA polymerase of pea chloroplasts. Biochem. Soc. Trans. 1, 892-894 (1973)

Blair, G.E., Ellis, R.J.: Protein synthesis in chloroplasts. I. Light-driven synthesis of the large subunit of fraction I protein by isolated pea chloroplasts. Biochim. Biophys. Acta 319, 223-234 (1973)

Blamire, J., Flechtner, V.R., Sager, R.: Regulation of nuclear DNA replication by the chloroplast in Chlamydomonas. Proc. Natl. Acad. Sci. USA 71, 2867-2871 (1974)

Blobel, G., Dobberstein, B.: Transfer of proteins across membranes. I. Presence of proteolytically processed and unprocessed nascent immunoglobulin light chains on membrane-bound ribosomes of murine myeloma. J. Cell Biol. 67, 835-851 (1975)

Boardman, N.K.: Ribosome composition and chloroplast development in Phaseolus vulgaris. Exp. Cell Res. 43, 474-482 (1966)

Bogorad, L., Mets, L.J., Mullinix, K.P., Smith, H.J., Strain, G.C.: Possibilities for intracellular integration: the ribonucleic acid polymerases of chloroplasts and nuclei, and genes specifying chloroplast ribosomal proteins. Biochem. Soc. Symp. 38, 17-41 (1973)

Bogorad, L., Davidson, J.N., Hanson, M.R., Mets, L.: Intergenomic cooperation in the synthesis of chloroplast ribosomes of Chlamydomonas. In: Proc. 12th Int. Bot. Congr., Abstr., p. 398. Leningrad 1975

Bogorad, L., Davidson, J.N., Hanson, M.R.: The genetics of the chloroplast ribosome in Chlamydomonas reinhardii. In: Nucleic Acids and Protein Synthesis in Plants, pp. 135-154; Bogorad, L., Weil, J.H. (eds.). New York-London: Plenum Press 1977

Bohnert, H.J., Schmitt, J.M.: Synthesis of high-molecular weight RNA in isolated chloroplasts. Hoppe Seylers Z. Physiol. Chem. 355, 1179 (1974)

Bohnert, H.J., Driesel, A.J., Herrmann, R.G.: Characterization of the RNA compounds synthesized by isolated chloroplasts, pp. 629-636. In: see ref. Bücher et al. (1976)

Börner, T.: Struktur und Funktion der genetischen Information in den Plastiden. Biol. Zentralbl. 92, 545-561 (1973)

Börner, T., Knoth, R., Herrmann, F., Hagemann, R.: Struktur und Funktion der genetischen Information in den Plastiden. V. Das Fehlen von ribosomaler RNS in den Plastiden der Plastommutante "Mrs. Parker" von Pelargonium zonale. Theor. Appl. Genet. 42, 3-11 (1972)

Bottomley, W.: Some effects of Triton X-100 on pea chloroplasts. Plant Physiol. 46, 437-441 (1970)

Bottomley, W., Whitfeld, P.R.: The products of in vitro transcription and translation of spinach chloroplast DNA, pp. 657-662. In: see ref. Akoyunoglou and Argyroudi-Akoyunoglou (1978)

Bottomley, W., Smith, H.J., Bogorad, L.: RNA polymerase of maize: partial purification and properties of the chloroplast enzyme. Proc. Natl. Acad. Sci. USA 68, 2412-2416 (1971a)

Bottomley, W., Spencer, D., Wheeler, A.M., Whitfeld, P.R.: The effect of a range of RNA polymerase inhibitors on RNA synthesis in higher plant chloroplasts and nuclei. Arch. Biochem. Biophys. 143, 269-275 (1971b)

Bottomley, W., Spencer, D., Whitfeld, P.R.: Protein synthesis in isolated spinach chloroplasts: comparison of light-driven and ATP-driven synthesis. Arch. Biochem. Biophys. 164, 106-117 (1974)

Bottomley, W., Higgins, T.J.V., Whitfeld, P.R.: Differential recognition of chloroplast and cytoplasmic messenger RNA by 70S and 80S ribosomal systems. FEBS Lett. 63, 120-124 (1976)

Boulter, D., Ellis, R.J., Yarwood, A.: Biochemistry of protein synthesis in plants. Biol. Rev. 47, 113-175 (1972)

Bourque, D.P., Wildman, S.G.: Evidence that nuclear genes code for several chloroplast ribosomal proteins. Biochem. Biophys. Res. Commun. 50, 532-537 (1973)

Brandt, P., Wiessner, W.: Unterschiedliche Temperaturoptima der DNA-abhängigen RNA-Polymerasen von Euglena gracilis, Stamm Z und ihre Bedeutung für die experimentelle Erzeugung der permanenten Apochlorose durch höhere Temperatur. Z. Pflanzenphysiol. 85, 53-60 (1977)

Brantner, J.H., Dure, L.S.: The developmental biochemistry of cotton and seed embryogenesis and germination. VI. Levels of cytosol and chloroplast aminoacyl-tRNA synthetases during cotyledon development. Biochim. Biophys. Acta 414, 99-114 (1975)

Brown, R.D., Haselkorn, R.: Chloroplast RNA population in dark-grown, light-grown and greening Euglena gracilis. Proc. Natl. Acad. Sci. USA 68, 2536-2539 (1971)

Brown, R.D., Bastia, D., Haselkorn, R.: Effect of rifampicin on transcription in chloroplasts of Euglena. In: RNA Polymerase and Transcription, pp. 309-327; Silvester, L. (ed.). Amsterdam: North Holland 1970

Bücher, Th., Neupert, W., Sebald, W., Werner, S.: Genetics and Biogenesis of Chloroplasts and Mitochondria. Amsterdam: Elsevier/North-Holland 1976

Buetow, D.E., Kissel, M.S., Zablen, L.: Phylogenetic origin of chloroplast 16S ribosomal RNA, pp. 641-644. In: see ref. Bücher et al. (1976)

Burkard, G., Keller, E.B.: Poly(A) polymerase and poly(G) polymerase in wheat chloroplasts. Proc. Natl. Acad. Sci. USA 71, 389-393 (1974)

Burkard, G., Eclancher, B., Weil, J.H.: Presence of N-formyl-methionyl tRNA in bean chloroplasts. FEBS Lett. 4, 285-287 (1969)

Burkard, G., Guillemaut, P., Weil, J.H.: Comparative studies of the tRNAs and the aminoacyl-tRNA synthetases from the cytoplasm and the chloroplasts of Phaseolus vulgaris. Biochim. Biophys. Acta 224, 184-198 (1970)

Burkard, G., Vaultier, J.P., Weil, J.H.: Differences in the level of plastid-specific tRNAs in chloroplasts and etioplasts of Phaseoclus vulgaris. Phytochemistry 11, 1351-1353 (1972)

Burkard, G., Guillemaut, P., Steinmetz, A., Weil, J.H.: Transfer ribonucleic acid and transfer ribonucleic acid recognizing enzymes in bean cytoplasm, chloroplasts, etioplasts and mitochondria. Biochem. Soc. Symp. 38, 43-56 (1973)

Carritt, B., Eisenstadt, J.M.: RNA synthesis in isolated chloroplasts: Characterization of the newly synthesized RNA. FEBS Lett. 36, 116-120 (1973)

Cattolico, R.A., Senner, J.W., Jones, R.F.: Changes in cytoplasmic and chloroplast ribosomal ribonucleic acid during the cell cycle of Chlamydomonas reinhardii. Arch. Biochem. Biophys. 156, 58-65 (1973)

Chan, P., Wildman, S.G.: Chloroplast DNA codes for the primary structure of the large subunit of fraction I protein. Biochim. Biophys. Acta 277, 677-680 (1972)

Chang, S.H., Hecker, L.J., Silberklang, M., Brum, C.K., Rajbhandary, U.L., Barnett, W.E.: Nucleotide sequence of phenylalanine transfer RNA from the chloroplasts of Euglena gracilis. Cell 9, 717-723 (1976)

Chelm, B.K., Hallick, R.B.: Changes in the expression of the chloroplast genome of Euglena gracilis during chloroplast development. Biochemistry 15, 593-599 (1976)

Chen, J.L., Wildman, S.G.: "Free" and membrane-bound ribosomes, and nature of products formed by isolated tobacco chloroplasts incubated for protein synthesis. Biochim. Biophys. Acta 209, 207-219 (1970)

Chua, N.H., Blobel, G., Siekevitz, P.: Isolation of cytoplasmic and chloroplast ribosomes and their dissociation into active subunits from Chlamydomonas reinhardii. J. Cell Biol. 57, 798-814 (1973a)

Chua, N.H., Blobel, G., Siekevitz, P., Palade, G.E.: Attachment of chloroplast polysomes to thylakoid membranes in Chlamydomonas reinhardii. Proc. Natl. Acad. Sci. USA 70, 1554-1559 (1973b)

Chua, M., Blobel, G., Siekevitz, P., Palade, G.: Periodic variations in the ratio of free to thylakoid-bound chloroplast ribosomes during the cell cycle of Chlamydomonas reinhardii. J. Cell Biol. 71, 497-514 (1976)

Ciferri, O.: Mechanism of protein synthesis in higher plants. Phytochem. Soc. Symp. 11, 113-136 (1975)

Ciferri, O., Tiboni, O.: Elongation factors for chloroplast and mitochondrial protein synthesis in Chlorella vulgaris. Nature New Biol. 245, 209-211 (1973)

Ciferri, O., Tiboni, O.: Evidence for the synthesis in the chloroplast of elongation factor G. Plant Sci. Lett. 7, 455-466 (1976)

Clark, M.F.: Polyribosomes from chloroplasts. Biochim. Biophys. Acta 91, 671-674 (1964)

Coen, D.M., Bedbrook, J.R., Bogorad, L., Rich, A.: Maize chloroplast DNA fragment encoding the large subunit of ribulose bisphosphate carboxylase. Proc. Natl. Acad. Sci. USA 74, 5487-5491 (1977)

Cohen, D., Schiff, J.A.: Events surrounding the early development of *Euglena* chloroplasts. Photoregulation of the transcription of chloroplastic and cytoplasmic ribosomal RNA. Arch. Biochem. Biophys. 177, 201-216 (1976)

Curtis, S., Rawson, J.: Transcription of nuclear unique DNA during chloroplast development in *Euglena gracilis*. J. Cell Biol. 75, XXXa (1977)

Dennis, P.P.: Stable ribonucleic acid synthesis during the cell division cycle in slowly growing *Escherichia coli* B/r. J. Biol. Chem. 247, 204-208 (1972)

Detchon, P., Possingham, J.V.: Ribosomal RNA distribution during leaf development in spinach. Phytochemistry 11, 943-947 (1972)

Detchon, P., Possingham, J.V.: Chloroplast ribosomal ribonucleic acid synthesis in cultured spinach leaf tissue. Biochem. J. 136, 829-836 (1973)

Dobberstein, B., Blobel, G., Chua, N.H.: In vitro synthesis and processing of a putative precursor for the small subunit of ribulose-1,5-bisphosphate carboxylase of *Chlamydomonas reinhardii*. Proc. Natl. Acad. Sci. USA 74, 1082-1085 (1977)

Drumm, H.E., Margulies, M.M.: In vitro protein synthesis by plastids of *Phaseolus vulgaris*. Plant Physiol. 45, 435-442 (1970)

Dubois, E.G., Dirheimer, G., Weil, J.H.: Methylation of yeast tRNA^{Asp} by enzymes from cytoplasm, chloroplasts and mitochondria of *Phaseolus vulgaris*. Biochim. Biophys. Acta 374, 332-341 (1975)

Dyer, T.A., Bowman, C.M.: A sequence analysis of low-molecular weight rRNA from chloroplasts of flowering plants, pp. 645-651. In: see ref. Bücher et al. (1976)

Dyer, T.A., Miller, R.H., Greenwood, A.D.: Leaf nucleic acids. I. Characteristics and role in the differentiation of plastids. J. Exp. Bot. 22, 125-136 (1971)

Dyer, T.A., Bowman, C.M., Payne, P.I.: The low-molecular weight RNAs of plant ribosomes: Their structure, function and evolution. In: Nucleic Acids and Protein Synthesis in Plants, pp. 121-133; Bogorad, L., Weil, J.H. (eds.). New York-London: Plenum Press 1977

Eaglesham, A.R.J., Ellis, R.J.: Protein synthesis in chloroplasts. II. Light-driven synthesis of membrane protein by isolated pea chloroplasts. Biochim. Biophys. Acta 335, 396-407 (1974)

Edelman, M., Reisfeld, A.: Identification, in vitro translation, and control of the main chloroplast membrane protein synthesized in *Spirodela*, pp. 641-652. In: see ref. Akoyunoglou and Argyroudi-Akoyunoglou (1978)

Ellis, R.J.: The synthesis of chloroplast membranes in *Pisum sativum*. In: Membrane Biogenesis, pp. 247-278; Tsagaloff, A. (ed.). New York: Plenum Publ. 1975a

Ellis, R.J.: Inhibition of chloroplast protein synthesis by lincomycin and 2-(4-methyl-2,6-dinitroanilino)-N-methyl-propionamide. Phytochemistry 14, 89-93 (1975b)

Ellis, R.J.: Protein synthesis by isolated chloroplasts. Biochim. Biophys. Acta 463, 185-215 (1977)

Ellis, R.J., Hartley, M.R.: Sites of synthesis of chloroplast proteins. Nature (New Biol.) 233, 193-196 (1971)

Ellis, R.J., Hartley, M.R.: Nucleic acids of chloroplasts. In: Biochemistry of Nucleic Acids; Burton U. (ed.). Butterworth and University Park Press 1974

Ellis, R.J., Blair, G.E., Hartley, M.R.: The nature and function of chloroplast protein synthesis. Biochem. Soc. Symp. 38, 157-162 (1973)

Fairfield, S.A., Barnett, W.E.: On the similarity between the tRNAs of organelles and prokaryotes. Proc. Natl. Acad. Sci. USA 68, 2972-2976 (1971)

Falk, H.: Rough thylakoids: Polysomes attached to chloroplast membranes. J. Cell Biol. 42, 582-587 (1969)

Feierabend, J., De Boer, J.: Comparative analysis of the action of cytokinin and light on the formation of ribulosebisphosphate carboxylase and plastid biogenesis. Planta 142, 75-82 (1978)

Feierabend, J., Schrader-Reichardt, U.: Biochemical differentiation of plastids and other organelles in rye leaves with a high-temperature-induced deficiency of plastid ribosomes. Planta 129, 133-145 (1976)

Feierabend, J., Wildner, G.: Formation of the small subunit in the absence of the large subunit of ribulose 1,5-bisphosphate carboxylase in 70S ribosome-deficient rye leaves. Arch. Biochem. Biophys. 186, 283-291 (1978)

Freyssinet, G.: Changes in chloroplast ribosomal protein in a streptomycin-resistant mutant of *Euglena gracilis*. Plant Sci. Lett. 5, 305-311 (1975)

Freyssinet, G.: The protein synthesizing system of *Euglena:* Synthesis of ribosomal proteins in vivo and their characterization. Physiol. Veg. 15, 519-550 (1977)

Freyssinet, G., Schiff, J.A.: The chloroplast and cytoplasmic ribosomes of *Euglena*. II. Characterization of ribosomal proteins. Plant Physiol. 53, 543-554 (1974)

Freyssinet, G., Morle, F., Nigon, V.: Chloroplast ribosomal proteins of *Euglena gracilis* immunological studies, pp. 653-656. In: see ref. Bücher et al. (1976)

Galling, G.: Der Einfluß von Rifampicin, Chloramphenicol und Cycloheximid auf den Uridin-Einbau in chloroplastidäre Ribosomenvorstufen von *Chlorella*. Planta 98, 50-62 (1971)

Galling, G.: Synthese von Plastidenribosomen im Entwicklungszyklus von *Chlorella*. Biochem. Physiol. Pflanz. 164, 575-581 (1973)

Galling, G.: RNA synthesis in synchronous cultures of unicellular algae. In: Les Sycles Cellulaires et leur Blocage chez Plusieurs Protistes, pp. 225-231. Paris: Ed. CNRS 1975

Galling, G., Salzmann, Ch., Spiess, E.: Synthese von Chlorophyll und Strukturelementen des Plastiden in *Chlorella* ohne Beteiligung der Chloroplasten-Ribosomen. Planta 114, 269-284 (1973)

Galling, G., Rössler-Hedenskog, M., Lorenzen, H.: Endogene Produktionsschwankungen bei *Scenedesmus acutus* und ihre Beziehung zum Nukleinsäure-Stoffwechsel. Planta 124, 219-229 (1975)

Gelvin, S., Heizmann, P., Howell, S.H.: Identification and cloning of the chloroplast gene coding for the large subunit of ribulose-1,5-bisphosphate carboxylase from *Chlamydomonas reinhardii*. Proc. Natl. Acad. Sci. USA 74, 3193-3197 (1977)

Goins, D.J., Reynolds, R.J., Schiff, J.A., Barnett, W.E.: A cytoplasmic regulatory mutant of *Euglena:* Constitutivity for the light-inducible chloroplast transfer RNAs. Proc. Natl. Acad. Sci. USA 70, 1749-1752 (1973)

Grierson, D., Loening, U.: Ribosomal RNA precursors and the synthesis of chloroplast and cytoplasmic ribosomal ribonucleic acid in leaves of *Phaseolus aureus*.. Eur. J. Biochem. 44, 501-507 (1974)

Grierson, D., Chambers, S.E., Penniket, L.P.: Nucleic acid and protein synthesis in discs cut from mature leaves of *Nicotiana tabacum L.* and cultured on nutrient agar with and without kinetin. Planta 134, 24-34 (1977)

Gualerzi, C., Cammarano, P.: Species specificity of ribosomal proteins from chloroplasts and cytoplasmic ribosomes of higher plants. Biochim. Biophys. Acta 199, 203-213 (1970)

Gualerzi, C., Janda, H.G., Passow, H., Stöffler, G.: Studies on protein moiety of plant ribosomes. Enumeration of the proteins of the ribosomal subunits and determination of the degree of evolutionary conservation by electrophoretic and immunochemical methods. J. Biol. Chem. 249, 3347-3355 (1974)

Guderian, R.H., Pulliam, R.L., Gordon, M.P.: Characterization and fractionation of tobacco leaf transfer RNA. Biochim. Biophys. Acta 262, 50-65 (1972)

Guignery, G., Duranton, J.: Origin of mRNAs directing chloroplast protein synthesis in vitro, pp. 663-668. In: see ref. Akoyunoglou and Argyroudi-Akoyunoglou (1978)

Guillemaut, P., Keith, G.: Primary structure of bean chloroplastic tRNA^{Phe}. Comparison with *Euglena* chloroplastic tRNA^{Phe}. FEBS Lett. 84, 352-356 (1977)

Guillemaut, P., Weil, J.H.: Aminoacylation of *Phaseolus vulgaris* cytoplasmic, chloroplastic and mitochondrial tRNAS^{Met} and of *E.coli* tRNAs^{Met} by homologous and heterologous enzymes. Biochim. Biophys. Acta 407, 240-248 (1975)

Guillemaut, P., Burkard, G., Steinmetz, A., Weil, J.H.: Comparative studies on the tRNAs^{Met} from the cytoplasm, chloroplasts and mitochondria of *Phaseolus vulgaris*. Plant Sci. Lett. 1, 141-146 (1973)

Guillemaut, P., Steinmetz, A., Burkard, G., Weil, J.H.: Aminoacylation of tRNA^{Leu} species from *Escherichia coli* and from the cytoplasm, chloroplasts and mitochondria of *Phaseolus vulgaris* by homologous and heterologous enzymes. Biochim. Biophys. Acta 378, 64-72 (1975)

Guillemaut, P., Martin, R., Weil, J.H.: Purification and base composition of a chloroplastic tRNA^{Phe} from *Phaseolus vulgaris*. FEBS Lett. 63, 273-277 (1976)

Haff, L.A., Bogorad, L.: Hybridization of maize chloroplast DNA with transfer ribonucleic acids. Biochemistry 15, 4105-4109 (1976a)

Haff, L.A., Bogorad, L.: Poly(adenylic acid)-containing RNA from plastids of maize. Biochemistry 15, 4110-4115 (1976b)

Hallick, R.B., Lipper, C., Richards, O.C., Rutter, W.J.: Isolation of a transcriptionally active chromosome from chloroplasts of Euglena gracilis. Biochemistry 15, 3039-3045 (1976)

Hanson, M.R., Davidson, J.N., Mets, L.J., Bogorad, L.: Characterization of chloroplast and cytoplasmic ribosomal proteins of Chlamydomonas reinhardii by two-dimensional gel electrophoresis. Mol. Gen. Genet. 132, 105-118 (1974)

Harel, E., Bogorad, L.: Effect of light on ribonucleic acid metabolism in greening maize leaves. Plant Physiol. 51, 10-16 (1973)

Harris, E.H., Eisenstadt, J.M.: Initiation of polysome formation in chloroplasts isolated from Euglena gracilis. Biochim. Biophys. Acta 232, 167-170 (1971)

Harris, E.H., Preston, J.F., Eisenstadt, J.M.: Amino acid incorporation and products of protein synthesis in isolated chloroplasts of Euglena gracilis. Biochemistry 12, 1227-1233 (1973)

Hartley, M.R., Ellis, R.J.: Ribonucleic acid synthesis in chloroplasts. Biochem. J. 134, 249-262 (1973)

Hartley, M.R., Wheeler, A., Ellis, R.J.: Protein synthesis in chloroplasts. V. Translation of messenger RNA for the large subunit of fraction I protein in a heterologous cell-free system. J. Mol. Biol. 91, 67-77 (1975)

Hartley, M.R., Head, C.W., Gardiner, J.: The synthesis of chloroplast RNA. In: Acides Nucléiques et Synthése des Protéines chez les Végétaux, pp. 419-423; Bogorad, L., Weil, J.H. (eds.). Paris: Ed. CNRS 1977

Hearing, V.J.: Protein synthesis in isolated etioplasts after light stimulation. Phytochemistry 12, 227-282 (1973)

Heber, U.: Metabolite exchange between chloroplasts and cytoplasm. Annu. Rev. Plant Physiol. 25, 28-55 (1974)

Hecker, L.J., Egan, J., Reynolds, R.J., Nix, C.E., Schiff, J.A., Barnett, W.E.: The sites of transcription and translation for Euglena chloroplastic aminoacyl-tRNA synthetases. Proc. Natl. Acad. Sci. USA 71, 1910-1914 (1974)

Hecker, L.J., Uziel, M., Barnett, W.E.: Comparative base composition of chloroplast and cytoplasmic tRNAs[Phe] from Euglena gracilis. Nucleic Acids Res. 3, 371-380 (1976).

Heizmann, P.: Maturation of chloroplast rRNA in Euglena gracilis. Biochem. Biophys. Res. Commun. 56, 112-118 (1974a)

Heizmann, P.: La synthese des RNA ribosomiques an cours de l'eclairement d'Euglenes etiolees. Biochim. Biophys. Acta 353, 301-312 (1974b)

Heizmann, P., Verdier, G.: Expression of nuclear and chloroplast genomes in Euglena during greening and in bleached mutants, pp. 623-628. In: see ref. Akoyunoglou and Argyroudi-Akoyunoglou (1978)

Heizmann, P., Trabuchet, G., Verdier, G., Freyssinet, G., Nigon, V.: Influence de l'eclairement sur l'evolution des polysomes dans des cultures d'Euglena gracilis étiolées. Biochim. Biophys. Acta 227, 149-160 (1972)

Heizmann, P., Verdier, G., Nigon, V.: Light induction of RNA synthesis in Euglena. In: Les Cycles Cellulaires et leur Blocage chez Plusieurs Protistes, pp. 247-254. Paris: Ed. CNRS 1975

Heizmann, P., Salvador, G., Nigon, V.: Occurence of plastidal rRNAs and plastidal structures in bleached mutants of Euglena gracilis. Exp. Cell Res. 99, 253-260 (1976)

Herrmann, R.G., Bohnert, H.J., Driesel, A., Hobom, B.: The location of rRNA genes on the restriction endonuclease map of the Spinacia oleracea chloroplast DNA, pp. 351-359. In: see ref. Bücher et al. (1976)

Highfield, P.E., Ellis, R.J.: Synthesis and transport of the small subunit of chloroplast ribulose bisphosphate carboxylase. Nature (London) 271, 420-424 (1978)

Hirvonen, A.P., Price, C.A.: Chloroplast ribosomes in the proplastids of Euglena gracilis. Biochim. Biophys. Acta 232, 696-704 (1971)

Honeycutt, R.C., Margulies, M.M.: Protein synthesis in *Chlamydomonas reinhardii*: Evidence for synthesis of proteins of chloroplastic ribosomes on cytoplasmic ribosomes. J. Biol. Chem. 248, 6145-6153 (1973)

Hoober, J.K., Blobel, G.: Characterization of the chloroplastic and cytoplasmic ribosomes of *Chlamydomonas reinhardii*. J. Mol. Biol. 41, 121-138 (1969)

Howell, S.H.: Cell cycle regulation of messenger RNAs coding for chloroplast proteins in *Chlamydomonas reinhardii*, pp. 679-686. In: see ref. Akoyunoglou and Argyroudi-Akoyunoglou (1978)

Howell, S., Heizmann, P., Gelvin, S.: Localization of the gene coding for the large subunit of ribulose bisphosphate carboxylase on the chloroplast genome of *Chlamydomonas reinhardii*, pp. 625-628. In: see ref. Bücher et al. (1976)

Ingle, J.: Synthesis and stability of chloroplast ribosomal RNAs. Plant Physiol. 43, 1448-1454 (1968a)

Ingle, J.: The effect of light and inhibitors on chloroplast and cytoplasmic RNA synthesis. Plant Physiol. 43, 1850-1854 (1968b)

Ingle, J., Possingham, J.V., Wells, R., Leaver, C.J., Loening, U.E.: The properties of chloroplast ribosomal RNA. Symp. Soc. Exp. Biol. 24, 303-325 (1970)

Ingle, J., Wells, R., Possingham, J.V., Leaver, C.J.: The origins of chloroplast ribosomal RNA. In: Autonomy and Biogenesis of Mitochondria and Chloroplasts, pp. 393-401; Boardman, N.K., Linnane, A.W., Smillie, R.M. (eds.). Amsterdam: North-Holland 1971

Jacobson, A.B., Swift, J., Bogorad, L.: Cytochemical studies concerning the occurence and distribution of RNA in plastids of *Zea mays*. J. Cell Biol. 17, 557-570 (1963)

Jeannin, G., Burkard, G., Weil, J.H.: Aminoacylation of *Phaseolus vulgaris* cytoplasmic, chloroplastic and mitochondrial tRNAs^{Pro} and tRNAs^{Lys} by homologous and heterologous enzymes. Biochim. Biophys. Acta 442, 24-31 (1976)

Jeannin, G., Burkard, G., Weil, J.H.: Characterization of *Phaseolus vulgaris* cytoplasmic, chloroplastic and mitochondrial tRNAs^{Phe}; aminoacylation by homologous and heterologous enzymes. Plant Sci. Lett. 13, 75-81 (1978)

Jennings, R.C., Ohad, I.: Biogenesis of chloroplast membranes XI. Evidence for the translation of extra-chloroplast RNA on chloroplast ribosomes in a mutant of *Chlamydomonas reinhardii*, y-1. Arch. Biochem. Biophys. 153, 79-87 (1972)

Jones, B.L., Nagabhushan, N., Gulyas, A., Zalik, S.: Two-dimensional acrylamide gel electrophoresis of wheat leaf cytoplasmic and chloroplast ribosomal proteins. FEBS Lett. 23, 167-170 (1972)

Joussaume, M.: Mise en évidence de deux formes de RNA polymérase dépendante du DNA dans les chloroplastes isolés de feuilles de pois. Physiol. Veg. 11, 69-82 (1973)

Kanabus, J., Cherry, J.H.: Isolation of an organ-specific leucyl-tRNA synthetase from soybean seedlings. Proc. Natl. Acad. Sci. USA 68, 873-876 (1971)

Kaveh, D., Harel, E.: Light-induced change in the pattern of protein synthesis during the early stages of greening of etiolated maize leaves. Plant Physiol. 51, 671-676 (1973)

Kislev, N., Selsky, M.J., Norton, C., Eisenstadt, J.M.: tRNA and tRNA aminoacyl synthetases of chloroplasts, mitochondria and cytoplasm from *Euglena gracilis*. Biochim. Biophys. Acta 287, 256-269 (1972)

Kloppstech, K., Schweiger, H.G.: Nuclear genome codes for chloroplast ribosomal proteins in *Acetabularia*. II. Nuclear transplantation experiments. Exp. Cell Res. 80, 69-78 (1973)

Kloppstech, K., Schweiger, H.G.: The site of synthesis of chloroplast ribosomal proteins. Plant Sci. Lett. 2, 101-105 (1974)

Krauspe, R., Parthier, B.: Chloroplast- and cytoplasmic specific aminoacyl transfer ribonucleic acid synthetases of *Euglena gracilis*: Separation, characterization and site of synthesis. Biochem. Soc. Symp. 38, 111-135 (1973)

Krauspe, R., Parthier, B.: Chloroplast and cytoplasmic aminoacyl-tRNA synthetases of *Euglena gracilis*. Biochem. Physiol. Pflanz. 165, 18-36 (1974)

Kung, S.: Expression of chloroplast genomes in higher plants. Annu. Rev. Plant Physiol. 28, 401-437 (1977)

Leis, J.P., Keller, E.B.: N-formylmethionyl-tRNA$_F$ of wheat chloroplasts. Its synthesis by a wheat transformylase. Biochemistry 10, 889-894 (1971)

Lesiewicz, J.L., Herson, D.S.: A reinvestigation of the sites of transcription and translation of *Euglena* chloroplastic phenylalanyl-tRNA synthetase. Arch. Mikrobiol. 105, 117-121 (1975)

Locy, R.D., Cherry, J.H.: Purification and characterization of two tyrosyl-tRNA synthetase activities from soybenan cotyledons. Phytochemistry 17, 19-27 (1978)

Lyttleton, J.W.: Isolation of ribosomes from spinach chloroplasts. Exp. Cell Res. 26, 312-317 (1962)

Mache, R., Jallifier-Verne, M., Rozier, C., Loiseaux, S.: Molecular weight determination of precursor, mature and postmature plastid ribosomal RNA from spinach using fully denaturing conditions. Biochim. Biophys. Acta 517, 390-399 (1978)

Margulies, M.M., Michaels, A.: Ribosomes bound to chloroplast membranes in *Chlamydomonas reinhardii*. J. Cell Biol. 60, 65-77 (1974)

Margulies, M.M., Michaels, A.: Free and membrane-bound chloroplast polyribosomes in *Chlamydomonas reinhardii*. Biochim. Biophys. Acta 402, 297-308 (1975)

Margulies, M.M., Weistrop, J.: A chloroplast membrane fraction enriched in chloroplast ribosomes, pp. 657-660. In: see ref. Bücher et al. (1976)

Matsuda, Y.: Studies on chloroplast development in *Chlamydomonas reinhardii*. III. Three phases in the greening process of y-1 cells and the stages of chloroplast ribosomal RNA synthesis. Biochim. Biophys. Acta 366, 45-55 (1974)

McCrea, J.M., Hershberger, C.L.: Chloroplast DNA codes for transfer RNA. Nucleic Acid Res. 3, 2005-2018 (1976)

McCrea, J.M., Hershberger, C.L.: Chloroplast codes for tRNA from cytoplasmic polysomes. Nature (London) 274, 717-719 (1978)

Mendiola-Morgenthaler, L.R., Morgenthaler, J.J., Price, C.A.: Synthesis of coupling factor CF protein by isolated spinach chloroplasts. FEBS Lett. 62, 96-100 (1976)

Merrick, W.C., Dure III, L.S.: Specific transformylation of one methionyl-tRNA from cotton seedling chloroplasts by endogenous and *Escherichia coli* transformylases. Proc. Natl. Acad. Sci. USA 68, 641-644 (1971)

Merrick, W.C, Dure, L.S.: The developmental biochemistry of cotton seed embryogenesis and germination. IV. Levels of cytoplasmic and chloroplastic transfer ribonucleic acid species. J. Biol. Chem. 247, 7988-7999 (1972)

Michaels, A., Margulies, M.M.: Amino acid incorporation into protein by ribosomes bound to chloroplast thylakoid membranes. Formation of discrete products. Biochim. Biophys. Acta 390, 352-362 (1975)

Mikulovich, T.P.: Synthesis of plastid and cytoplasmic ribosomal RNAs in isolated pumpkin cotyledons. I. Effect of detachment and light. Biochem. Physiol. Pflanz. 172, 93-100 (1978)

Mikulovich, T.P., Wollgiehn, R., Khokhlova, W.A., Neumann, D., Kulaeva, O.N.: Synthesis of plastid and cytoplasmic ribsomal RNAs in isolated pumpkin cotyledons. II. Effect of cytokinin and light. Biochem. Physiol. Pflanz. 172, 101-110 (1978)

Miller, M.J., McMahon, D.: Synthesis and maturation of chloroplast and cytoplasmic ribosomal RNA in *Chlamydomonas reinhardii*. Biochim. Biophys. Acta 366, 35-44 (1974)

Mohr, H.: Phytochrome and chloroplast development. Endeavour N.S. 1, 107-114 (1977)

Morgenthaler, J.J., Mendiola-Morgenthaler, L.: Synthesis of soluble, thylakoid, and envelope membrane proteins by spinach chloroplasts purified from gradients. Arch. Biochem. Biophys. 172, 51-58 (1976)

Munns, R., Scott, N.S., Smillie, R.M.: RNA synthesis during chloroplast development in *Euglena gracilis*. Phytochemistry 11, 45-52 (1972)

Munsche, D., Wollgiehn, R.: Die Synthese von ribosomaler RNA in Chloroplasten von *Nicotiana rustica*. Biochim. Biophys. Acta 249, 106-117 (1973)

Munsche, D., Wollgiehn, R.: Altersabhängige Labilität der ribosomalen RNA aus Chloroplasten von *Nicotiana rustica*. Biochim. Biophys. Acta 340, 437-445 (1974)

Nigon, V., Verdier, G., Salvador, G., Heizmann, P., Ravel-Chapuis, P., Freyssinet, G.: Biochemical sequences during the greening of dark-grown *Euglena gracilis*, pp. 629-640. In: see ref. Akoyunoglou and Argyroudi-Akoyunoglou (1978)

Nikolaev, N., Schlessinger, D., Wellauer, P.K.: 30S pre-ribosomal RNA of *E.coli* and products of cleavage by ribonuclease III: Length and molecular weight. J. Mol. Biol. 86, 741-747 (1974)

Nover, L.: Density labeling of chloroplast-specific leucyl-tRNA synthetase in greening cells of *Euglena gracilis*. Plant Sci. Lett. 7, 403-407 (1976)

Ohad, I.: Biogenesis of chloroplast membranes. In: Membrane Biogenesis, pp. 279-350; Tzagoloff, A. (ed.). New York-London: Plenum Press 1975

Ohad, I., Siekevitz, P., Palade, G.E.: Biogenesis of chloroplast membranes. I. Plastid differentiation in a dark-grown algal mutant *(Chlamydomonas reinhardii)*. J. Cell Biol. 35, 521-552 (1967a)

Ohad, I., Siekevitz, P., Palade, G.E.: Biogenesis of chloroplast membranes. II. Plastid differentiation during greening of a dark-grown algal mutant *(Chlamydomonas reinhardii)*. J. Cell Biol. 35, 553-584 (1967b)

Oparin, A.D., Odintsova, M.S., Yurina, N.P.: Chloroplast ribosomes as ribosomes of the prokaryotic type. Biochem. Physiol. Pflanz. 168, 175-183 (1975)

Parthier, B.: Existenz und Realisierung extrachromosomaler genetischer Information in Plastiden und Mitochondrien. Biol. Rundsch. 8, 289-306 (1970)

Parthier, B.: Cytoplasmic site of synthesis of chloroplast aminoacyl-tRNA synthetases in *Euglena gracilis*. FEBS Lett. 38, 70-74 (1973)

Parthier, B.: Lichtabhängige Transformation von Proplastiden zu Chloroplasten. Acta Histochem. Suppl. 17, 77-93 (1976)

Parthier, B., Krauspe, R.: Assignement to chloroplast and cytoplasm of three *Euglena gracilis* aminoacyl-tRNA synthetases with ambiguous specificity for transfer RNA. Plant Sci. Lett. 1, 221-227 (1973)

Parthier, B., Krauspe, R.: Chloroplast and cytoplasmic transfer RNA of *Euglena gracilis*. Biochem. Physiol. Pflanz. 165, 1-17 (1974)

Parthier, B., Neumann, D.: Structural and functional analysis of some plastid mutants of *Euglena gracilis*. Biochem. Physiol. Pflanz. 171, 547-562 (1977)

Parthier, B., Wollgiehn, R.: Nukleinsäuren und Proteinsynthese in Plastiden. In: Probleme der biologischen Reduplikation. pp. 244-272; Sitte, P. (ed.). Berlin-Heidelberg-New York: Springer 1966

Parthier, B., Krauspe, R., Samtleben, S.: Light-stimulated synthesis of aminoacyl-tRNA synthetases in greening *Euglena gracilis*. Biochim. Biophys. Acta 277, 335-341 (1972)

Parthier, B., Mueller-Uri, F., Krauspe, R.: The aminoacyl-tRNA synthetases in *Euglena* chloroplasts, pp. 687-693. In: see ref. Akoyunoglou and Argyroudi-Akoyunoglou (1978)

Parthier, B., Krauspe, R., Munsche, D., Wollgiehn, R.: The biogenesis of chloroplasts. In: The Chemistry and Biochemistry of Plant Proteins, pp. 167-210; Harborne, J.B., van Sumere, C.F. (eds.). London-New York: Academic Press 1975

Paterson, B.D., Smillie, R.M.: Developmental changes in ribosomal ribonucleic acid and fraction I protein in wheat leaves. Plant Physiol. 47, 196-198 (1971)

Philippovich, I.I., Bezsmertnaya, I.N., Oparin, A.I.: On the localization of polyribosomes in the system of chloroplast lamellae. Exp. Cell Res. 79, 159-168 (1973)

Pine, K., Klein, A.O.: Regulation of polysome formation in etiolated bean leaves by light. Dev. Biol. 28, 280-289 (1972)

Polya, G.M., Jagendorf, A.T.: Wheat leaf RNA polymerase. I, II. Arch. Biochem. Biophys. 146, 635-657 (1971)

Posner, H.B., Rosner, A.: Effect of chloramphenicol on RNA synthesis in *Spirodela*. Plant Cell Physiol. 16, 361-365 (1975)

Poulson, R., Beevers, L.: Nucleic acid metabolism during greening and unrolling of barley leaf segments. Plant Physiol. 46, 315-319 (1970)

Ramiasa, J., Guillemaut, P., Weil, J.H.: Codon recognition pattern of *Phaseolus vulgaris* cytoplasmic and chloroplastidic tRNAs. FEBS Lett. 75, 128-132 (1977)

Rawson, J.R.Y.: A measurement of the fraction of chloroplast DNA transcribed in *Euglena*. Biochem. Biophys. Res. Commun. 62, 539-545 (1975)

Rawson, J.R.Y., Boerma, C.L.: A measurement of the fraction of chloroplast DNA transcribed during chloroplast development in *Euglena gracilis*. Biochemistry 15, 588-592 (1976)

Rawson, J.R.Y., Boerma, C.L., Curtis, S.: Transcription of EcoRI-chloroplast DNA fragments in *Euglena gracilis* during chloroplast development. J. Cell Biol. 75, XXXa (1977)

Reger, B.J., Fairfield, S.A., Epler, J.L., Barnett, W.E.: Identification and origin of some chloroplast aminoacyl-tRNA synthetases and tRNA. Proc. Natl. Acad. Sci. USA 67, 1207-1213 (1970)

Reger, B.J., Smillie, R.M., Fuller, R.C.: Light-stimulated production of a chloroplast-localized system for protein synthesis in *Euglena gracilis*. Plant Physiol. 50, 24-27 (1972)

Reisfeld, A., Jacob, K.M., Edelman, M.: Molecular weight of chloroplast messenger RNA translating full-size RuDP carboxylase large subunit, pp. 669-674. In: see ref. Akoyunoglou and Argyroudi-Akoyunoglou (1978)

Richter, G., Dirks, W.: Blue-light induced development of chloroplasts in isolated seedling roots. Preferential synthesis of chloroplast ribosomal RNA species. Photochem. Photobiol. 27, 155-160 (1978)

Rosner, A., Porath, D., Gressel, J.: The distribution of plastid ribosomes and the integrity of plastid ribosomal RNA during the greening and maturation of *Spirodela* fronds. Plant Cell Physiol. 15, 891-902 (1974)

Rosner, A., Jacob, K.M., Gressel, J., Sagher, D.: The early synthesis and possible function of a 0.5×10^6 M_r RNA after transfer of dark-grown Spirodela plants to light. Biochem. Biophys. Res. Commun. 67, 383-391 (1975)

Rosner, A., Reisfeld, A., Jacob, K.M., Gressel, J., Edelmann, M.: Shifts in the RNA and protein metabolism of *Spirodela* (duckweed). In: Acides Nucléiques et Synthése des Protéines chez les Végétaux, pp. 561-568; Bogorad, L., Weil, J.H. (eds.). Paris: Ed. CNRS (1977)

Roussaux, J., Hoffelt, M., Farineau, N.: Evolution des RNA ribosomaux an cours des verdissement de cotylédons de concombre en presénce de 6-benzylaminopurine. Can. J. Bot. 54, 2328-2336 (1976)

Sager, R., Palade, G.E.: Chloroplast structure in green and yellow strains of *Chlamydomonas*. Exp. Cell Res. 7, 584-588 (1954)

Sagher, D., Crosfeld, H., Edelman, M.: Large subunit ribulose bisphosphate carboxylase messenger RNA from *Euglena* chloroplasts. Proc. Natl. Acad. Sci. 73, 722-728 (1976)

Sarantoglou, V., Imbault, P., Weil, J.H.: Partial purification and properties of *Euglena* cytoplasmic and chloroplastic valyl-tRNA synthetases, pp. 695-700. In: see ref. Akoyunoglou and Argyroudi-Akoyunoglou (1978)

Schiemann, J., Wollgiehn, R., Parthier, B.: Isolation of a transcription-active RNA polymerase-DNA complex from *Euglena* chloroplasts. Biochem. Physiol. Pflanz. 171, 474-478 (1977)

Schiemann, J., Wollgiehn, R., Parthier, B.: DNA-dependent RNA polymerase in *Euglena gracilis* chloroplasts. Biochem. Physiol. Pflanz. 172, 507-519 (1978)

Schiff, J.A.: The control of chloroplast differentiation in *Euglena*. In: Proc. 3rd Int. Congr. Photosynth. Rehovot, pp. 1691-1717; Avron, M. (ed.). Amsterdam: Elsevier 1974

Schwartzbach, S.D., Hecker, L.J., Barnett, W.E.: Transcriptional origin of *Euglena* chloroplast tRNAs. Proc. Natl. Acad. Sci. USA 73, 1984-1988 (1976)

Schweiger, H.G.: Synthesis of RNA in Acetabularia. Symp. Soc. Exp. Biol. 24, 327-344 (1970)

Scott, N.S.: Precursors of chloroplast ribosomal RNA in *Euglena gracilis*. Phytochemistry 15, 1207-1213 (1976)

Scott, N.S., Munns, R., Graham, D., Smillie, R.M.: Origin and synthesis of chloroplast ribosomal RNA and photoregulation during chloroplast biogenesis. In: Autonomy and Biogenesis of Mitochondiria and Chloroplasts, pp. 383-392; Boardman, N.K., Linnane, A.W., Smillie, R.M. (eds.). Amsterdam: North-Holland 1971a

Scott, N.S., Nair, H., Smillie, R.M.: The effect of red irradiation on plastid ribosomal RNA synthesis in dark-grown pea seedlings. Plant Physiol. 47, 385-388 (1971b)

Siddell, S.G., Ellis, R.J.: Protein synthesis in chloroplasts. Characteristics and products of protein synthesis in vitro in etioplasts and developing chloroplasts from pea leaves. Biochem. J. 146, 675-685 (1975)

Siersma, P.W., Chiang, K.S.: Conservation and degradation of cytoplasmic and chloroplast ribosomes in *Chlamydomonas reinhardii*. J. Mol. Biol. 58, 167-185 (1971)

Sirevag, R., Levine, R.P.: Fatty acid synthetase from *Chlamydomonas reinhardii*. Sites of transcription and translation. J. Biol. Chem. 247, 2586-2591 (1972)

Smith, H.: Changes in plastid ribosomal RNA and enzymes during the growth of barley leaves in darkness. Phytochemistry 9, 965-975 (1970)

Smith, H.: Phytochrome-mediated assembly of polyribosomes in etiolated bean leaves. Eur. J. Biochem. 65, 161-170 (1976)

Smith, H.J., Bogorad, L.: The polypeptide subunit structure of the DNA-dependent RNA polymerase of Zea mays chloroplasts. Proc. Natl. Acad. Sci. USA 71, 4839-4842 (1974)

Smith, H., Steward, G.R., Berry, D.R.: The effect of light on plastid ribosome RNA and enzymes at different stages of barley etioplast development. Phytochemistry 9, 977-983 (1970)

Spiess, H.: Analysis of the chloroplast ribosomal proteins from Chlamydomonas reinhardii streptomycin-resistant and dependent mutants by two-dimensional gel electrophoresis. Plant Sci. Lett. 10, 103-113 (1977)

Sprey, B.: Ribosomale RNA and Thylakoidmembranen in Plastiden von Chlorophylldefekt-mutanten der Gerste. Z. Pflanzenphysiol. 67, 223-243 (1972)

Steup, M.: Die Wirkung von blauem und rotem Licht auf die Synthese ribosomaler RNA bei Chlorella. Arch. Mikrobiol. 105, 143-151 (1975)

Surzycki, S.J., Shellenbarger, D.L.: Purification and characterization of a putative sigma factor from Chlamydomonas reinhardii. Proc. Natl. Acad. Sci. USA 73, 3961-3965 (1976)

Surzycki, S.J., Goodenough, U.W., Levine, R.P., Armstrong, J.J.: Nuclear and chloroplast control of chloroplast structure and function in Chlamydomonas reinhardii. Symp. Soc. Exp. Biol. 24, 13-24 (1970)

Suyama, Y., Hamada, J.: Imported tRNA: Its synthetase as a probable transport protein, pp. 763-770. In: see ref. Bücher et al. (1976)

Tewari, K.K., Wildman, S.G.: Function of Chloroplast DNA. II. Studies in DNA-dependent RNA polymerase activity of tobacco chloroplasts. Biochem. Biophys. Acta 186, 358-372 (1969)

Tewari, K.K., Wildman, S.G.: Information content in the chloroplast DNA. Symp. Soc. Exp. Biol. 24, 147-179 (1970)

Tewari, K.K., Kolodner, R., Chu, N.M., Mecker, R.: Structure of chloroplast DNA. In: Nucleic Acids and Protein Synthesis in Plants, pp. 15-36; Bogorad, L., Weil, J.H. (eds.). New York-London: Plenum Press 1977

Thien, W., Schopfer, P.: Control by phytochrome of cytoplasmic and plastid rRNA accumulation in cotyledons of mustard seedlings in the absence of photosynthesis. Plant Physiol. 56, 660-664 (1975)

Tiboni, O., Di Pasquale, G., Ciferri, O.: Ribosomes and translation factors from isolated spinach chloroplasts. Plant Sci. Lett. 6, 419-429 (1976)

Travis, R.L., Lin, C.Y., Key, J.L.: Enhancement by light of the in vitro protein synthetic activity of cytoplasmic ribosomes isolated from dark-grown maize seedlings. Biochim. Biophys. Acta 277, 606-614 (1972)

Trewavas, A.: The turnover of nucleic acids in Lemna minor. Plant Physiol. 45, 742-751 (1970)

Vasconcelos, A.C.: Synthesis of proteins by isolated Euglena gracilis chloroplasts. Plant Physiol. 58, 719-721 (1976)

Vasconcelos, A.C., Bogorad, L.: Proteins of cytoplasmic, chloroplast and mitochondrial ribosomes of some plants. Biochim. Biophys. Acta 228, 492-502 (1971)

Vasconcelos, A.C., Mendiola-Morgenthaler, L.R., Floyd, G.L., Salisbury, J.L.: Fractionation and analysis of polypeptides of Euglena gracilis chloroplasts. Plant Physiol. 58, 87-90 (1976)

Vedel, F., D'Aoust, M.J.: Polyacrylamid gel analysis of high molecular weight ribonucleic acid from etiolated and green cucumber cotyledons. Plant Physiol. 46, 81-85 (1970)

Verdier, G.: Synthesis and translation site of light-induced mRNAs in etiolated Euglena gracilis. Biochim. Biophys. Acta 407, 91-98 (1975)

Verdier, G., Trabuchet, G., Heizmann, P., Nigon, V.: Effet de l'eclairement sur les synthéses de DNA et de sequences polyadenyliques dans des cultures d'Euglena gracilis etiolées. Biochim. Biophys. Acta 312, 528-539 (1973)

Vreman, H., Thomas, R., Corse, J., Swaminathan, S., Murai, N.: Cytokinins in tRNA obtained from Spinacia oleracea leaves and isolated chloroplasts. Plant Physiol. 61, 296-306 (1978)

Walden, R., Leaver, C.J.: Regulation of chloroplast protein synthesis during germination and early development of cucumber (*Cucumis sativus*), pp. 251-256. In: see ref. Akoyunoglou and Argyroudi-Akoyunoglou (1978)

Weil, J.H.: Cytoplasmic and organellar tRNAs in plants. In: Nucleic Acids in Plants. West Palm Beach: CRC Press 1978

Weil, J.H., Burkard, G., Guillemaut, P., Jeannin, G., Martin, R., Steinmetz, A.: tRNAs and aminoacyl-tRNA synthetases in plant cytoplasm, chloroplasts and mitochondria. In: Nucleic Acids and Protein Synthesis in Plants, pp. 97-120; Bogorad, L., Weil, J.H. (eds). New York-London: Plenum Press 1977

Wheeler, A.M., Hartley, M.R.: Major mRNA species from spinach chloroplasts do not contain poly(A). Nature (London) 257, 66-67 (1975)

Whitfeld, P.R.: Chloroplast RNA. In: The Ribonucleic Acids, 2nd ed., pp. 297-332; Stewart, P.R., Letham, D.S. (eds.). Berlin-Heidelberg-New York: Springer 1977

Whitfeld, P.R., Spencer, D., Bottomley, W.: Products of chloroplast DNA-directed transcription and translation. In: The Biochemistry of Gene Expression in Higher Organisms, pp. 504-522; Pollak, J.K., Lee, W.J. (eds.). Sidney: Australian and New Zealand Book Corp. 1973

Whitfeld, P.R., Leaver, C.J., Bottomley, W., Atchison, B.A.: A new species of chloroplast RNA. In: Acides Nucléiques et Synthése des Protéines chez les Végétaux, pp. 235-241; Bogorad, L., Weil, J.H. (eds.). Paris: Ed. CNRS 1977

Wildner, G.F.: The kinetics of appearance of chloroplast proteins and the effect of cycloheximide and chloramphenicol on their synthesis. (The greening process in *Euglena gracilis* I). Z. Naturforsch. 31c, 157-162 (1976).

Williams, G.R., Novelli, G.D.: Ribosome changes following illumination of dark-grown plants. Biochim. Biophys. Acta 155, 183-192 (1968)

Williams, G.R., Williams, A.S., George, S.A.: Hybridization of leucyl-transfer ribonucleic acid isoacceptor from green leaves with nuclear and chloroplast deoxyribonucleic acid. Proc. Natl. Acaud. Sci. USA 70, 3498-3501 (1973)

Wilson, R., Chiang, K.: Temporal programming of chloroplast and cytoplasmic rRNA transcription in the synchronous cell cycle of *Chlamydomonas reinhardii*. J. Cell Biol. 72, 470-481 (1977)

Wittmann, H.G.: Structure and function of 70S ribosomal proteins. Biochem. J. 129, 30-31 (1972)

Wollgiehn, R., Munsche, D.: RNS-Synthese in isolierten Chloroplasten von *Nicotiana rustica*. Biochem. Physiol. Pflanz. 163, 137-155 (197a)

Wollgiehn, R., Ruess, M., Munsche, D.: Ribonucleinsäuren in Chloroplasten. Flora A157, 92-108 (1966)

Wollgiehn, R., Lerbs, S., Munsche, D.: Synthesis of ribosomal RNA in chloroplasts from tobacco leaves of different age. Biochem. Physiol. Pflanz. 170, 381-387 (1976)

Wollgiehn, R., Lerbs, S., Munsche, D.: Eigenschaften einer Chloroplasten-RNA vom Molekulargewicht 0.5×10^6 aus *Nicotiana rustica*. Biochem. Physiol. Pflanz. 173, 60-69 (1978)

Woodcock, C.L.F., Bogorad, L.: Nucleic acids and information processing in chloroplasts. In: Structure and Function of Chloroplasts, pp. 89-129; Gibbs, M. (ed.). Berlin-Heidelberg-New York: Springer 1971

Zablen, L.B., Kissel, M.S., Woese, C.R., Buetow, D.E.: Phylogenetic origin of the chloroplast and prokaryotic nature of its ribosomal RNA. Proc. Natl. Acad. Sci. USA 72, 2418-2422 (1975)

Biosynthesis of Thylakoids and the Membrane-Bound Enzyme Systems of Photosynthesis

F.H. HERRMANN, TH. BÖRNER, and R. HAGEMANN

Wissenschaftsbereich Genetik, Sektion Biowissenschaften
Martin-Luther-Universität, Halle/Saale, GDR

I. Introduction

The chloroplasts are the sites of photosynthesis in eukaryotic plant cells. Within these organelles the processes of photosynthetic energy uptake, transfer, and transformation take place in the thylakoids. In this article we wish to describe characteristics of the thylakoids and of the membrane-bound enzyme systems, their biosynthesis, and especially the cooperation of nuclear DNA and plastid DNA and of the protein-synthesizing systems in the cytoplasm and in the plastid during the ontogenetic development of the chloroplast.

In green algae and higher plants the photosynthetic electron transport chain is associated with the thylakoid membranes and consists of two photosystems (= PS) connected by a series of electron carriers and membrane-bound enzymes (cf. Trebst 1974; Krogmann 1976). The thylakoid membranes can be fractionated by differential centrifugation in the presence of nonionic detergents into subchloroplast fragments which are different in photochemical activities and chemical composition (cf. Boardman 1970). The light fraction has a high ratio of chlorophyll (= chl) a to chlorophyll b and possesses mainly PS I properties, whereas the heavy fraction has a low chl a to b ratio, and is enriched in PS II activity. A more complete solubilization of thylakoid membranes is obtained with anionic detergents such as sodium dodecyl sulfate (SDS), lithium dodecyl sulfate (LDS) or sodium dodecyl benzene sulfonate (SDBS). The photochemical activities are inactivated by SDS, LDS or SDBS (Ogawa et al. 1966; Thornber et al. 1967a,b; Delepelaire and Chua 1979). The SDS-soluble chloroplast membranes are separable into chlorophyll-containing bands and numerous polypeptides by SDS-polyacrylamide gel electrophoresis (SDS-PAGE) (cf. Anderson 1975; Thornber 1975). With these methods the chloroplast membranes of higher plants have been separated into at least 15 to 26 polypeptides. Chua and Bennoun (1975) separated the thylakoid membranes of *Chlamydomonas reinhardii* by SDS-PAGE into at least 33 polypeptides ranging in molecular weight from 68,000 to less than 10,000. More than 40 bands could be resolved in the patterns of thylakoid membranes of *Hordeum* and *Spinacea* obtained from thylakoid preparations after preceeding lipid extraction by slab gel SDS-PAGE and discontinuous buffer systems (Henriques and Park 1975; Henriques et al. 1975).

Several approaches have been used for the identification of the chloroplast membrane polypeptides and membrane-bound enzyme systems:

a) Fractionation of membranes into small fragments enriched in either PS I or PS II activities and the analysis of the polypeptide composition of these subchloroplast fragments.

b) Characterization of membrane polypeptides of mutants, which are either pigment-deficient or have specific defects in their electron transport pathway.

c) Comparison of polypeptides of developing chloroplasts of algae and higher plants, especially during the process of greening.

d) Purification and functional characterization of membrane-bound enzymes.

e) Use of antibodies against particular thylakoid polypeptides in studying their function and localization.

When seeds of angiosperms are germinated in the dark, the seedlings contain proplastids (or leucoplasts) which possess no chlorophyll; the same is true for *Euglena* cells grown in the dark. After exposure of these seedlings or cells to light, the proplastids undergo intense internal biochemical and ultrastructural changes. The synthesis of chlorophyll begins, and the proplastids perform a metamorphosis into chloroplasts. This differentiation is connected with the biogenesis of the typical thylakoid membrane systems of the mature chloroplast. In the following we will describe the characteristics of the thylakoid and the membrane-bound enzymes and afterwards characterize the processes of their biogenesis.

II. The Chlorophyll-Protein Complexes of Chloroplasts

When chloroplast membranes are solubilized with SDS or SDBS and separated with SDS-PAGE, several chlorophyll-containing bands are visible: Two major chlorophyll-protein complexes, termed CP I (= P-700 chlorophyll a protein, Dietrich and Thronber 1971; = CP I chlorophyllin, Thornber et al. 1979) and CP II or LHCP (= light-harvesting chlorophyll a/b protein, Thornber and Highkin 1974; = CP II chlorophyllin, Thornber et al. 1979) and several minor chlorophyll-containing bands.

Thornber et al. (1967a) observed sporadically the appearance of a minor CP between CP I and CP II. By variation of the ratio of SDS/chl for solubilization of chloroplast lamellae Herrmann and Meister (1972) found as a reproducible phenomenon seven pigment protein complexes: In addition to the major components CP I and CP II a bluish-green minor pigment complex migrating more slowly than CP I (designated CP Ia) and four yellow-green minor components (II a-d) between CP I and CP II. These minor components have been characterized spectroscopically (visible spectra and their second derivative; chl a/b ratio). Hiller et al. (1974) found a 69 kD CP migrating between CP I and CP II.

Since 1977 several research groups confirmed the appearence of additional CP's. Anderson et al. (1978) and Boardman et al. (1978) reported the separation and characterization of six complexes: CP 1a (an oligomer of CP 1); CP 1; LHCP[1] and LHCP[2] (oligomers of the LHCP); CP a (a chlorophyll a protein complex which may represent the PS II reaction centre complex) and LHCP[3] (the monomer of LHCP). Henriques and Park (1978a,b) described five chlorophyll-protein complexes: Two chl a and three chl a/b protein bands in the order CP I (= chl a),

IIb, IIa, A (= chl a), LHCP. Remy et al. (1977, 1978) reported the isolation of some minor bands in addition to CP I and CP II: CP I, IIa (= LHCP dimer), IIb (containing only chl a and no chl b), and IIc (= CP II, LHCP). Markwell et al. (1978) and Miles (1979) resolved and characterized five chlorophyll-protein complexes: A-1 (= CP I); AB-1 (LHCP a/b 1); AB-2; A-2 (= CP with only chl a, reaction center of PS 2) and AB-3. The absence of all LHCP a/b complexes (AB-1, AB-2, AB-3) in chl-b-lacking mutants of barley and corn indicates that these components are monomers (AB-3) and oligomers (AB-1, AB-2) of the LHCP.

One of these minor CP's between CP I and CP II contains primarily only chl a (and no chl b). This component has been characterized by Hayden and Hopkins (1977), Anderson et al. (1978), Boardman et al. (1978), Henriques and Park (1978a,b), Remy et al. (1978), Wessels and Borchert (1978) and Miles et al. (1979). The apoprotein of this complex has a molecular weight of 47 kD (Wessels and Borchert 1978). This complex has been suggested to be a component of the PS II reaction center. Delepelaire and Chua (1979), using lithium dodecyl sulfate polyacrylamide gel electrophoresis of thylakoid membranes of *Chlamydomonas* at 4° C, isolated between CP I and CP II two minor complexes (CP III and IV), which only contain chl a and β carotene; the apoprotein of CP III is polypeptide 5 (M_r 50,000) and that of CP IV is polypeptide 6 (M_r 47,000). This observation together with results on photosystem II mutants provides indirect evidence that CP III and CP IV may be involved in the primary photochemistry of PS II.

The identification of the reaction center of PS II seems possible using different mutants. There are two mutants, one in barley (viridis c[12]) and one in *Chlamydomonas reinhardii* which contain the polypeptides of CP II (LHCP), but lack significant PS II activity (Machold and Hoyer-Hansen 1976; Chua and Bennoun 1975). These mutants are deficient in a band with an apparent molecular weight of 45,000 or 47,000, resp. Herrmann (1972) analyzed the plastome mutant en:viridis-1 of *Antirrhinum majus* which is deficient in photosystem II and found that this mutant still contains CP II, but lacks a band with an apparent molecular weigth in the range between 40,000 and 50,000. These results implicate that (a) membrane polypeptide(s) in the range between 45,000 and 50,000 are components of the reaction center of photosystem II (PS II). Vernon et al. (1971) and Vernon and Klein (1975) have isolated particles (TSF 2a) containing the reaction center of PS II and enriched in a polypeptide with an apparent molecular weight of 44,000 daltons.

A chlorophyll a-containing complex that migrates more slowly than CP I in gels was at first described in extracts from higher plants by Herrmann and Meister (1972). Its occurrence was confirmed recently by other laboratories (Hoarau et al. 1977; Anderson et al. 1978; Boardman et al. 1978; Reinman and Thornber 1979). The spectrum of this complex is similar to that of CP I. This additional complex may represent a supramolecular complex of CP I.

The progress in separation and analysis of thylakoid complexes is to a very great extent dependent upon the particular techniques used. The progress outlined in the preceeding paragraphs shows how much methodic improvements led to further new insights. This progress is obviously continuous. Markwell et al. (1979a,b) have described a new technique which may initiate new insights. The major difference of this new technique from previous gel electrophoresis systems is the replacement of

SDS in the electrophoresis buffer by a zwitterionic detergent: Deriphat 160 (disodium N-lauryl-β-iminopropionate) or Miranol S2M-SF (Markwell et al. 1979a,b). The use of Deriphat 160 has led so far to the following conclusions: No significant amount of free chlorophyll exists in the chloroplast thylakoid membranes in vivo. Most of the free pigment seen previously in SDS gels was generated during electrophoresis and was not the result of the solubilization technique (with SDS). The chlorophyll-protein complexes resolved with this method have very different characteristics from those observed in former systems. The complexes isolated have been designated: N (MW: 250,000), O (150,000), P (between 60,000 and 100,000) and Q (50,000). Deriphat 160 and Miranol S2M-SF thylakoid extracts retained the photochemical activity of photosystem II.

III. The Major Chlorophyll-Protein Complexes

During the last 10 years most studies concentrated on the major CP I and CP II (isolated after SDS, or LDS electrophoresis). These CP's contain the main thylakoid intrinsic proteins. Thornber et al. (1967a) calculated that 28% and 49% of the total protein in the detergent solubilized extract is contained in CP I and CP II respectively. Detailed information of the characteristics and functions of the CP's I and II is available from the excellent reviews of Anderson (1975), Thornber (1975), Thornber and Alberte (1976), Boardman et al. (1978), Gillham et al. (1978), and Thornber et al. (1979).

A. The Chlorophyll-Protein Complex I (CP I)

CP I accounts for 25–30% of the total chlorophyll of the chloroplast (Remy et al. 1977; Anderson et al. 1978; Henriques and Park 1978a,b; Markwell et al. 1978). Using the older techniques for isolation 18% had been estimated (Ogawa et al. 1966; Thornber et al. 1967a; Thornber 1969; Genge et al. 1974; Shiozawa et al. 1974; Brown et al. 1975). The chl a/b ratio was determined to range from 5 to 12 (Ogawa et al. 1966; Thornber et al. 1967a; Machold et al. 1971; Herrmann and Meister 1972; Chua et al. 1975; Nakamura et al. 1976). In some preparations chl b was found to be absent (Genge et al. 1974; Brown et al. 1975; Thornber 1975).

CP I contains probably only chl a and β carotene in a molar ratio of 20–30 chl a/1 β carotene (Ogawa et al. 1966; Thornber 1969; Shiozawa et al. 1974; Thornber 1975). The ubiquitous distribution of CP I in many bacteria and all eukaryotic plants (Brown et al. 1975) makes it — on a theoretical basis — rather improbable, that chl b, which is not present throughout the plant kingdom, is an essential pigment of CP I.

The CP I contains P-700 (Largett-Barley and Kreutz 1969); therefore Dietrich and Thornber (1971) introduced the name P-700 chlorophyll a-protein for the previously described CP I. The chl a/P-700 ratio of the isolated complex I is 40 (Shiozawa et al. 1974; Ke et al. 1975; Malkin 1975). Recently Bengis and Nelson (1977) prepared a P-700 chlorophyll a-protein complex, which was active in the light-induced reversible bleaching of P-700. P-700 is associated with a 70,000 dalton polypeptide. Two copies of this polypeptide are probably required to form a P-700 pigment and each one of them contains about 20 chl a molecules as an

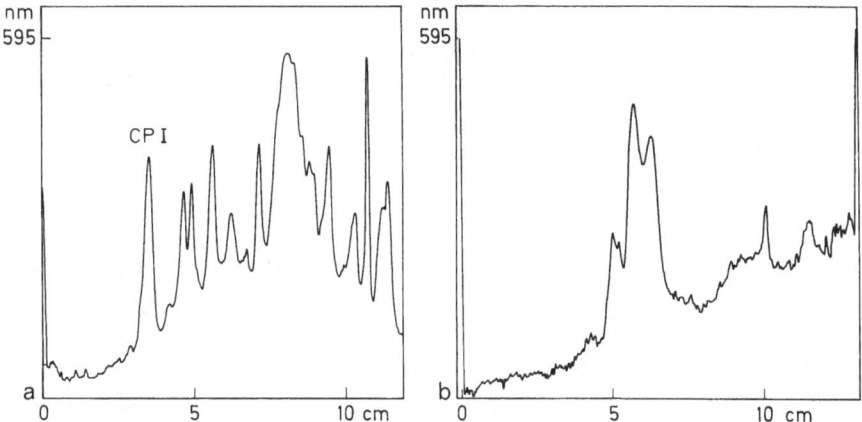

Fig. 1a,b. Electrophoretic pattern of chloroplast lamellar proteins of *Antirrhinum majus* (**a**) and of the isolated CP I (**b**) in 11% polyacrylamide gels containing 5 M urea

intrinsic part of the light-harvesting antenna of PS I. Galacto- and phospholipids occur in trace amounts (Thornber et al. 1967b; Dietrich and Thornber 1971). The amino acid composition of the two complexes is different (Thornber et al. 1967a,b; Thornber and Alberte 1976). Thornber et al. (1967b) determined a sedimentation coefficient of 9S for CPI.

The reports on molecular weight and polypeptide composition of CP I are still controversial. An apparent molecular weight of about 110,000 daltons was determined in several higher plants for CP I as a whole by calibrated SDS-PAGE. In some experiments an apparent molecular weight of 70,000 (Guigneri et al. 1974; Hayden and Hopkins 1976 for *Zea mays*; Picaud and Acker 1976 for *Chlamydomonas*) and 60,000 (Lürtz 1975 for *Avena*) was determined. Guigneri et al. (1974), Lürtz (1975), and Picaud and Acker (1975) used a buffer system with dithiotheitol.

The difficulties for the exact determination of the molecular weight of CPs in SDS-systems have been discussed intensively (Felgenhauer 1974; Machold 1974; Takagi et al. 1975). Chua et al. (1975) showed that the CPI has an abnormal migration rate; accordingly, its molecular weight cannot be measured accurately by calibrated SDS gels. Removal of the lipids gives a single polypeptide with an apparent molecular weight of 66,000. Lipids play an important role in the stabilization of the complexes in the membrane. The lipid extraction of membranes preceeding SDS-electrophoresis may remove the whole complexes or parts of them (cf. Boardman et al. 1977). In order to determine the polypeptide components of the complexes it is preferable to use SDS-PAGE with thylakoids which have not been lipid-extracted, and then to perform re-electrophoresis of an isolated complex after removal of pigments by heat denaturation or lipid extraction or by adding SH-reagents (such as mercaptoethanol) or reducing agents (such as $NABH_4$). Removal of lipids from CPI usually increases the electrophoretic mobility of the polypeptide(s). Machold (1975) and Bengis and Nelson (1975, 1977) expressed the view that the CPI or P-700 reaction center consists of one 70,000-dalton polypeptide. Chua et al. (1975) determined also only one component of the CPI from *Chlamydomonas* (M.W. 66,000), spinach and chinese cabbage (M.W. 64,000).

On the other hand Anderson and Levine (1974) reported that the complex I in *C. reinhardii* contains two polypeptides in the molecular weight range of 60,000. According to Picaud and Acker (1975) the complex consists of 56,000- and 44,000-dalton polypeptides. CP I of *Lactuca* is reported to consist of several subunits with a major one of 67,000 daltons (Henriques and Park 1977). Schumann et al. (1975) and Herrmann (1978) separated the CP I of several higher plants (*Antirrhinum, Pelargonium, Lycopersicon, Hordeum*, and seven species of *Nicotiana*) into several different components (Fig. 1).

Recently Bar-Nun et al. (1977) demonstrated that the CP I in *C. reinhardii* consists of a major polypeptide(s) of molecular weight of 64,000 which is synthesized within the chloroplasts; in addition to the 64,000-dalton major polypeptide, several minor bands are released from the isolated complex I following complete denaturation. These polypeptides have a molecular weight range of 46,000 to 50,000. The 46,000 molecular weight region can be further resolved into several polypeptides.

The chlorophyll/protein ratio of the CP I in higher plants and in blue-green algae is 14 mol chl/110,000 g complex (Thornber et al. 1967b; Thornber 1969). In *C. reinhardii* Bar-Nun et al. (1977) calculated a molar ratio of about 29 chlorophyll molecules per one polypeptide of 64,000 daltons, or, if the complex contains two polypeptides (63,000 and 65,000), the molar ratio could be about 15 chl/polypeptide, Chua et al. (1975) calculated that each constituent polypeptide chain (64,000 daltons) is associated with 8 or 9 chlorophyll molecules.

The spectroscopical similarities of the separated major pigment protein complexes with those of PS I- and PS II-enriched subchloroplast fractions (Ogawa et al. 1975) and the demonstration that SDS-treated PS I and PS II particles were enriched in CP I and II, respectively (Sironval et al. 1967; Thornber et al. 1967a) led initially to the suggestion that CP I and CP II might be associated with PS I and PS II, respectively.

This suggestion has been confirmed with regard to CP I and PS I. There are reports showing that mutants of *C. reinhardii* (Chua et al. 1975; Picaud and Acker 1975; Bennoun et al. 1977), of *Scenedesmus* (Gregory et al. 1971), and of *Antirrhinum* and *Pelargonium* (Herrmann 1971; Herrmann et al. 1976) which are impaired in PS I reaction do not contain CP I. However, Bar-Nun et al. (1977) suggested that one should not simply identify CP I with PS I, because they could not detect CP I in the temperature-sensitive mutant T 4 of *C. reinhardii* (grown at 37° C), which has a normal PS I activity (Kretzer et al. 1976).

B. The Chlorophyll-Protein Complex II (CP II)

In most chloroplasts of higher plants the chlorophyll-protein complex II contains 40–60% of the total chlorophyll. This complex has equimolar quantities of chl a and chl b (Kung and Thornber 1971; Thornber and Olson 1971; Argyroudi-Akoyunoglou and Akoyunoglou 1973; Thornber and Highkin 1974; Kan and Thornber 1976) or a chl a/chl b ratio greater than 1 (cf. Boardman et al. 1977). Every carotenoid (besides β-carotene) present in whole chloroplasts is found in the CP II (Ogawa et al. 1966; Thornber et al. 1967; Thornber and Highkin 1974). The galacto- and phospholipid content of the CP II is so small and variable that it is not known whether both components are integral parts of the complex (Ogawa et al. 1966; Thornber et al. 1967b; Thornber 1975; Thornber and Alberte 1976). The sediment coefficient of CP II was determined to be 2.3–3.1 S (Thornber et al. 1967b; Kung and Thornber 1971; Kan and Thornber 1976).

The CP II represents about 50% of the total protein in the membrane detergent extract. The molecular weight has been reported to be between 23,000 and 25,000 daltons (Kung and Thornber 1971; Eaglesham and Ellis 1974; Hiller et al. 1974; Machold 1974; Herrmann et al. 1976; Kan and Thornber 1976; Vasconcelos et al. 1976). The polypeptide composition of CP II is not yet clear: some authors observed only one band, others two bands; but it appears that higher plant CP II contains two polypeptides. Bar-Nun et al. (1977) showed that CP II of *Chlamydomonas* after denaturation by heat or after the action of S-S reducing agents is composed of two polypeptides with a molecular weight of 24,000 and 22,000 daltons, which bind 12 chlorophylls (a and b).

The CP II has been found to contain two polypeptides in *Acetabularia* (21,500 and 23,000 daltons; Apel et al. 1975) and also in several higher plants (26,000 and 24,000; Anderson and Levine 1974; Herrmann et al. 1976). The major polypeptide of about 25,000 daltons, which is observed by electrophoretical separation of lipid-free chloroplast membranes, is correlated with the CP II. This major 25,000-dalton zone is composed of two polypeptides (Hoober 1970; Levine et al. 1972; Klein and Vernon 1974a-c; Wessels and Borchert 1975). Süss (1976) reported, that four polypeptides (M1–M4) correspond to CP II.

The chlorophyll-protein complex II does not contain the reaction center of photosystem II. This conclusion was drawn because chl b-less mutants (Anderson and Levine 1974; Genge et al. 1974; Thornber and Highkin 1974; Henriques and Park 1975; Machold et al. 1976) with PS I and PS II activities do not contain CP II. Henriques and Park (1975) and Machold et al. (1976) showed that one of the polypeptides of CP II is present in the chl b-less mutants of barley. On the other hand, in greenish mutant plastids of *Antirrhinum* having impaired photosystem II (Herrmann 1972; Herrmann and Knoth 1977) CP II is nevertheless present, whereas a protein band with the MG between 40,000 and 50,000 is absent.

IV. Membrane-Bound Enzymes

The main functions of chloroplasts are the photosynthetic production of ATP and NADPH and the use of this "assimilatory power" in the generation of carbohydrates. These processes are catalyzed by the enzymes of the photosynthetic electron transport chain and of the Calvin cycle. In addition the chloroplast harbors a large number of enzymes performing the replication, transcription, and translation of the chloroplast-specific genetic information (see Wollgiehn and Parthier; Bottomley, this vol.) and enzymes involved in the synthesis of small molecules such as amino acids, acyl lipids, and pigments (for review see Givan and Harwood 1976). Some of these functions are known to take place in the stroma of the chloroplast, and the respective enzymes are readily released on rupture of the chloroplast envelope. This is true, for example, for most enzymes of the Calvin cycle, though one of them, the ribulosebisphosphate carboxylase, is associated with the outer surface of the lamellar membrane and visualized as 14–15 nm particles by freeze-etching (Miller and Staehelin 1976). On the other hand, the processes of ATP and NADPH generation are intimately associated with the internal chloroplast membrane structures (for review see Krogmann 1976), the DNA-dependent RNA polymerase of chloroplasts was found to be tightly bound to membranes (Bennett and Ellis 1973; Bottomley et al. 1971; Schiemann et al. 1977), and chloroplast protein synthesis is performed in part by ribosomes connected with thylakoids (Philippovich et al. 1973; Margulies and Michaels 1974).

Thus, a great part of the chloroplast membrane polypeptides separated by polyacrylamide gel electrophoresis may be identical with enzymes of the electron transport chain or other processes. The functional identification of the chloroplast membrane polypeptides found after electrophoretic separation is an urgent necessity for further work on chloroplast differentiation. In the following the first results in this field are discussed.

Recent studies revealed that the most prominent membrane polypeptides do not directly relate to the components of the photosynthetic electron transport chain. These polypeptides are (1) the protein moieties of the chlorophyll-protein complexes I and II (see above), (2) the large and small subunits of fraction I protein (ribulosebisphosphate carboxylase) (Nielsen 1975; Henriques and Park 1976), and (3) the α and β subunits of coupling factor CF_1 (McEvoy and Lynn 1973; Börner et al. 1976; Henriques and Park 1976). The subunits of fraction I protein appear as major components only in the polypeptide pattern if whole chloroplasts were treated with SDS. Preparations of buffer-washed chloroplast membranes from barley contain no or only traces of fraction I protein as shown by immunological reaction, and the isolated large subunit of fraction I protein migrates in the gel to the same position as a membrane polypeptide which is obviously not identical with this subunit (Börner, Herrmann and Reichenbächer, unpubl.). CF_1 has been identified with particles of 14–15 nm on the outer membrane surface by the deep-etching technique (Garber and Steponkus 1974; Miller and Staehelin 1976). The CF_1 particles seem to be excluded from the partition region of grana thylakoids and from the lamellar body of etioplasts, but are capable of movement during grana unstacking and restacking (Oleszko and Moudrianakis 1974; Miller and Staehelin 1976; Wellburn 1977). CF_1 has a molecular weight of about 325,000 and consists of (at least) five subunits, designated as α, β, γ, δ and ε, with molecular weights of 59,000, 56,000, 37,000, 17,500 and 13,000, respectively (for review see Nelson 1976). In many structural respects CF_1 and its mitochondrial counterpart F_1 are very similar (cf. Senior 1973). However, whereas all subunits of F_1 are synthesized on cytoplasmic ribosomes (cf. Schatz 1976), the subunits of CF_1 are synthesized partly on cytoplasmic, partly on chloroplast ribosomes (see below). Washing chloroplast membranes with EDTA releases the CF_1 and at the same time reduces two polypeptides in the membrane polypeptide pattern which could be identified to be the α and β subunit of CF_1 (Börner et al. 1976; Henriques and Park 1976). Süss (1976) demonstrated the same mobility of membrane polypeptides and of four CF_1 subunits by one-dimensional electrophoresis and identified five CF_1 subunits with membrane polypeptides by two-dimensional electrophoresis (Süss 1978).

Also the subunits of cytochrome f, of cytochrome b-563, and cytochrome b-559 migrate in gels to the same position as certain minor membrane polypeptides (Süss 1976; unpubl. results). Immunological studies indicate that cytochrome f, which is involved in linear electron transport, is located on the outer surface of the membrane, whereas cytochrome f involved in cyclic phosphorylation seems to reside inside the membrane (Schmid et al. 1977). Cytochrome b-559 is on or near the surface of the membrane, as indicated by the oxidation of the reduced cytochrome by external oxidants (Cramer and Horton 1974).

A number of further components of the photosynthetic electron transport chain has been shown to be associated with the chloroplast thylakoid system mainly by immunological techniques: ferredoxin, ferredoxin-NADP$^+$ reductase, plastocyanine, plastoquinone, polypeptides belonging to the reaction centers of PS I and PS II (cf. Berzborn and Lockau 1977; Schmid et al. 1978). Until now these enzymes could not be detected in the pattern of membrane polypeptides after electrophoresis. It should be emphasized, however, that the immunological data concerning the position of the proteins within the membrane or on its inner or outer surface are still in part controversial. This is possibly due to artifacts caused by the preparation of the membrane material or by the mobility of the proteins in the membrane, and certainly the data are in many cases insufficient (for a critical review about the possiblities and limitations of the immunological approach cf. Berzborn and Lockau 1977). Certainly more membrane polypeptides will be identified with certain enzymes in the next future. However, the identification only on the basis of electrophoretic mobility needs further proof by other techniques, and some of the minor "chloroplast" membrane polypeptides may be unmasked as contaminations by other cell components such as nuclei or mitochondria.

V. Sites of Transcription and Translation
of the Genetic Information for Plastid Membrane Polypeptides

Plastid biogenesis in general needs the cooperation of (at least) two genetic systems: the machinery for transcription and translation of the plastid itself, and that of the nuclear-cytoplasmic compartment. Papers dealing with the site of synthesis and the coding site of membrane polypeptides published since 1970 are listed in Table 1. The table clearly demonstrates that the division of labor between the two compartments is also true for the biogenesis of the plastid membranes.

Several approaches were used in order to find the sites of synthesis of the membrane polypeptides and to clarify whether their genetic information is encoded in plastid DNA or in the DNA of the nucleus. Since the results obtained vary according to the approach applied we will deal with the approaches and the main results in the following section.

A. Effects of Specifically Acting Antibiotics

The most common approach to the site of synthesis of plastid components is the use of specific inhibitors of translation on cytoplasmic 80S ribosomes (cycloheximide) or on plastidic 70S ribosomes (chloramphenicol, lincomycin,

Table 1. Papers dealing with the sites of transcription and/or translation of genetic information for chloroplast membrane components

Membrane component	Genus	N-DNA	C-Rib	P-DNA	P-Rib	Reference	Method
Chloroplast membrane polypeptides (not identified)	*Acetabularia*	—				Apel and Schweiger (1972)	NT
	Acetabularia		—		—	Apel and Schweiger (1973)	IH
	Antirrhinum			—		Herrmann (1972)	M
	Chlamydomonas		—		—	Hoober (1970)	IH
	Chlamydomonas		—		—	Eytan and Ohad (1970)	IH
	Chlamydomonas				—	Boynton et al. (1972)	LPR
	Chlamydomonas				—	Goodenough (1971)	IH
	Chlamydomonas		—			Hoober and Stegeman (1973)	IH
	Chlamydomonas				—	Chua et al. (1973)	MR
	Chlamydomonas				—	Margulies et al. (1975)	MR
	Chlorella		—			Galling et al. (1973)	IH
	Euglena		—		—	Bishop et al. (1973)	IH
	Euglena		—		—	Harris et al. (1973)	IP
	Euglena		—		—	Liebers and Parthier (1974)	IH
	Euglena			—		Bingham and Schiff (1976)	MU
	Gossypium	—				Benedict and Kohel (1970)	M
	Gossypium			—		Kohel and Benedict (1972)	MU
	Hordeum		—		—	Sprey (1972)	LPR
	Hordeum		—		—	Börner et al. (1976)	LPR
	Hordeum		—		—	Knoth and Hagemann (1977)	LPR
	Lycopersicon		—		—	Toyama (1972)	IH
	Ochromonas				—	Smith-Johannsen and Gibbs (1972)	IH
	Oenothera	—		—		Schötz (1970)	GPD
	Pelargonium		—		—	Börner et al. (1972)	LPR
	Phaseolus				—	Bradbeer et al. (1974)	IH
	Pisum				—	Eaglesham and Ellis (1974)	IP
	Pisum				—	Cashmore (1976)	IH
	Secale		—			Schäfers and Feierabend (1976)	LPR
	Spinacia				—	Bottomley et al. (1974)	IP
	Spinacia				—	Morgenthaler and Mendiola-Morgenthaler (1976)	IP
	Vicia		—		—	Dyer at al. (1971)	LPR
	Vicia		—		—	Machold and Aurich (1972)	IH
	Vicia				—	Hachtel (1976)	IP
Etioplast membrane polypeptides (not identified)	*Hordeum*		—		—	Börner et al. (1976)	LPR
	Pisum				—	Sidell and Ellis (1975)	IP
	Zea				—	Hearing (1973)	IP
Polypeptides of prolamellar bodies	*Euglena*				—	Hovenkamp-Obbema (1974)	IH
	Pisum				—	Srivastava et al. (1971)	IH
	Hordeum				—	Börner and Knoth (unpubl.)	LPR

Table 1 (continued)

Membrane component	Genus	N-DNA	C-Rib	P-DNA	P-Rib	Reference	Method
Envelope poly-peptides	*Euglena*	—				Michaels and Gibor (1972)	MU
	Euglena	—				Neumann and Parthier (1973)	IH
	Hordeum		—			Sprey (1972)	LPR
	Hordeum		—			Knoth and Hagemann (1977)	LPR
	Pelargonium		—			Börner et al. (1972)	LPR
	Pisum				—	Joy and Ellis (1975)	IP
	Secale		—			Schäfers and Feierabend (1976)	LPR
	Spinacia				—	Morgenthaler and Mendiola-Morgenthaler (1976)	IP
	Vicia		—			Dyer et al. (1971)	LPR
CP I	*Antirrhinum*			—		Herrmann (1971)	M
	Chlamydomonas			—		Bennoun et al. (1977)	M
	Chlamydomonas				—	Bar-Nun et al. (1977)	IH
	Hordeum				—	Morgan and Griffith (1975)	IH
	Pisum				—	Cashmore (1976)	IH
	Vicia				—	Machold and Aurich (1972)	IH
	Vicia				—	Hachtel (1976)	IP
Polypeptide of PS I reaction center	*Chlamydomonas*				—	Chua and Gillham (1977)	IH
PS I poly-peptides	*Chlamydomonas*		—		—	Bar-Nun and Ohad (1977)	IH
CP II	*Nicotiana*	—				Kung et al. (1972)	CR
	Pisum		—			Cashmore (1976)	IH
	Pisum		—			Ellis (1975)	IH
	Vicia		—			Machold and Aurich (1972)	IH
Polypeptide of PS II reaction center	*Chlamydomonas*				—	Chua and Gillham (1977)	IH
PS II poly-peptides	*Chlamydomonas*		—		—	Bar-Nun and Ohad (1977)	IH
CF_1	*Phaseolus*		—		—	Horak and Hill (1972)	IH
	Spinacia		—		—	Mendiola-Morgenthaler et al. (1976)	IP
Ferredoxin	*Chlamydomonas*				—	Surzycki et al. (1970)	LPR
	Chlamydomonas	—			—	Armstrong et al. (1971)	IH
	Nicotiana	—				Kwanyuen and Wildman (1975)	CR
	Phaseolus				—	Haslett et al. (1973)	IH
Ferredoxin-NADP-reductase	*Chlamydomonas*				—	Surzycki et al. (1970)	LPR
	Chlamydomonas	—			—	Armstrong et al. (1971)	IH
	Euglena	—			—	Vaisberg et al. (1976)	IH

Table 1 (continued)

Membrane component	Genus	N-DNA	C-Rib	P-DNA	P-Rib	Reference	Method
Cytochrom-552	*Euglena*		—		—	Wildner (1976)	IH
Cytochrom-553	*Chlamydomonas*		—	—	—	Armstrong et al. (1971)	IH
	Chlamydomonas				—	Levine and Armstrong (1972)	LPR
Cytochrom-559	*Chlamydomonas*				—	Surzycki et al. (1970)	LPR
Cytochrom-563	*Chlamydomonas*		—	—	—	Armstrong et al. (1971)	IH
	Chlamydomonas				—	Levine and Armstrong (1972)	LPR
	Euglena		—			Wildner (1976)	IH
Cytochrom-564	*Chlamydomonas*		—			Surzycki et al. (1970)	LPR

Abbreviations: *N-DNA*, *P-DNA* = Nuclear DNA, plastid DNA (site of transcription); *C-Rib*, *P-Rib* = Cytoplasmic ribosomes, plastid ribosomes (site of translation); *CR* = Biochemical analysis of hybrids resulting from crosses between plant species differing in properties of membrane polypeptides; *GPD* = Genome-plastom disharmony (in *Oenothera*); *IH* = Use of inhibitors of transcription or translation; *IP* = Protein synthesis in isolated plastids; *NT* = Transplantation of nucleus (in *Acetabularia*); *LPR* = Use of plants lacking plastid ribosomes; *M* = Use of mutants with altered membrane polypeptide pattern; *MR* = Protein synthesis on plastid ribosomes attached to isolated membranes; *MU* = Use of mutants lacking plastid DNA and plastid ribosomes (in *Euglena*).

streptomycin). Using this method one has to exclude secondary effects of the translation inhibitors on transport processes, energy supply, regulatory events etc. (cf. Ellis and MacDonald 1970; Schlender et al. 1972; McMahon 1975). Therefore, results obtained by the inhibition of transcription and translation can provide only limited information about the sites of coding and synthesis of chloroplast polypeptides. Nevertheless, taken together, studies with inhibitors demonstrate unanimously that cytoplasmic and plastid ribosomes provide proteins to the biogenesis of chloroplast membranes. Prolamellar bodies are found in etioplasts after inhibition of plastid protein synthesis in algae and higher plants, suggesting that their formation is independent of proteins made by plastidic ribosomes (Srivastava et al. 1971; Hovenkamp-Obbema 1974). From experiments with inhibitors it has been concluded that the protein moiety of CP I is synthesized on plastid ribosomes, whereas the protein moiety of CP II is made in the cytoplasm (Machold and Aurich 1972; Ellis 1975; Morgan and Griffiths 1975; Cashmore 1976; Bar-Nun et al. 1977). However, because the complexes may contain more than one polypeptide (see above) and because the appearance of the complexes may need a certain stage of development of membrane structures and a simultaneous synthesis of chlorophylls, the results have to be interpreted cautiously (cf. Börner et al. 1975).

B. In Vitro Protein Synthesis in Isolated Chloroplasts

Direct evidence that some membrane polypeptides are synthesiszed inside the plastids is provided by studies on protein synthesis in isolated chloroplasts (Bottomley et al. 1974; Eaglesham and Ellis 1974; Chua 1976; Hachtel 1976; Morgenthaler and Mendiola-Morgenthaler 1976; Ellis 1977; Ellis and Barraclough 1978; Highfield and Ellis 1978; Reardon et al. 1978) and in isolated etioplasts (Sidell and Ellis 1975, 1977). A similar approach is to use isolated plastid membranes with attached ribosomes for in vitro protein synthesis (Margulies et al. 1975).

After separation by one-dimensional electrophoresis, the observed number of polypeptides synthesized by isolated chloroplasts is only between 5 and 9. Using electrophoresis in one dimension and isolectric focusing in the second dimension for separation of chloroplast polypeptides, Ellis (1977) found more than 15 membrane polypeptides among the products of protein synthesis in isolated plastids. Radioactivity was incorporated by chloroplasts and etioplasts into the same polypeptides (Sidell and Ellis 1975). Among the labeled proteins were CP I (Hachtel 1976, but see Eaglesham and Ellis 1974) and three subunits of the coupling factor (Mendiola-Morgenthaler et al. 1976; Ellis 1977).

C. Gene Mapping

Since the introduction of restriction enzymes into molecular biology rapid progress was achieved also concerning structure and function of chloroplast DNA. Two ways are followed to elucidate the function of defined chloroplast DNA regions: (1) Plastid mRNA species are isolated, translated in vitro and the translation product is characterized, if possible identified with a known protein. After that one has to try whether this mRNA specifically hybridizes with a certain fragment of chloroplast DNA obtained by digestion with restriction enzymes. In this way it has been shown that a 32,000-dalton polypeptide of chloroplast membranes is encoded in chloroplast DNA (Coen et al. 1978). (2) Total chloroplast DNA or restriction fragments of this DNA are transcribed and translated in a cell-free system and the products of the coupled transcription-translation are analyzed. Using this approach, the large subunit of fraction I protein could be synthesized, but hitherto no chloroplast membrane polypeptide was identified (Bottomley and Whitfeld 1978).

D. Genetic-Biochemical Approach

One of the most promising methods for further analysis is a combined genetic-biochemical approach. Wildman and co-workers introduced the biochemical analysis of hybrids resulting from crosses between plants differing in the peptide pattern of a certain protein into chloroplast research. In this way it was shown that differences in the protein moiety of CP II between *Nicotiana* species are inherited in a Mendelian manner and therefore the genetic information of this protein should be encoded in the nuclear DNA (Kung et al. 1972).

E. Mutants with Altered Polypeptide Pattern

Theoretically this approach (D.) can also be applied to the analysis of Mendelian or non-Mendelian (plastom) mutants. A number of nuclear gene and plastom mutants have been described which lack certain membrane polypeptides of the chloroplast (Gregory et al. 1971; Herrmann 1971, 1972; Thornber and Highkin 1974; Börner et al. 1975; Henriques and Park 1975; Herrmann et al. 1976; Machold and Høyer-Hansen 1976; Bennoun et al. 1977, 1978; Simpson et al. 1978). But in the case of these mutants it cannot be ruled out that the altered thylakoid membrane phenotype is a secondary effect of the mutation, and no exact conclusion can be drawn with respect to the localization of the genetic information for the respective polypeptide. Nevertheless, a control of formation of the lacking polypeptides by the respective DNA can be postulated (Herrmann 1971, 1972; Herrmann et al. 1976; Bennoun et al. 1977). The disadvantage of this approach can be overcome by the search for mutants making variant polypeptides (Chua 1976).

F. Mutants with Blocked Plastid Protein Synthesis

Another type of mutant is used in studying the role of cytoplasmic ribosomes in plastid biogenesis: mutants with deficient plastid protein synthesis. The rationale of studies on such mutants is similar to that of the "DNA-lacking" *Euglena* mutants.

All plastid components found in the mutants in spite of the deficiency in plastid protein synthesis have to be synthesized outside the plastids on cytoplasmic ribosomes (cf. Börner et al. 1976). In the same way plants can be used which lost their plastid ribosomes due to a temperature-sensitivity of ribosome formation (cf. Feierabend 1976). Studies on plastid ribosome-deficient plants also demonstrate the dependency of plastid membrane biogenesis upon cytoplasmic and plastid protein synthesis (Dyer et al. 1971; Börner et al. 1972, 1976; Boynton et al. 1972; Sprey 1972; Schäfers and Feierabend 1976; Knoth and Hagemann 1977). If it can be demonstrated that plastid protein synthesis is totally blocked in the mutant (cf. Börner et al. 1976; Hagemann and Börner 1978), this approach will provide exact data about the function of cytoplasmic ribosomes in the biogenesis of chloroplasts.

In plastid ribosome-deficient plants it was clearly shown that the plastid envelope and the prolamellar body are mainly composed of polypeptides synthesized in the cytoplasm. Ribosome-deficient plastids lack fraction I protein and CF_1 (Börner et al.1974, 1976; Börner et al. 1979), supporting the results obtained with isolated chloroplasts that subunits of these proteins are synthesized on chloroplast ribosomes.

G. Mutants Lacking Plastid DNA

In *Euglena* there exists a group of mutants (the most famous among them is W_3BUL; Schiff et al. 1971) which are reported to lack plastid DNA and plastid ribosomes. All plastid remnants in these mutants must have their origin outside the plastids, provided that there is not a nondetectable trace of plastid DNA in the mutants. However, there is good evidence for the presence of plastid structures and indirect evidence for the presence of plastid DNA in W₃BUL and comparable mutants (Walles and Kronestedt 1974; Heizmann et al. 1976, 1978; Parthier and Neumann 1977). Bingham and Schiff (1976) found a number of plastid membrane polypeptides in W₃BUL using labeled sulfolipid as a marker and concluded that these polypeptides were synthesized by cytoplasmic ribosomes. But at least five membrane components may be encoded in plastid DNA. They were only present in the light-grown wild-type, appear during greening, and were lacking in the mutant. However, it cannot be excluded that they are of nuclear origin and that their absence is due to a regulatory defect.

H. Transplantation of the Nucleus and Genome–Plastome Disharmony

Certain methods are restricted to one species only. In *Acetabularia* it is easily possible to transplant the nucleus from one cell into another. If this is done with cells differing, for example, in the pattern of membrane polypeptides, and if these differences are "transplanted" together with the nucleus, it has to be concluded that the genetic information of the nucleus determines these differences (Apel and Schweiger 1972).

In *Oenothera* the phenomenon of genome-plastom disharmonies was observed. The disharmony results beside other effects in irregularities in plastid membrane structure, suggesting an influence of both genome and plastom on membrane biogenesis in plastids (Schötz 1970). These effects of genome–plastom disharmony were rather unspecific, and therefore only of limited value for studies on the problem discussed.

Table 1 contains a list of papers dealing with sites of translation and transcription of plastid membrane polypeptides. The significance of the results published in these papers clearly varies according to the method applied. Especially the results obtained with inhibitors of RNA and protein synthesis are only of restricted value. This cannot be pointed out in the table, but if a group of polypeptides is concerned, e.g., "unidentified membrane polypeptides", apparently contradictory results have been reported. These contradictions can be explained by the fact that one paper deals, for example, only with the function of plastid DNA, the other only with the function of nuclear DNA; i.e., if it is reported that the formation of certain plastid membrane polypeptides is controlled by nuclear DNA (cf. Apel and Schweiger 1972), it does not mean that the formation of others may not be under control of plastid DNA (cf. Herrmann 1972).

In spite of the vast number of publications dealing with this matter, we have to the present only very limited information about the sites of transcription and translation of chloroplast polypeptides. This is mainly due to the fact that only a small number of chloroplast membrane polypeptides have been identified and functionally characterized. In conclusion, the majority of the data collected indicate that most of the plastid membrane proteins are synthesized on cytoplasmic ribosomes. A rather limited number of membrane polypeptides was found to be synthesized on plastid ribosomes, among them some envelope proteins and three subunits of the coupling factor CF_1. Most probably the polypeptides which are synthesized on chloroplast ribosomes are also encoded in chloroplast DNA.

VI. Biogenesis of Chloroplast Membranes in Higher Plants

Meristematic cells of higher plants contain plastids in form of proplastids, undifferentiated double-membrane-bound organelles 0.5–1 μm in diameter. In the light they develop into chloroplasts, a process which has been only poorly studied (cf. Leech 1977). In the dark, proplastids develop into etioplasts (Weier and Brown 1970; Bradbeer et al. 1974b; Robertson and Laetsch 1974). The overwhelming majority of studies on plastid biogenesis in higher plants deals with the con version of etioplasts into chloroplasts under the influence of light. Several photoreceptors affect this development (see Hagemann, this vol.) which is accompanied by drastic ultrastructural changes (see Schnepf, this vol.).

Etioplasts are regarded as chloroplasts whose normal development has been stopped by the absence of light. However, as conducted from ultrastructural and biochemical analyses, etioplasts, as compared with proplastids, have reached a considerable level of development. Dark-grown leaves have been found to contain all the enzymes of the Calvin cycle (Bradbeer et al. 1974b) and a large amount of membrane protein, including membrane-bound enzymes. In spite of the large quantity of work done on this problem, there is no consensus whether the differentiation of internal membranes of the etioplast into the stroma and grana thylakoids of the chloroplast is only a rearrangement of preexisting membrane material, or whether and to what extent a de novo synthesis of polypeptides not present in the etioplast takes place during greening. There is especially a considerable debate whether the protein moieties of the two chlorophyll-protein complexes found in the chloroplast are already components of the etioplast (see below). In *Chlamydomonas* there is first information available which indicates that there may be some specific sequence of addition of polypeptides to developing membranes in the light-requiring mutant, *y-l* (see below). Such information does not yet exist for higher plants.

There is no doubt that etioplasts possess the components of the electron transport system, including plastocyanin, cytochromes f, b-563 and b-559 (low potential form) (Plesničar and Bendall 1973). The high potential form of cytochrome b-559 is not detectable in etioplasts, and the possibility is discussed that it may be present in modified form with a much lower redox potential (cf. Bendall 1977). No or little further synthesis of plastocyanin and cytochromes occurs during greening of etiolated barley leaves (Plesničar and Bendall 1973) and also comparable amounts of ferredoxin-NADP$^+$ reductase and of coupling factor of photophosphorylation (CF$_1$, chloroplast ATPase) are detectable in dark-grown barley by immunological methods (Börner, unpubl.). In contrast, a high rate of synthesis of cytochromes and of ferredoxin are reported for greening bean etioplasts (Bradbeer 1973; Haslett et al. 1973). From measurements of the activity of trypsin-activated Ca^{2+}-dependent chloroplast ATPase in *Zea* (Lockshin et al. 1971) and of light-triggered dithiothreitol-activated Mg^{2+}-dependent chloroplast ATPase in *Phaseolus* (Gregory and Bradbeer 1975) is concluded, that most or all of the chloroplast ATPase is already present in etioplasts. This is in agreement with the results of Wellburn (1977), who in addition found that the distance between the individual CF$_1$ particles attached to the lamellar membranes progressively increases with the length of illumination.

Etioplasts contain protochlorophyllide as precursor of chlorophyll. Protochlorophyllide can be extracted from etioplasts as a protochlorophyllide-protein complex, the so-called protochlorophyllide holochrome (see review by Boardman et al. 1977). Protochlorophyllide holochrome is extracted from bean leaves in a soluble form with a molecular weight of about 600,000 (Boardman 1962). Molecular weight determinations by gel filtration resulted in two particles with apparent molecular weights of 300,000 and 550,000 (Schopfer and Siegelman 1968). A protochlorophyllide holochrome with a molecular weight of 63,000 was obtained by extraction of barley leaves with saponine (Henningsen and Kahn 1971). It was suggested that this holochrome represents an active subunit, and a holochrome extracted with saponine from bean leaves with a molecular weight of 170,000 may represent a dimer or trimer (Henningsen et al. 1974). The first process observed in the transition of etioplast membranes into chloroplast membranes is the conversion of protochlorophyllide in chlorophyllide a. The newly formed chlorophyllide a is still bound to protein in a complex with a molecular weight of 63,000, is then reduced in size to 29,000, possibly by dissociation into a colorless activating enzyme and a chlorophyllide a-carrier protein, followed by an apparent increase in molecular weight to 100,000 (Henningsen et al. 1974). This increase in molecular weight may be the result of the incorporation of chlorophyll a into the developing membrane (cf. Griffiths 1977). Until now one can only speculate about the fate of the protein moiety of the (proto-)chlorophyllide holochrome during the conversion into chlorophyll a and b, and about its relation to the protein moieties of the chlorophyll-protein complexes present in the photosynthetically active membranes of chloroplasts. A model for the organization of protochlorophyllide in the prolamellar body has been proposed by Boardman and Anderson (1978).

As reported for algae (see above), in higher plants etioplast and chloroplast membranes differ in their polypeptide patterns after separation in polyacrylamide gels (Lürssen 1971; Cobb and Wellburn 1973; Nielsen 1975; Høyer-Hansen et al. 1976; Schumann and Börner 1976; Hiller et al. 1978) and at early stages of greening newly synthesized polypeptides are inserted into existing membranes (Cobb and Wellburn 1973; Bogorad 1975; Nielsen 1975). This is in disagreement with the results of other investigators who contend that all membrane polypeptides of the chloroplast are present in the etioplast and that only quantitative, not qualitative, changes occur during greening (Lagoutte and Duranton 1972; Argyroudi-Akoyunoglou and Akoyunoglou 1973; Remy 1973; Duval and Duranton 1974; Lütz 1975). Particularly the preexistence of the protein moieties of the two major chlorophyll-protein complexes is emphasized. This difference in the results and in the interpretation is mainly due to insufficiency of the method applied. The first results of immunological studies also need further proof. The apoprotein of photosystem I was immunologically detected in barley etioplasts by Griffiths et al. (1976). Its incorporation into a functional photosystem I is dependent on plastid protein synthesis. Volodarsky et al. (1976) report that barley etioplasts contain one antigen to both chlorophyll-protein complexes I and II of the chloroplast. Several attempts have been made to correlate the appearance of the chlorophyll-protein complexes with the development of the activities of photosystem I and II during greening (cf. Thornber and Alberte 1976; Boardman et al. 1977); but it should be emphasized in this context again, that at least the chlorophyll-protein complex II does not belong to the reaction center of photosystem II (see above). However, there is general coincidence that the protein moieties of chlorophyll-protein complexes I and II, preexisting in the etioplast in the same form or another or not at all, increase in amount during greening, and become the dominating protein bands in the polypeptide pattern of chloroplast membranes.

There is good evidence that the synthesis of mRNA for the precursor of the apoprotein of chlorophyll-protein complex II is light-induced and under the control of phytochrome (Apel and Kloppstech 1978). Also the mRNA for a 32,000-dalton poylpeptide of chloroplast membranes was shown to be accumulated during light-induced greening (Coen et al. 1978).

VII. The Biogenesis of Chloroplast Membranes in Chlamydomonas

The biogenesis of chloroplast membranes in algae has been intensively investigated in *Chlamydomonas*. Similar studies have also been performed with several other algae (e.g., *Euglena, Chlorella, Scenedesmus*), but on a smaller scale. This part deals with *Chlamydomonas* only; the results of other algae are summarized from the point of view of the genetic control of biogenesis of membrane polypeptides in Table 1.

The wild-type *Chlamydomonas reinhardii* cells are able to synthesize chloroplasts and chloroplast membranes when continuously grown in the dark, the membranes so formed are photosynthetically active.

The *y-1* mutant cells have apparently lost the ability to form chlorophyll in the dark, but are able to synthesize protochlorophyll(ide), which can be converted photochemically to chlorophyll(ide), as in higher plants. If dark-grown *y-1* mutant cells are exposed to light, chloroplast membranes are synthesized preferentially, and reach the level of organization and functionality similar to that present in light-grown cells. The greening process of *y-1* cell has been extensively studied in experiments in which the composition of lipids and pigments, proteins, structural changes and photosynthetic activity have been analyzed during the normal greening process, as well as during greening with alternative light and dark exposure and use of protein and RNA synthesis inhibitors. The results of these works have been reviewed by Ohad (1972, 1975) and are shown schematically in Fig. 2 (Ohad 1975).

The analysis of the membrane proteins by (1) gel electrophoresis in an acidic medium after solubilization of the membranes with a mixture of phenol, acetic acid, and water, and (2) SDS-PAGE (Hoober 1970; Hoober 1972; Levine and Duram 1973; Regitz and Ohad 1974; Bar-Nun and Ohad 1975; Chua and Bennoun 1975; Hoober and Stegeman 1976; Chua and Gillham 1977) shows that only some of the polypeptides are synthesized in the chloroplast, and others on the cytoplasmic ribosomes. Several major membrane polypeptides of M.W. ranging between 20,000 and 40,000 are the translation products of cytoplasmic ribosomes (cf. Ohad 1975). These polypeptides were found to be associated with the photosystem II complex, prepared by treatment with digitonin (Levine et al. 1972) or Triton X-100 (Jennings and Eytan 1973). The polypeptide composition of PS II complex analyzed by acidic gel electrophoresis consisted of mostly L-proteins (light-induced, lamellar protein; Eytan and Ohad 1970; Jennings and Eytan 1973). The L-proteins correspond to the two predominant polypeptides of about 28,000 and 24,000 daltons (designated

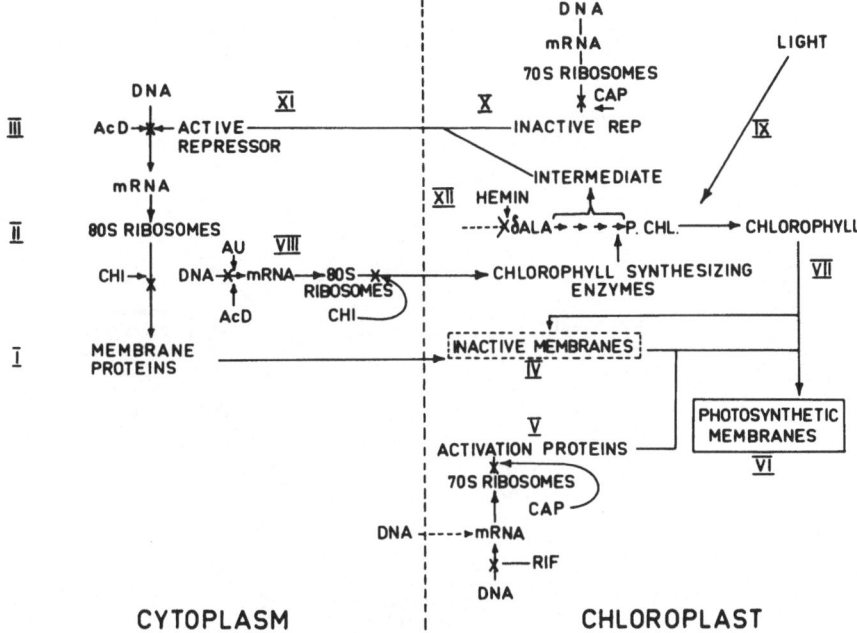

Fig. 2. Schematic representation of the processes of chlorophyll and protein synthesis in the cytoplasm and chloroplast of *C. reinhardii* leading to the formation of functional chloroplast membranes. Membrane proteins (1) (peptides a, b, c in Hoober's nomenclature, *L-protein*, or peptides Va, Vb) are synthesized by cytoplasmic 80S ribosomes (II) and coded for by nuclear DNA (III). The transcription process can be inhibited by actinomycin D, but not by rifampicin. The translation process can occur in the dark and is inhibited by cycloheximide (II). The proteins are transferred to the chloroplast and become incorporated into chloroplast membrane (IV), which as such, are photosynthetically inactive. Activation proteins synthesized in the chloroplast (V), coded for by chloroplast—as well as nuclear DNA—and translated on 70S ribosomes inhibited by chloramphenicol are added to the inactive membrane which becomes photosynthetically functional (VI). The activation proteins include components which are essential for the activity of the photosystem II and I reaction centers and can be made and inserted into the growing membrane in absence of concomitant chlorophyll synthesis or protein synthesis in the cytoplasm, provided that the inactive membranes containing the cytoplasmic proteins are already formed. The synthesis of protochlorophyll(ide) (VII) form δALA is mediated by a series of enzymatic reactions carried by several enzymes, including an enzyme which is turning over rapidly. The synthesis of this enzyme(s) is light independent. The transcription process is rifampicin-resistant and, hence, depends on nuclear DNA. The translation process occurs in the cytoplasm (VIII), since it is not sensitive to chloramphenicol. Light is continuously required for the photoconversion of the protochlorophyll(ide) into chlorophyll(ide) in the y-1 mutant, but not in the wild type (IX). In absence of illumination, chlorophyll precursors or intermediates accumulate in small amounts and combine with an aporepressor protein (X) synthesized on 70S ribosomes (the origin of the genetic information for this protein is not yet clear). The active repressor inhibits the transcription process for the formation of the cytoplasmic membrane proteins (XI), and, thus, all membrane components synthesis is blocked. However, membrane proteins can be synthesized in the dark if chloramphenicol, which will inhibit the formation of the aporepressor, is added to the cells. The synthesis of chlorophyll can be dissociated from that of the membrane proteins by use of hemin, which apparently inhibit the formation of δALA (XII). For the possible sequence of addition of chlorophyll and cytoplasmatic and activation proteins to the growing membrane, see scheme in Fig. 3. Symbols: - - - intermediate steps; — × — point of inhibition. The term *cytoplasm* is used to define the cell minus the chloroplast (Ohad 1975)

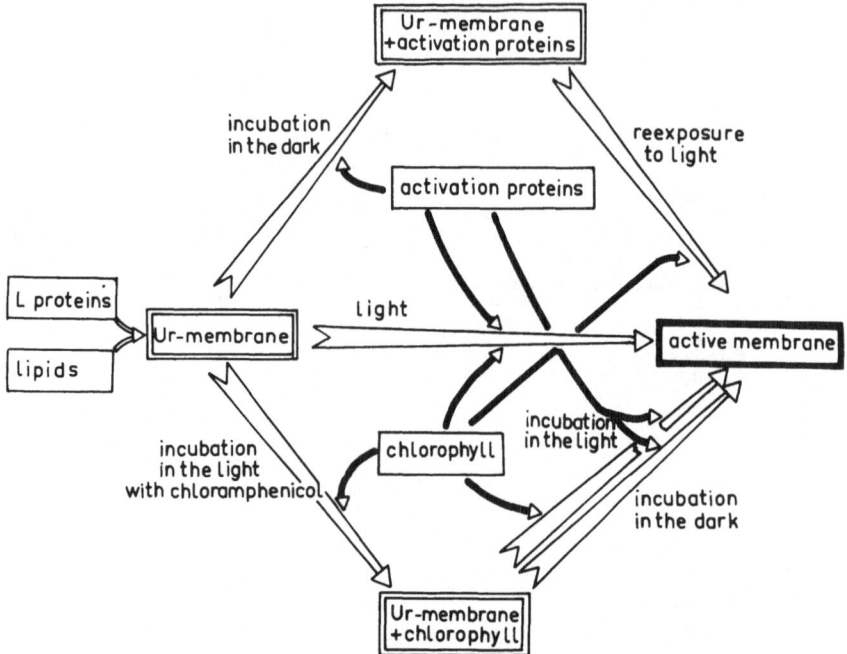

Fig. 3. Schematic representation of the modulation of membrane composition and activity during membrane formation under different experimental conditions. The term *Ur-membrane* designates the basic membrane structure composed of *L proteins* of cytoplasmic origin and lipids which might serve as an obligatory acceptor frame for the additions of chlorophyll and activation proteins of chloroplast origin (Ohad 1975)

b and c, respectively), which Hoober (1970, 1972) separated by SDS-PAGE of *Chlamydomonas* thylakoid membranes. Kan and Thornber (1976) found that these polypeptides are subunits of the apoprotein of the CP II. Cycloheximid completely blocks the synthesis of these polypeptides (Hoober 1976), indicating that polypeptides b and c are of cytoplasmic origin. The modification of the SDS-gel electrophoresis method according to Laemmli (1970) yields still another pattern in which the L-protein(s) are separated into four bands III, IV, Va and Vb (Bar-Nun and Ohad 1974; Regitz and Ohad 1974), translated in the cytoplasma (Bar-Nun and Ohad 1977).

Extraction procedures and PAGE methods used by several authors vary widely. Therefore the comparison is very difficult. According to Bar-Nun's nomenclature, polypeptide III and IV coincide with polypeptides IIa and IIb, respectively, in Hoober's (1970) and Levine's (1972, 1973) and 11–12 in Chua and Bennoun's (1975). The single polypeptide IIc (Hoober 1970; Anderson and Levine 1974) corresponds to Va and Vb (Bar-Nun et al. 1977) and 15–17 (Chua and Bennoun 1975).

Eytan and Ohad (1970) have proposed that the polypeptides made in the cytoplasma are essential for the formation of the membrane structure, but not sufficient to allow photosynthetic electron flow. Greening of *Chlamydomonas y-1* cells in the presence of CAP, which inhibits synthesis of chloroplast-made proteins, results in the formation of nonphotoactive membranes. Activity can be regained by reactivation of chloroplast ribosomes (Eytan and Ohad 1970; Hoober 1970). Chua (1976) showed that at least nine polypeptides are made inside the chloroplast.

The products of chloroplast protein synthesis include polypeptides 2 and 6, which associated with the reaction center activity of PS I and PS II respectively (Chua et al. 1975; Bennoun and Chua 1976, Chua and Bennoun 1975), polypeptide 5 and several other polypeptides whose functions are unknown. The function of polypeptide 5 has not yet been identified, but on the basis of genetic criteria (a molecular weight variant of this poylypeptide is synthesized by a mutant, *thm-u-1*, whose defect is inherited uniparentally; Adams et al. 1976) appears to be coded by chloroplast DNA (Chua 1976).

Recently Bar-Nun and Ohad (1977) showed that both PS I and PS II photoactive preparations contain polypeptides of chloroplastic and cytoplasmic origin (Figs. 2 and 3). Two of the polypeptides of cytoplasmic origin (III and IV) are common to both preparations. Polypeptide III is apparently not required for PS I and PS II electron flow, since it can be removed by trypsin digestion of the membrane without loss of these activities (Regitz and Ohad 1974). The T_4 temperature mutant of *Chlamydomonas reinhardii* (Chua and Bennoun 1975) grown at nonpermissive temperature (37°C) lacked two polypeptides of chloroplastic origin. The PS II activity was absent, while PS I activity was not affected. The PS II activity could be restored following synthesis and insertion into membrane of the two missing polypeptides at permissive temperature (25°; Kretzer et al. 1976). It could additionally be postulated that the polypeptides of chloroplastic origin are required for the formation of photosynthetic reaction centers, since other components of photosynthetic electron transfer, such as cyt. f, ferredoxin, and plastocyanin are already present in the nonactive membranes (cf. Ohad 1975).

From the results of greening experiments Bar-Nun and Ohad (1977) reported that the complexes can be formed only when chlorophyll is synthesized simultaneously with the polypeptide moiety. These authors speculate that the association between chlorophyll and the polypeptide may occur prior to or during the insertion of the chlorophyll-protein complexes into the developing membrane, resulting in a thermodynamically more stable molecular conformation of the light-dependent polypeptides which are synthesized in the cytoplasm under nuclear control. The association of chlorophyll with the corresponding polypeptides may play a role in the transport across the chloroplast envelope of the polypeptides synthesized light-dependent under nuclear control in the cytoplasm (Bar-Nun et al. 1977).

References

Adams, G.M.W., Van Winkle-Swift, K.P., Gillham, N.W., Boynton, J.E.: Plastid inheritance in *Chlamydomonas reinhardii*. In: The Genetics of Algae, pp. 69-118; Lewin, R.A. (ed.). Berkeley: Univ. Calfifornia Press 1976

Akoyunoglou, G., Argyroudi-Akoyunoglou, J.H. (eds.): Chloroplast Development, 888 p. Amsterdam: Elsevier/North-Holland 1978

Anderson, J.M.: The molecular organization of chloroplast thylakoids. Biochim. Biophys. Acta 416, 191-235 (1975)

Anderson, J.M., Levine, R.P.: The relationship between chlorophyll protein complexes and chloroplast membrane polypeptides. Biochim. Biophys. Acta 357, 118-126 (1974)

Anderson, J.M., Waldron, J.C., Thorne, S.W.: Chlorophyll-protein complexes of spinach and barley thylakoids. Spectral characterization of six complexes resolved by an improved electrophoretic procedure. FEBS Lett. 92, 227-233 (1978)

Apel, K., Bogorad, L., Woodcock, C.L.F.: Chloroplast membranes of the green alga *Acetabularia mediterranea*. I. Isolation of the photosystem II. Biochim. Biophys. Acta 387, 568-579 (1975)

Apel, K., Kloppstech, K.: Light-induced appearance of mRNA coding a/b protein, pp. 653-656. In: see ref. Akoyunoglou and Argyroudi-Akoyunoglou (1978)

Apel, K., Schweiger, H.-G.: Nuclear dependency of chloroplast proteins in *Acetabularia*. Eur. J. Biochem. 25, 229-238 (1972)

Apel, K., Schweiger, H.-G.: Sites of synthesis of chloroplast-membrane proteins. Evidence of three types of ribosomes engaged in chloroplast protein synthesis. Eur. J. Biochem. 38, 373-383 (1973)

Argyroudi-Akoyunoglou, J.H., Akoyunoglou, G.: On the formation of photosynthetic membrane in bean plants. Photochem. Photobiol. 18, 219-228 (1973)

Armstrong, J.J., Surzycki, S.J., Moll, B., Levine, R.P.: Genetic transcription and translation specifying chloroplast components in *Chlamydomonas reinhardii*. Biochemistry 10, 692-701 (1971)

Bar-Nun, S., Ohad, I.: Cytoplasmic and chloroplastic origin of chloroplast membrane proteins associated with PS II and PS I active center in *Chlamydomonas reinhardii y-1*. In: Proc. 3rd Int. Congr. Photosynth; Avron, M. (ed.), pp. 1627-1638 Amsterdam: Elsevier 1975

Bar-Nun, S., Ohad, I.: Presence of polypeptides of cytoplasmic and chloroplastic origin in isolated photoactive preparations of photosystem I and II in *Chlamydomonas reinhardii y-1*. Plant Physiol. 59, 161-166 (1977)

Bar-Nun, S., Schantz, R., Ohad, I.: Appearance and composition of chlorophyll-protein complexes I and II during chloroplast membrane biogenesis in *Chlamydomonas reinhardii y-1*. Biochim. Biophys. Acta 459, 451-467 (1977)

Bendall, D.S.: Development of photosynthetic electron transport in greening barley. Biochem. Soc. Trans. 5, 84-88 (1977)

Benedict, C.R., Kohel, R.J.: Photosynthetic rate of a virescent cotton mutant lacking chloroplast grana. Plant Physiol. 45, 519-521 (1970)

Bengis, C., Nelson, N.: Purification and properties of the photosystem I reaction center from chloroplast. J. Biol. Chem. 250, 2783-2788 (1975)

Bengis, C., Nelson, N.: Subunit structure of chloroplast photosystem I reaction center. J. Biol. Chem. 252, 4564-4569 (1977)

Bennett, J., Ellis, R.J.: Solubilization of the membrane-bound deoxyribonucleic acid-dependent ribonucleic acid polymerase of pea chloroplasts. Biochem. Soc. Trans. 1, 892-894 (1973)

Bennoun, P., Chua, N.-H.: Methods for the detection and characterization of photosynthetic mutants in *Chlamydomonas reinhardii*, pp. 33-39. In: see ref. Bücher et al. (1976)

Bennoun, P., Girard, J., Chua, N.-H.: A uniparental mutant of *Chlamydomonas reinhardii* deficient in the chlorophyll-protein complex CP I. Mol. Gen. Genet. 153, 343-348 (1977)

Bennoun, P., Masson, A., Piccioni, R., Chua, N.-H.: Uniparental mutants of *Chlamydomonas reinhardii* defective in photosynthesis, pp.721-726. In: see ref. Akoyunoglou and Argyroudi-Akoyunoglou (1978)

Berzborn, R.J., Lockau, W.: Antibodies. In: Encyclopedia of Plant Physiology, New Series, Vol. V, Photosynthesis I; Trebst, A., Avron, M. (eds.), pp. 283-296. Berlin-Heidelberg-New York: Springer 1977

Bingham, S., Schiff, J.A.: Cellular origin of plastid membrane polypeptides in *Euglena*, pp. 79-86. In: see ref. Bücher et al. (1976)

Bishop, D.G., Bain, J.M., Smillie, R.M.: The effect of antibiotics on the ultrastructure and photochemical activity of a developing chloroplast. J. Exp. Bot. 24, 361-375 (1973)

Boardman, N.K.: Studies on a protochlorophyll-protein complex. I. Purification and molecular-weight determination. Biochim. Biophys. Acta 62, 63-79 (1962)

Boardman, N.K.: Physical separation of the photosynthetic photochemical systems. Annu. Rev. Plant Physiol. 21, 115-140 (1970)

Boardman, N.K., Anderson, J.M.: Composition, structure and photochemical activity of developing and mature chloroplasts, pp. 1-14. In: see ref. Akoyunoglou and Argyroudi-Akoyunoglou (1978)

Boardman, N.K., Anderson, J.M., Goodchild, D.J.: Chlorophyll-protein complexes and structure of mature and developing chloroplasts. Curr. Top. Bioenerg. 8, 35-109 (1978)

Börner, Th.: Untersuchungen zur Chloroplastenbiogenese an Plastidenribosomen-defizienten Plastommutanten von Hordeum vulgare L. und Pelargonium zonale hort. Thesis B, Universität Halle (1979)

Börner, Th., Knoth, R., Herrmann, F., Hagemann, R.: Struktur und Funktion der genetischen Information in den Plastiden. V. Das Fehlen von ribosomaler RNS in den Plastiden der Plastommutante "Mrs. Parker" von Pelargonium zonale Ait. Theor. Appl. Genet. 42, 3-11 (1972)

Börner, Th., Knoth, R., Herrmann, F., Hagemann, R.: Struktur und Funktion der genetischen Information in den Plastiden. X. Das Fehlen von Fraktion-I-Protein in den weißen Plastiden einiger Sorten von Pelargonium zonale. Biochem. Physiol. Pflanz. 165, 429-432 (1974)

Börner, Th., Schumann, B., Krahnert, S., Pechauf, M., Herrmann, F.H., Knoth, R., Hagemann, R.: Struktur und Funktion der genetischen Information in den Plastiden. XIII. Lamellarproteine bleicher Plastiden von Plastom- und Genmutanten von Hordeum und Lycopersicon. Biochem. Physiol. Pflanz. 168, 185-193 (1975)

Börner, Th., Schumann, B., Hagemann, R.: Biochemical studies on a plastid ribosome-deficient mutant of Hordeum vulgare, pp. 41-48. In: see ref. Bücher et al. (1976)

Börner, Th., Manteuffel, R., Wellburn, A.R.: Enzymes of plastid ribosome-deficient mutants. Chloroplast ATPase (CF₁). Protoplasma 98, 153-161 (1979)

Bogorad, L.: Eukaryotic intracellular relationships. In: Membrane Biogenesis; Tzagoloff, A. (ed.), pp. 201-245. New York-London: Plenum Press 1975

Bottomley, W., Smith, H.J., Bogorad, L.: RNA polymerases of maize: partial purification and properties of the chloroplast enzyme. Proc. Natl. Acad. Sci. USA 68, 2415-2416 (1971)

Bottomley, W., Spencer, A., Whitfeld, P.R.: Protein synthesis in isolated spinach chloroplasts: comparison of light-driven and ATP-driven synthesis. Arch. Biochem. Biophys. 164, 106-117 (1974)

Bottomley, W., Whitfeld, P.R.: The products of cell-free transcription and translation of total spinach chloroplast DNA, pp. 657-662. In: see ref. Akoyunoglou and Argyroudi-Akouyunoglou (1978)

Boynton, J.E., Gillham, N.W., Chabot, J.F.: Chloroplast ribosome deficient mutants in the green alga Chlamydomonas reinhardii and the question of chloroplast ribosome function. J. Cell Sci. 10, 267-305 (1972)

Bradbeer, J.W.: The synthesis of chloroplast enzymes. In: Biosynthesis and its Control in Plants; Milborrow, B.V. (ed.), pp. 279-302. London-New York: Academic Press 1973

Bradbeer, J.W., Gyldenholm, A.O., Ireland, H.M.M., Smith, J.W., Rest, J., Edge, H.J.W.: Plastid development in primary leaves of Phaseolus vulgaris. VIII. The effects of the transfer of dark-grown plants to continuous illumination. New Phytol. 73, 271-279 (1974a)

Bradbeer, J.W., Ireland, H.M.M., Smith, J.W., Rest, J., Edge, J.H.W.: Plastid development in primary leaves of Phaseolus vulgaris. VII. Development during growth in continuous darkness. New Phytol. 73, 263-270 (1974b)

Brown, J.S., Alberte, R.S., Thornber, J.P.: Comparative studies on the occurrence and spectral composition of chlorophyll-protein complexes in a wide variety of plant material. In: Proc. 3rd Int. Congr. Photosynth.; Avron M. (ed.), pp. 1951-1962. Amsterdam: Elsevier 1975

Bücher, T., Neupert, W., Sebald, W., Werner, S.: Genetics and Biogenesis of Chloroplasts and Mitochondria. Int. Conf. Munich, Germany, August 1976. Amsterdam-New York-Oxford: North-Holland 1976

Cashmore, A.R.: Protein synthesis in plant leaf tissue. The site of synthesis of the major proteins. J. Biol. Chem. 251, 2848-2853 (1976)

Chua, N.-H.: A uniparental mutant of *Chlamydomonas reinhardii* with a variant thylakoid membrane polypeptide, pp. 323-330. In: see ref. Bücher et al. (1976)

Chua, N.-H., Bennoun, P.: Thylakoid membrane polypeptides of *Chlamydomonas reinhardii:* wild-type and mutant strains deficient in photosystem II reaction center. Proc. Natl. Acad. Sci. USA 72, 2175-2179 (1975)

Chua, N.-H., Blobel, G., Siekevitz, P., Palade, G.E.: Attachment of chloroplast polysomes to thylakoid membranes in *Chlamydomonas reinhardii*. Proc. Natl. Acad. Sci. USA 70, 1554-1558 (1973)

Chua, N.-H., Gillham, N.W.: The sites of synthesis of the principal thylakoid membrane polypeptides in *Chlamydomonas reinhardii*. J. Cell Biol. 74, 441-452 (1977)

Chua, N.-H., Matlin, K., Bennoun, P.: A chlorophyll-protein complex lacking in photosystem I mutants of *Chlamydomonas reinhardii*. J. Cell Biol. 67, 361-377 (1975)

Cobb, A.H., Wellburn, A.R.: Developmental changes in the leaves of SDS extractable polypeptides during plastid morphogenesis. Planta 114, 131-142 (1973)

Coen, D.M., Bedbrook, J.R., Link, G., Grebanier, A., Steinback, K., Beaton, A., Rich, A., Bogorad, L.: Genes and mRNAs for maize chloroplast proteins: changes during light-induced chloroplast development, pp. 559-564. In: see ref. Akyounoglou and Argyroudi-Akoyunoglou (1978)

Cramer, W.A., Horton, P.: On the oxidation mechanism of cytochrome b^{559} by photosystem I. Abstr. 3rd Int. Congr. Photosynth., p. 23, Rehovot 1974

Delepelaire, P., Chua, N.-H., Lithium dodecyl sulfate/polyacrylamide gel electrophoresis of thylakoid membranes at 4 °C: Characterization of two additional chlorophyll a-protein complexes. Proc. Natl. Acad. Sci. USA 76, 111-115 (1979)

Dietrich, W.E., Jr., Thornber, J.P.: The P700-chlorophyll a-protein of a blue-green alga. Biochim. Biophys. Acta 245, 482-493 (1971)

Duval, D., Duranton, J.: Cation localization on plastid membrane proteins from *Zea mays* L. Photosynthetica 8, 1-8 (1974)

Dyer, T.A., Miller, R.H., Greenwood, A.D.: Leaf nucleic acids. I. Characteristics and role in the differentiation of plastids. J. Exp. Bot. 22, 125-136 (1971)

Eaglesham, A.R.J., Ellis, R.J.: Protein synthesis in chloroplasts. II. Light-driven synthesis of membrane proteins by isolated pea chloroplasts. Biochim. Biophys. Acta 335, 396-407 (1974)

Ellis, R.J.: The synthesis of chloroplast membranes in *Pisum sativum*. In: Membrane Biogenesis; Tzagoloff, A. (ed.), pp. 247-278. New York-London: Plenum Press 1975

Ellis, R.J.: Protein synthesis by isolated chloroplasts. Biochim. Biophys. Acta 463, 185-215 (1977)

Ellis, R.J., MacDonald, I.R.: Specificity of cycloheximide in higher plant systems. Plant Physiol. 46, 227-232 (1970)

Ellis, R.J., Barraclough, R.: Synthesis and transport of chloroplast proteins inside and outside the cell, pp. 185-194. In: see ref. Akoyunoglou and Argyroudi-Akoyunoglou (1978)

Eytan, G., Ohad, I.: Biogenesis of chloroplast membranes. VI. Cooperation between cytoplasmic and chloroplast ribosomes in the synthesis of photosynthetic lamellar proteins during the greening process in a mutant of *Chlamydomonas reinhardii y-1*. J. Biol. Chem. 245, 4297-4307 (1970)

Feierabend, J.: Temperature-sensitivity of chloroplast ribosome formation in higher plant; pp. 99-102. In: see ref. Bücher et al. (1976)

Feierabend, J.: Cooperation of cytoplasmic and plastidic protein synthesis in rye leaves, pp. 207-214. In: see ref. Akoyunoglou and Argyroudi-Akoyunoglou (1978)

Felgenhauer, K.: Evaluation of molecular size by gel electrophoretic techniques. Hoppe Seylers Z. Physiol. Chem. 355, 1281-1290 (1974)

Galling, G., Salzmann, C., Spiess, E.: Synthese von Chlorophyll und Strukturelementen der Plastiden in *Chlorella* ohne Beteiligung der Chloroplasten-Ribosomen. Planta 114, 269-284 (1973)

Garber, M.P., Steponkus, P.L.: Identification of chloroplast coupling factor by freeze-etching and negative staining techniques. J. Cell Biol. 63, 24-34 (1974)

Genge, S., Pilger, D., Hiller, R.G.: The relationship between chlorophyll b and pigment-protein complex II. Biochim. Biophys. Acta 347, 22-30 (1974)

Gillham, N.W., Boynton, J.E., Chua, N.-H.: Genetic control of chloroplast proteins. Curr. Top. Bioenerg. 8 (Part B), 211-260 (1978)

Givan, C.V., Harwood, J.L.: Biosynthesis of small molecules in chloroplasts of higher plants. Biol. Rev. 51, 365-406 (1976)

Goodenough, U.W.: The effect of inhibition of RNA and protein synthesis on chloroplast structure and function in wild-type *Chlamydomonas reinhardii*. J. Cell Biol. 50, 35-49 (1971)

Goodenough, U.W., Togasaki, R.K., Paszewski, A., Levine, R.P.: Inhibition of chloroplast ribosome formation by gene mutation in *Chlamydomonas reinhardii*. In: Autonomy and Biogenesis of Mitochondria and Chloroplasts; Boardman, N.K., Linnane, A.V., Smillie, R.M. (eds.), pp. 224-232. Amsterdam: North-Holland 1971

Gregory, P., Bradbeer, J.W.: Plastid development in the primary leaves of *Phaseolus vulgaris*. Development of plastid adenosine triphosphatase activity during greening. Biochem. J. 148, 433-438 (1975)

Gregory, R.P.F., Raps, S., Bartsch, W.: Are specific chlorophyll-protein complexes required for photosynthesis? Biochim. Biophys. Acta 234, 330-334 (1971)

Griffiths, W.T.: Studies in vitro on plastid development. Biochem. Soc. Trans. 5, 88-90 (1977)

Griffiths, W.T., Morgan, N.L., Mapleston, R.E.: Chlorophyll synthesis and the development of photosynthetic activity, pp. 111-118. In: see ref. Bücher et al. (1976)

Guignery, G., Luzzati, A., Duranton, J.: On the specific binding of protochlorophyllide and chlorophyll to different peptide chains. Planta 115, 227-243 (1974)

Hachtel, W.: In vitro synthesis of membrane proteins by isolated chloroplasts of *Vicia faba*. Ber. Dtsch. Bot. Ges. 89, 185-192 (1976)

Hagemann, R., Börner, Th.: Plastid ribosome-deficient mutants of higher plants as a tool in studying chloroplast biogenesis, pp. 709-720. In: see ref. Akoyunoglou and Argyroudi-Akoyunoglou (1978)

Harris, E.H., Preston, J.F., Eisenstadt, J.M.: Amino acid incorporation and products of protein synthesis in isolated chloroplasts of *Euglena gracilis*. Biochemistry 13, 1227-1234 (1973)

Haslett, B.G., Cammack, R., Whatleg, F.R.: Quantitative studies on ferredoxin in greening bean leaves. Biochem. J. 136, 697-703 (1973)

Hayden, D.B., Hopkins, W.G.: Membrane polypeptides and chlorophyll-protein complexes on maize mesophyll chloroplasts. Can. J. Bot. 54, 1584-1689 (1976)

Hayden, D.B., Hopkins, W.G.: A second distinct chlorophyll-protein complex in maize mesophyll chloroplasts. Can. J. Bot. 55, 2525-2529 (1977)

Hearing, V.J.: Protein synthesis in isolated etioplasts after light stimulation. Phytochemistry 12, 277-282 (1973)

Heizmann, Ph., Salvador, G.F., Nigon, V.: Occurrence of plastidal rRNAs and plastidal structures in bleached mutants of *Euglena gracilis*. Exp. Cell Res. 99, 253-260 (1976)

Heizmann, P., Verdier, G., Younis, H.: Transcription of nuclear and chloroplast genomes in *Euglena* during greening and in bleached mutants, pp. 623-628. In: see ref. Akoyunoglou and Argyroudi-Akoyunoglou (1978)

Henningsen, K.W., Kahn, A.: Photoactive subunits of protochlorophyll(ide) holochrome. Plant Physiol. 47, 685-690 (1971)

Henningsen, K.W., Thorne, S.W., Boardman, N.K.: Properties of protochlorophyllide and chlorophyll(ide) holochromes from etiolated and greening leaves. Plant Physiol. 53, 419-425 (1974)

Henriques, F., Park, R.: Further chemical and morphological characterization of chloroplast membranes from a chlorophyll b-less mutant of *Hordeum vulgare*. Plant Physiol. 55, 763-767 (1975)

Henriques, F., Park, R.: Identification of chloroplast membrane peptides with subunits of coupling factor and ribulose-1,5 disphosphate carboxylase. Arch. Biochem. Biopyhys. 176, 472-478 (1976)

Henriques, F., Park, R.: Polypeptide composition of chlorophyll-protein complexes from romaine lettuce. Plant Physiol. 60, 64-68 (1977)

Henriques, F., Park, R.B.: Characterization of three new chlorophyll-protein complexes. Biochem. Biophys. Res. Commun. 81, 1113-1118 (1978a)

Henriques, F., Park, R.B.: Spectral characterization of five chlorophyll-protein complexes. Plant Physiol. 62, 856-860 (1978b)

Henriques, F., Vaughan, W., Park, R.: High resolution gel electrophoresis of chloroplast membrane polypeptides. Plant Physiol. 55, 338-339 (1975)

Herrmann, F.H.: Genetic control of pigment-protein complexes I and Ia of the plastid mutant en:alba-1 of Antirrhinum majus. FEBS Lett. 19, 267-269 (1971)

Herrmann, F.H.: Chloroplast lamellar proteins of the plastid mutant en:viridis-1 of Antirrhinum majus having impaired photosystem II. Exp. Cell Res. 70, 452-453 (1972)

Herrmann, F.H.: Polypeptide composition of chlorophyll protein complex I from several plants, pp. 221-227. In: see ref. Akoyunoglou and Argyroudi-Akoyunoglou (1978)

Herrmann, F.H., Knoth, R.: Photosystem II deficiency in a plastid mutant of Antirrhinum majus: Biochemical and ultrastructural studies. Abstr. Int. Conf. Regul. Dev. Process. Plants, p. 64, Halle 1977

Herrmann, F.H., Meister, A.: Separation and spectroscopical properties of pigment-protein complexes in Antirrhinum chloroplasts. Photosynthetica 6, 177-182 (1972)

Herrmann, F.H., Schumann, B., Börner, Th., Knoth, R.: Struktur und Funktion der genetischen Information in den Plastiden. XII. Die plastidalen Lamellarproteine der photosynthesedefekten Plastommutanten en:gil-1 ("Mrs. Pollock") und der Genmutante "Cloth of Gold" von Pelargonium zonale Ait. Photosynthetica 10, 164-171 (1976)

Highfield, P.E., Ellis, R.J.: Synthesis and transport of the small subunit of chloroplast ribulose bisphosphate carboxylase. Nature (London) 271, 420-424 (1978)

Hiller, R.G., Genge, S., Pilger, D.: Evidence for a dimer of the light-harvesting chlorophyll-protein complex II. Plant Sci. Lett. 2, 239-242 (1974)

Hiller, R.G., Pilger, T.B.G., Genge, S.: Formation of chlorophyll protein complexes during greening of etiolated barley leaves, pp. 215-220. In: see ref. Akoyunoglou and Argyroudi-Akoyunoglou (1978)

Hoarau, J., Remy, R., Leclerc, J.C.: Hétérgéneité des variations spectrales photoinduites vers 700 nm observées sur les membranes chlorophylliennes et les complexes chlorophylle-protéines isolés de divers organismes photosynthetiques. Biochim. Biophys. Acta 462, 659-670 (1977)

Hoober, J.K.: Sites of synthesis of chloroplast membrane polypeptides in Chlamydomonas reinhardii y-1. J. Biol. Chem. 245, 4327-4334 (1970)

Hoober, J.K.: A major polypeptide of chloroplast membranes of Chlamydomonas reinhardii. J. Cell Biol. 52, 84-96 (1972)

Hoober, J.K.: Synthesis of the major thylakoid polypeptides during greening of Chlamydomonas reinhardii y-1, pp. 87-94. In: see ref. Bücher et al. (1976)

Hoober, J.K., Stegeman, W.J.: Control of the synthesis of a major polypeptide of chloroplast membranes in Chlamydomonas reinhardii. J. Cell Biol. 56, 1-12 (1973)

Hoober, J.K., Stegeman, W.J.: Kinetics and regulation of synthesis of the major polypeptides of thylakoid membranes in Chlamydomonas reinhardii y-1 at elevated temperatures. J. Cell Biol. 70, 326-337 (1976)

Horak, A., Hill, R.D.: Adenosine triphosphatase of bean plastids. Its properties and site of formation. Plant Physiol. 49, 365-370 (1972)

Hovenkamp-Obbema, R.: Effect of chloramphenicol on the development of proplastids in Euglena gracilis. II. The synthesis of carotenoids. Z. Pflanzenphysiol. 73, 439-447 (1974)

Høyer-Hansen, G., Machold, O., Kahn, A.: Polypeptide composition of internal membranes from barley etioplasts. Carlsberg Res. Commun. 41, 349-357 (1976)

Jennings, R.C., Eytan, G.: Biogenesis of chloroplast membranes XIV. Inhomogeneity of membrane protein distribution in photosystem particles obtained from Chlamydomonas reinhardii y-1. Arch. Biochem. Biophys. 159, 813-820 (1973)

Joy, K.W., Ellis, R.J.: Protein synthesis in chloroplasts. IV. Polypeptides of the chloroplast envelope. Biochim. Biophys. Acta 378, 143-151 (1975)

Kan, K., Thornber, J.P.: Light-harvesting chlorophyll a/b-protein of Chlamydomonas reinhardii. Plant Physiol. 57, 47-52 (1976)

Ke, B., Sugahara, K., Shaw, E.R.: Further purification of "Triton subchloroplast fraction I" (TSF-I) particles. Isolation of a cytochrome-free high-P-700 particle and a complex containing cytochrome f and b₆, plasto-cyanin and iron-sulfur protein(s). Biochim. Biophys. Acta 408, 12-25 (1975)

Klein, S.M., Vernon, L.P.: Protein composition of spinach chloroplasts and their photosystem I and photosystem II subfragments. Photochem. Photobiol. 19, 43-49 (1974a)

Klein, S.M., Vernon, L.P.: Polypeptide composition of photosynthetic membranes from Chlamydomonas reinhardii and Anabena variabilis. Plant Physiol. 53, 777-778 (1974b)

Klein, S.M., Vernon, L.P.: Protein arrangement in chloroplast membranes: Studies with P-diazonium-benzene-S^{35}-sulfonic acid. Ann. N.Y. Acad. Sci. 227, 568-579 (1974c)

Knoth, R., Hagemann, R.: Struktur und Funktion der genetischen Information in den Plastiden. XVI. Die Feinstruktur der Plastiden und der elektronenmikroskopische Nachweis echter Mischzellen in Blättern der Plastommutationen auslösenden Genmutante albostrians von Hordeum vulgare. Biol. Zentralbl. 69, 141-150 (1977)

Kohel, R.J., Benedict, C.R.: Plastom control of chloroplast development in cotton, Gossypium hirsutum L. Crop Sci. 12, 343-346 (1972)

Kretzer, F., Ohad, I., Bennoun, P.: Ontogeny, insertion and activation of two thylakoid peptides required for photosystem II activity in the nuclear temperature-sensitive T_4 mutant of Chlamydomonas reinhardii, pp. 25-32. In: see ref. Bücher et al. (1976)

Krogmann, D.W.: The organization of photosynthetic enzymes on the chloroplast membrane. In: The Enzymes of Biological Membranes; Martonosi (ed.), pp. 143-166. New York: Plenum Publ. Corp. 1976

Kung, S.D., Thornber, J.P.: Photosystem I and II chlorophyll-protein complexes of higher plant chloroplasts. Biochim. Biophys. Acta 253, 285-289 (1971)

Kung, S.D., Thornber, J.P., Wildman, S.G.: Nuclear DNA codes for the photosystem II chlorophyll-protein of chloroplast membranes. FEBS Lett. 24, 185-188 (1972)

Kwanyuen, P. Wildman, S.G.: Nuclear DNA codes for Nicotiana ferredoxin. Biochim. Biophys. Acta 405, 167-174 (1975)

Laemmli, U.K.: Cleavage of structural protein during the assembly of the head of bacteriophage T_4. Nature (London) 227, 680-685 (1970)

Lagoutte, B., Duranton, J.: Action of light at the structural protein level on etiolated plastids from Zea mays. FEBS Lett. 28, 333-336 (1972)

Largett-Barley, I., Kreutz, W.: Characterization of pigment-protein complexes related to photosystem I and II. Prog. Photosynth. Res. 1, 149-158 (1969)

Leech, R.M.: Etioplast structure and its relevance to chloroplast development. Biochem. Soc. Trans. 5, 81-84 (1977)

Levine, R.P., Armstrong, J.: The site of synthesis of two chloroplast cytochromes in Chlamydomonas reinhardii. Plant Physiol. 49, 661-662 (1972)

Levine, R.P., Duram, H.A.: The polypeptides of stacked and unstacked Chlamydomonas reinhardii chloroplast membrane and relationship to photosystem II activity. Biochim. Biophys. Acta 325, 565-572 (1973)

Levine, R.P., Burton, W.G., Duram, H.A.: Membrane polypeptides associated with photochemical systems. Nature (London) 237, 176-177 (1972)

Liebers, H., Parthier, B.: Synthese der Lamellarproteine in Chloroplasten ergrünender Euglena gracilis. BPP 165, 517-530 (1974)

Lokshin, A., Falk, R.H., Bogorad, L., Woodcock, C.L.F.: A coupling factor for photosynthetic phosphorylation from plastids of light- and dark-grown maize. Biochim. Biophys. Acta 226, 366-382 (1971)

Lürssen, K.: Zur Lokalisierung der Proteinsynthese bei der Entwicklung des Lamellarsystems ergrünender Etioplasten. Z. Naturforsch. 26b, 725-729 (1971)

Lütz, C.: Biochemische und cytologische Untersuchungen zur Chloroplastenentwicklung. I. Die chemische Charakterisierung der Prolamellarkörper aus Etioplasten von Avena sativa L. Z. Pflanzenphysiol. 75, 346-359 (1975)

Machold, O.: Molekulargewichtsbestimmung von Chloroplasten-Proteinen durch Dodecylsulfat-Gelelektrophorese. Biochem. Physiol. Pflanz. 166, 149-162 (1974)

Machold, O.: On the molecular nature of chloroplast thylakoid membranes. Biochim. Biophys. Acta 382, 494-505 (1975)

Machold, O., Aurich, O.: Sites of synthesis of chloroplast lamellar proteins in Vicia faba. Biochim. Biophys. Acta 281, 103-112 (1972)

Machold, O., Høyer-Hansen, G.: Polypeptide composition of thylakoids from viridis and xantha mutants in barley. Carlsberg Res. Commun. 41, 359-366 (1976)

Machold, O., Meister, A., Adler, K.: Spektroskopische Eigenschaften von elektrophoretisch getrennten Chlorophyll-Protein-Komplexen der Photosysteme I und II aus *Vicia faba* and *Chlorella pyrenoidosa.* Photosynthetica 5, 160-165 (1971)

Machold, O., Meister, A., Sagromsky, H., Høyer-Hansen, G., von Wettstein, D.: Composition of photosynthetic membranes of wild type barley and chlorophyll b-less mutants. Photosynthetica 11, 200-206 (1977)

Malkin, R.: Photochemical properties of a photosystem I. subchloroplast fragment. Arch. Biochem. Biophys. 169, 77-83 (1975)

Margulies, M.M., Michaels, A.: Ribosomes bound to chloroplast membranes in *Chlamydomonas reinhardii.* J. Cell Biol. 60, 65-77 (1974)

Margulies, M.M., Tiffany, H.L., Michaels, A.: Vectorial discharge of nascent polypeptides attached to chloroplast thylakoid membranes. Biochem. Biophys. Res. Commun. 64, 735-739 (1975)

Markwell, J.P., Reinman, S., Thornber, J.P.: Chlorophyll-protein complexes from higher plants: a procedure for improved stability and fractionation. Arch. Biochem. Biophys. 190, 136-141 (1978)

Markwell, J.P., Thornber, J.Ph., Boggs, R.T.: Higher plant chloroplasts: evidence that all the chlorophyll exists as chlorophyll-protein complexes. Proc. Natl. Acad. Sci. USA 76, 1233-1235 (1979a)

Markwell, J.P., Miles, G.D., Boggs, R.T., Thornber, J.P.: Solubilization of chloroplast membranes by zwitterionic detergents. Effect on photosystem II activity. FEBS Lett. 99, 11-14 (1979b)

McEvoy, F.A., Lynn, W.S.: The peptides of chloroplast membranes. I. The soluble coupling factor (Ca^{2+}-ATPase). Arch. Biochem. Biophys. 156, 335-341 (1973)

McMahon, D.: Cycloheximide is not a specific inhibitor of protein synthesis in vivo. Plant Physiol. 55, 815-821 (1975)

Mendiola-Morgenthaler, C.R., Morgenthaler, J.J., Price, C.A.: Synthesis of coupling factor CF_1 protein by isolated spinach chloroplasts. FEBS Lett. 62, 96-100 (1976)

Michaels, A., Gibor, A.: Effects of ultraviolet irradiation on plastid continuity in *Euglena gracilis* (Abstr.) Am. J. Bot. 59, 654 (1972)

Miles, C.D., Markwell, J.P., Thornber, J.P.: Effect of nuclear mutation in maize on photosynthetic activity and content of chlorophyll-protein complexes. Plant Physiol. 64, 690-694 (1979)

Miller, K.R., Staehelin, L.A.: Analysis of the thylakoid outer surface. Coupling factor is limited to unstacked membrane regions. J. Cell Biol. 68, 30-47 (1976)

Morgan, N.L., Griffiths, W.T.: The development of photosystem I in greening barley. Biochem. Soc. Trans. 3, 391-392 (1975)

Morgenthaler, J.J., Mendiola-Morgenthaler, C.: Synthesis of soluble, thylakoid and envelope membrane proteins by spinach chloroplasts purified from gradients. Arch. Biochem. Biophys. 172, 51-58 (1976)

Nakamura, K., Ogawa, T., Shibata, K.: Chlorophyll and peptide compositions in the two photosystems of marine green algae. Biochim. Biophys. Acta 423, 227-236 (1976)

Nelson, N.: Structure and function of chloroplast ATPase. Biochim. Biophys. Acta 456, 314-338 (1976)

Neumann, D., Parthier, B.: Effects of nalidixic acid, chloramphenicol, cycloheximide, and anisomycin on structure and development of plastids and mitochondria in greening *Euglena gracilis.* Exp. Cell Res. 81, 255-268 (1973)

Nielsen, N.C.: Electrophoretic characterization of membrane proteins during chloroplast development in barley. Eur. J. Biochem. 50, 611-623 (1975)

Ogawa, T., Obata, F., Shibata, K.: Two pigment-proteins in spinach Chloroplasts. Biochim. Biophys. Acta 112, 223-234 (1966)

Ogawa, T., Bovey, F., Inoue, Y., Shibata, K.: Early stages of greening in etiolated bean leaves. In: Proc. 3rd Int. Congr. Photosynth; Avron, M. (ed.), pp. 1829-1832. Amsterdam: Elsevier 1975

Ohad, I.: Biogenesis and modulation of chloroplast membranes. In: Role of Membranes in Secretory Processes; Bolis, L., Keynes, R.D., Wilbrand, W. (eds.), pp. 24-51. Amsterdam: North-Holland 1972

Ohad, I.: Biogenesis of chloroplast membranes. In: Membrane Biogenesis; Mitochondria, Chloroplasts, and Bacteria; Tzagoloff (ed.), pp. 279-350. New York-London: Plenum Press 1975

Oleszko, S., Moudrianakis, E.N.: The visualization of the photosynthetic coupling factor in embedded spinach chloroplasts. J. Cell Biol. 63, 936-948 (1974)

Parthier, B., Neumann, D.: Structural and functional analysis of some plastid mutants of *Euglena gracilis*. Biochem. Physiol. Pflanz. 171, 547-562 (1977)

Philippovich, I.J., Bezsmertnaya, I.N., Oparin, A.I.: On the localization of polyribsomes in the system of chloroplast lamellae. Exp. Cell Res. 79, 159-168 (1973)

Picaud, A., Acker, S.: Etude de la structure de membranes chloroplastiques isolées de la souche sauvage et d'un mutant sans activité du photosystème I de *Chlamydomonas reinhardii*. FEBS Lett. 54, 13-17 (1975)

Plesničar, M., Bendall, D.S.: The photochemical activities and electron carriers of developing barley leaves. Biochem. J. 136, 803-812 (1973)

Reardon, E.M., Bartolf, M., Ortiz, W., Santoro, D., Zielinski, R., Price, C.A.: Isolation of chloroplasts active in protein synthesis from spinach and Euglena, pp. 277-282. In: see ref. Akoyunoglou and Argyroudi-Akoyunoglou (1978)

Regitz, G., Ohad, I.: Changes in the protein organization in developing thylakoid of *Chlamydomonas reinhardii y-1* as shown by sensitivity to trypsin. In: Proc. 3rd. Int. Congr. Photosynth.; Avron, M. (ed.), pp. 1615-1625. Amsterdam: Elsevier 1975

Reinman, S., Thornber, J.P.: The electrophoretic isolation and partial characterization of three chlorophyll-protein complexes from blue-green algae. Biochim. Biophys. Acta 547, 188-197 (1979)

Remy, R.: Pre-existence of chloroplast lamellar proteins in wheat etioplasts. Functional and protein changes during greening under continuous or intermitted light. FEBS Lett. 31, 308-312 (1973)

Remy, R., Hoarau, J.: New forms of chlorophyll-protein complexes from thylakoids of different photosynthesizing organisms, pp. 235-240. In: see ref. Akoyunoglou and Argyroudi-Akoyunoglou (1978)

Remy, R., Hoarau, J., Leclerc, J.C.: Electrophoretic and spectrophotometric studies of chlorophyll-protein complexes from tobacco chloroplasts. Isolation of a light-harvesting pigment protein complex with a molecular weight of 70.000. Photochem. Photobiol. 26, 151-158 (1977)

Robertson, D., Laetsch, W.M.: Structure and function of developing barley plastids. Plant Physiol. 54, 148-149 (1974)

Schäfers, H.-A., Feierabend, J.: Ultrastructural differentiation of plastids and other organelles in rye leaves with a high-temperature-induced deficiency of plastid ribosomes. Cytobiology 14, 75-90 (1976)

Schatz, G.: The biogenesis of mitochondria—a review. In: Genetics, Biogenesis and Bioenergetics of Mitochondria; Bandlow, W., Schweyen, R.J., Thomas, T.Y., Wolf, K., Kaudewitz, F. (eds.), pp. 163-177. Berlin-New York: de Gruyter 1976

Schiemann, J., Wollgiehn, R., Parthier, B.: Isolation of a transcription-active RNA polymerase-DNA complex from *Euglena* chloroplasts. Biochem. Physiol. Pflanz. 171, 474-478 (1977)

Schiff, J., Lyman, H., Russel, S.K.: Isolation of mutants from *Euglena gracilis*. In: Methods in Enzymology, Vol. XXIII; San Pietro, A. (ed.), Photosynthesis, Part A, pp. 143-162. New York-London: Academic Press 1971

Schlender, K.K., Sell, H.M., Bukovac, M.J.: Inhibition of selected plant systems by stereoisomers of chloramphenicol. Phytochemistry 11, 2949-2956 (1972)

Schmid, G.H., Radunz, A., Menke, W.: Localization and function of cytochrome f in the thylakoid membrane. Z. Naturforsch. 32c, 271-280 (1977)

Schmid, G.H., Menke, W., Radunz, A., Koenig, F.: Polypeptides of the thylakoid membrane and their functional characterization. Z. Naturforsch. 33c, 722-730 (1978)

Schötz, F.: Effects of the disharmony between genome and plastom on the differentiation of the thylakoid system in *Oenothera*. Symp. Soc. Exp. Biol. 24, 39-54 (1970)

Schopfer, P., Siegelman, H.W.: Purification of protochlorophyllide holochrome. Plant Physiol. 43, 990-996 (1968)

Schumann, B., Börner, Th.: Über den Zusammenhang zwischen Lamellarproteinen und Thylakoidmorphologie bei der Biogenese grüner, etiolierter und mutierter Plastiden von *Hordeum, Pelargonium* und *Lycopersicon*. Acta Histochim. Suppl. 17, 153-155 (1976)

Schumann, B., Herrmann, F., Börner, Th., Hagemann, R.: Separation of the photosystem I chlorophyll-protein 1 complex into several components. Photosynthetica 9, 410-411 (1975)

Senior, A.E.: The structure of mitochondrial ATPase. Biochim. Biophys. Acta 301, 249-277 (1973)

Shiozawa, J.A., Alberte, R.S., Thornber, J.F.: The P700-chlorophyll a-proteins. Arch. Biochem. Biophys. 165, 388-397 (1974)

Sidell, S.G., Ellis, R.J.: Protein synthesis in chloroplasts. Characteristics and products of protein synthesis in vitro in etioplasts and developing chloroplasts from pea leaves. Biochem. J. 146, 675-685 (1975)

Sidell, S.G., Ellis, R.J.: Protein synthesis by etioplasts. Biochem. Soc. Trans. 5, 98-102 (1977)

Simpson, D., Moller, B.L., Høyer-Hansen, G.: Freeze fracture structure and polypeptide composition of thylakoids of wild-type and mutant barley plastids, pp. 507-512. In: see ref. Akoyunoglou and Argyroudi-Akoyunoglou (1978)

Sironval, C., Clijsters, H., Michel, J.-M., Bronchart, R., Michel-Wodwert, M.-R.: Sur la séparation de deux fractions a partir des membranes des chloroplasts (systemes I and II), sur leurs propriétés sur l'organization et le functionnement de ces membranes. In: Le Chloroplaste; Sironval, C. (ed.), pp. 99-123. Paris: Masson 1967

Smith-Johannsen, H., Gibbs, S.P.: Effects of chloramphenicol on chloroplast and mitochondrial ultrastructure in *Ochromonas danica*. J. Cell Biol. 52, 598-614 (1972)

Spencer, D., Whitfeld, P.R., Bottomley, W., Wheeler, A.M.: The nature of the proteins and nucleic acids synthesized by isolated chloroplasts. In: Autonomy and Biogenesis of Mitochondria and Chloroplasts; Boardman, N.K., Linnane, A.W., Smillie, R.M. (eds.), pp. 372-382. Amsterdam-London: North-Holland 1971

Sprey, B.: Ribosomale RNA and Thylakoidmembranen in Plastiden von Chlorophylldefektmutanten der Gerste. Z. Pflanzenphysiol. 67, 223-243 (1972)

Srivastava, L.M., Vesk, M., Singh, A.P.: Effect of chloramphenicol on membrane transformation in plastids. Can. J. Bot. 49, 587-593 (1971)

Süss, K.H.: Identification of chloroplast thylakoid membrane polypeptides: coupling factor of photophosphorylation (CF_1) and cytochrome f. FEBS Lett. 70, 191-196 (1976)

Süss, K.H.: Untersuchungen über Struktur und Funktion von Proteinen der Thylakoidmembranen der Chloroplasten. Thesis A, Universität Halle, 1978

Süss, K.H., Damaschun, H., Damaschun, G., Zirwer, D.: Chloroplast coupling factor CF_1 in solution. Small-angle X-ray scattering and circular dichroism measurement. FEBS Lett. 87, 265-268 (1978)

Surzycki, S.J., Goodenough, U.W., Levine, R.P., Armstrong, J.J.: Nuclear and chloroplast control of chloroplast structure and function in *Chlamydomonas reinhardii*. Symp. Soc. Exp. Biol. 24, 13-17 (1970)

Takagi, T., Tsujii, K., Shirahama, K.: Binding isotherms of sodium dodecyl sulphate to protein polypeptides with special reference to SDS-polyacrylamide gel elctrophoreses. J. Biochem. 77, 939-947 (1975)

Thornber, J.P.: Comparison of a chlorophyll a-protein complex isolated from a blue-green bacteria and higher plants. Biochim. Biophys. Acta 172, 230-241 (1969)

Thornber, J.P.: Chlorophyll-proteins: Light-harvesting and reaction center components of plants. Annu. Rev. Plant Phys. 26, 127-158 (1975)

Thornber, J., Alberte, R.S.: Chlorophyll-proteins: membrane-bound photoreceptor complexes in plants. In: The Enzymes of Biological Membranes; Martonosi, A. (ed.), pp. 163-190. New York: Plenum Publ. Corp. 1976

Thornber, J.P., Highkin, H.R.: Composition of the photosynthetic apparatus of normal barley leaves and a mutant lacking chlorophyll b. Eur. J. Biochem. 41, 109-116 (1974)

Thornber, J.P., Olson, J.M.: Chlorophyll-proteins and reaction center preparations from photosynthetic bacteria, algae and higher plants. Photochem. Photobiol. 14, 329-341 (1971)

Thornber, J.P., Gregory, R.P.F., Smith, C.A., Bailey, G.L.: Studies on the nature of the chloroplast lamellae. I. Preparations and some properties of two chlorophyll-protein complexes. Biochemistry 6, 391-396 (1967a)

Thornber, J.P., Steward, J.C., Hatton, M.W.C., Bailey, J.L.: Nature of chloroplast lamellae. II. Chemical composition and further physical properties of two chlorophyll-protein complexes. Biochemistry 6, 2006-2014 (1967b)

Thornber, J.P., Markwell, J.P., Reinman, S.: Plant chlorophyll-protein complexes: Recent advances. Photochem. Photobiol., in press (1979)

Toyama, S.: Electron microscope studies on the morphogenesis in plastids. VI. Plastid development and fine structure in variegated leaves of tomato. Bot. Mag. 85, 1-10 (1972)

Trebst, A.: Energy conservation in photosynthetic electron transport of chloroplast. Annu. Rev. Plant Physiol. 25, 423-458 (1974)

Vaisberg, A.J., Schiff, J.A., Li, L., Freedman, Z.: Events surrounding the early development of Euglena chloroplasts. VIII. Photocontrol of the source of reducing power for chloramphenicol reduction by the ferredoxin-NADP reductase system. Plant Physiol. 57, 594-601 (1976)

Vasconcelos, A.C., Mendiola-Morgenthaler, L.R., Floyd, G.L., Salisbury, J.L.: Fractionation and analysis of polypeptides of Euglena gracilis chloroplast. Plant Physiol. 58, 87-90 (1976)

Vernon, L.P., Klein, S.M.: Nature of plant chlorophyll in vivo and their associated protein. Ann. N.Y. Acad. Sci. 244, 281-296 (1975)

Vernon, L.P., Shaw, E.R., Ogawa, T., Raveed, D.: Structure of photosystem I and photosystem II of plant chloroplasts. Photochem. Photobiol. 14, 343-357 (1971)

Volodarsky, A.D., Chaika, M.T., Abramchik, L.M., Chayanova, S.S., Tikhonovskaya, N.G., Savchenko, G.E.: Immunochemical characteristics of pigment-protein components from membranes of etioplasts and chloroplasts (russ.) Fiziol. Rast. 23, 1207-1213 (1976)

Walles, B., Kronestedt, E.: The autonomy of plastids in Euglena gracilis as elucidated by electron microscopy of bleached strains. Port. Acta Biol. A14, 201-214 (1974)

Weier, T.E., Brown, D.L.: Formation of the prolamellar body in 8-day, dark-grown seedlings. Am. J. Bot. 57, 267-275 (1970)

Wellburn, A.R.: Distribution of chloroplast coupling factor (CF_1) particles on plastid membranes during development. Planta 135, 191-198 (1977)

Wessels, J.S.G., Borchert, M.T.: Studies on subchloroplastic particles. Similarity of grana and stroma photosystem I and the protein composition of photosystem I and photosystem II particles. In: Proc. 3rd Int. Congr. Photosynth.; Avron, M. (ed.), pp. 473-484. Amsterdam: Elsevier 1975

Wessels, J.S.C., Borchert, M.T.: Polypeptide profiles of chlorophyll-protein complexes and thylakoid membranes of spinach chloroplasts. Biochim. Biophys. Acta (B) 503, 78-93 (1978)

Wildner, G.F.: The greening process in Euglena gracilis. I. The kinetics of appearance of chloroplast proteins and the effect of cycloheximide and chloramphenicol on their synthesis. Z. Naturforsch. 31c, 157-162 (1976)

Fraction I Protein

W. BOTTOMLEY

Division of Plant Industry, CSIRO, Canberra, Australia

I. Introduction

When examining the soluble protein from spinach leaves in the ultracentrifuge, Wildman and Bonner in 1947 coined the term Fraction I protein to designate a high molecular weight protein which occurred in very high yield. Subsequent studies in a number of laboratories have shown that this protein contributes up to 50% of the soluble protein of leaves and is probably the most abundant protein in nature (Kawashima and Wildman 1970a). It occurs with remarkably little variation in structure in all plants containing chlorophyll a, ranging from *Euglena* and the blue-green algae to all higher plants examined. Later Weissbach et al. (1956) succeeded in purifying the enzyme responsible for the first step in the Calvin photosynthetic cycle, the carboxylation of ribulosebisphosphate with CO_2 to form 3-phosphoglycerid acid, and it became apparent that the two proteins had very similar physical properties. Subsequently they have been shown to be identical. Hence Fraction I protein is responsible for catalyzing the initial step of photosynthesis, the process on which all life as we know it depends. More recently (Bowes et al. 1971), this enzyme has been found to be responsible for a second important activity in the plant, that of ribulosebisphosphate oxygenase, an enzyme activity which converts RuBP in the presence of oxygen to 3-phospho-D-glycerate and phosphoglycolate and is the first step of photorespiration.

The early work on Fraction I has already been covered by the extensive reviews of Akazawa (1970) and Kawashima and Wildman (1970a) while the importance of Fraction I as a marker in evolutionary genetics has recently been reviewed by Kung (1976).

II. Structure of Fraction I Protein

All Fraction I proteins so far isolated from higher plants have a sedimentation constant of around 18S corresponding to a molecular weight of 5.6×10^5 daltons (Kawashima and Wildman 1970a; Kung 1976). Treatment with sodium dodecylsulphate results in the dissociation of Fraction I proteins into two sizes of subunits, a large subunit of 56,000 daltons and a small subunit of 12,000 daltons (Kung et al. 1974). Earlier there was some confusion concerning the molecular

---

Abbreviation. *RuBP:* Ribulosebisphosphate

weight of the small subunit since gel filtration in alkaline solutions suggested a molecular weight of 24,000 daltons leading to the suggestion (Ellis 1973) that the peptide formed dimers in alkaline solutions. However, Roy et al. (1976a) measured the intrinsic viscosity of the small subunit and found this to be inconsistent with its existence as a dimer and suggested that, in alkaline solutions, the small subunit exists as a partially unfolded monomer. In an extensive study of crystalline Fraction I protein from tobacco (Baker et al. 1975, 1977), using X-ray diffraction, electron microscopy, and optical diffraction concluded that the holoprotein was made up of eight large subunits and eight small subunits arranged in two layers each with four subunit pairs. Recently, it has been demonstrated (Kung et al. 1974; Chen et al. 1976b) that when Fraction I protein from a variety of plants was examined by isoelectric focusing, the large subunits of each plant could be separated into three distinct polypeptides, while the small subunits yielded from one to four different polypeptides. Interspecific comparisons showed that the particular set of large and of small subunits could vary from one species to another.

Nicotiana tabacum and *N. pluboginifolia* each contain small subunits with different electrophoretic mobilities. The Fraction I protein from a hybrid of the two has been shown to be heterogeneous in charge when examined by two-dimensional agarose gel electrophoresis (Hirai 1977). This indicates that each Fraction I molecule is assembled using random small unit molecules rather than each holoprotein containing only one type of small subunit.

A comparison of the amino acid composition of the large and small subunits from a number of plants revealed that the amino acid composition of the large subunits from the higher plants (Kawashima and Wildman 1970a; Sugiyama and Akazawa 1970; Kawashima et al. 1971; Strobaek and Gibbons 1976), as well as those from *Chlorella ellipsoides* (Sugiyama et al. 1971) and *Euglena gracilis* (Rabinowitz et al. 1975), were very similar, whereas the small subunits vary widely in their amino acid composition. This similarity in the large subunits from different species is also apparent from their immunological properties, since antisera to the large subunits from different genera exhibit cross reactivity to one another, whereas the small subunit antisera show no common determinants with either small subunits from other species or the large subunit from the same species (Kawashima and Wildman 1971b; Kawashima et al. 1971; Gray and Kekwick 1974a; Brown et al. 1976).

It has been suggested that the lack of variability of the large subunit results from the cistrons being contained by the chloroplast DNA which is polyploid and therefore a barrier to mutation (Kirk 1972). The small subunit, which is coded for in the nuclear DNA is, as expected, more variable since this DNA is more responsive to mutational forces. Chen et al. (1976a) have calculated that the cistron coding for the large Fraction I subunits of *Nicotiana* spp. from Australia has not had one surviving mutation in 10^8 years, whereas the small subunit has a mutation rate similar to that of cytochrome c. However, examination of the tryptic peptide maps of four species of *Nicotiana* has led to the conclusion that the mutation rate is, in fact, higher than that predicted on the basis of the isoelectric points (Kung and Lee 1977).

While the structure of Fraction I proteins from higher plants and the green and most blue-green algae (Rabinowitz et al. 1975; Takabe et al. 1976) conform to the above generalizations, RuBP carboxylase from photosynthetic bacteria are more variable. Those from *Rhodospirillum rubrum Rhodopseudomonas* spp., and *Chlorobium* spp. are significantly smaller molecules (Anderson et al. 1968; Tabita and McFadden 1972; Tabita et al. 1974) while the enzyme from *Hydrogenomonas*, a chemoautotroph, contains RuBP carboxylase which was originally reported to consist of one subunit of 40,000 daltons (McFadden 1973). Recently Purohit and McFadden (1976) have shown that there are two large subunits of 52,000 and 56,000 daltons, occurring in a mol ratio of 3:5, as well as one small subunit of 15,000 daltons. Also Gibson and Tabita (1977) have isolated two forms of ribulose diphosphate carboxylase from *Rhodopseudomonas sphaeroides*. One form, like those from *Rhodospirillum* and *Chlorobium*, contains only the large subunit of about 54,000 daltons, the other form resembled higher plant enzymes by containing both a large and a small subunit. The RuBP carboxylase from the blue-green alga *Agmenellum quadriplicatium* of 456,000 daltons consists of only one subunit which has a M.W. of 56,000 daltons (Tabita et al. 1974).

Enzymes with oxygenase activity often contain cofactors such as copper, iron, or flavin nucleotides (Hayaishi 1974; Pistorius and Axelrod 1974). A report that RuBP carboxylase from spinach contained 1 g-atom of bound copper per mol of enzyme (Wishnick et al. 1969, 1970) was queried by Lorimer et al. (1973) who found less than 0.2 g-atom per mol. Chollet et al. (1975) have demonstrated the absence of bound copper, iron, or flavin nucleotide in crystalline Fraction I from tobacco.

III. The Isolation of Fraction I Protein

The most widely used procedures for the isolation and purification of Fraction I include deionization of crude leaf extracts by gel filtration followed by purification on Sephadex G 200 or Sepharose 6B and subsequent ion exchange chromatography with DEAE-cellulose (Trown 1965; Steer et al. 1968; Sugiyama et al. 1968; Kawashima 1969; Kawashima and Wildman 1971; Haslett et al. 1976; Strobaek and Gibbons 1976). Alternatively, density gradient centrifugation on gradients of sorbitol (Kleinkopf et al. 1970a) or sucrose (Goldthwaite and Bogorad 1971; Givan and Criddle 1972; McFadden et al. 1975; Takabe et al. 1976) takes advantage of the high molecular weight (560,000 daltons) of Fraction I to achieve considerable purification. Gray and Kekwick (1973b) have pointed out the desirability of the presence of di-isopropyl phosphofluoridate in the homogenization medium to prevent degradation, particularly of the large subunit.

An elegant method for the preparation and recrystallization of pure Fraction I protein from *Nicotiana* spp. has been described (Chan et al. 1972). This method takes advantage of the fact that the protein is very soluble in the presence of ribulosebisphosphate but insoluble in its absence when Mg^{2+} and HCO_3^- are present (Kwok et al. 1971). This method has been adapted to the isolation of large quantities of Fraction I protein from tobacco by Lowe (1977).

The subunits are readily separated by gel-filtration on Sephadex G-100 in the presence of urea (Moon and Thompson 1969), sodium dodecyl sulphate (Rutner and Lane 1967; Gray and Kekwick 1974a) or alkali (Kawashima and Wildman 1970b).

Recently Gray and Wildman (1976) have described a method of isolation of the mixed subunits by use of a column containing Fraction I antibodies linked covalently to Sepharose 4B, in which the crude extracts are passed through the column and the mixture of pure subunits eluted by dissociation of the antigen-antibody complexes with 8M urea.

IV. Localization of Fraction I Protein

Chloroplasts contain a highly complex internal structure within a surrounding double membrane envelope. Within this envelope is the stroma or mobile phase together with an extensive network of thylakoid membranes which may be appressed at intervals to form granal stacks. If the external membrane of the chloroplast is damaged to any extent by osmotic or physical shock, then Fraction I is readily lost to the surrounding medium. The ease with which this occurs, together with the fact that the chlorophyll-bearing membrane fraction of the chloroplast can readily be washed free of Fraction I, make it clear that this protein is localized in, and indeed must be the principal constituent of, the stroma of the chloroplast (Lyttleton and Ts'o 1958).

In 1975 Kagan-Zur and Lips found that, when the leaves of *Tropaeolum majus* were kept in the dark for extended periods prior to organelle isolation, the RuBP carboxylase activity appeared to be associated with a particulate fraction sedimenting ahead of the chloroplasts on sucrose gradients. Adding this particulate fraction back to chloroplasts from leaves kept in the dark, but not those from preilluminated leaves, stimulated CO_2 fixation. They suggest that these "microbodies" contain the Calvin cycle enzymes which become associated with the chloroplasts on exposure to light and, in the dark, dissociate and become distinct cytoplasmic entities. While it seems difficult to reconcile this proposal with the large amount of Fraction I protein which occurs in most chloroplasts, comprising about 50% of the chloroplast protein, it has already been suggested that Fraction I protein may occur in the form of a multi-enzmyme complex with other Calvin cycle enzymes (Kawashima and Wildman 1970a). In addition, it has been observed by phase contrast microscopy of living cells that parts of the mobile phase appear to dissociate from and reassociate with the main chloroplast structure (Wildman et al. 1962).

V. The Function of Fraction I Protein

It is now well established that Fraction I protein catalyzes the CO_2-fixing step in the Calvin cycle (Akazawa 1970; Kawashima and Wildman 1970a) in which one molecule of ribulosebisphosphate reacts with one molecule of CO_2 to yield two molecules of 3-phosphoglyceric acid. However, until recently, the observed Km [CO_2] of around $500\mu M$ for isolated Fraction I was far too high to account for the rate at which fixation occurs in vivo. In 1974, Bahr and Jensen, by careful rupture of spinach chloroplasts in hypotonic medium obtained a RuBP carboxylase activity with a Km [CO_2] of $11\text{-}18\mu M$ which is low enough to account for the in vivo rate.

However, this form underwent spontaneous transition to a high Km form. It has been observed that preincubation of the enzyme with CO_2 and Mg^{2+} caused an increase in the carboxylase activity (Pon et al. 1963; Murai and Akazawa 1972; Chu and Bassham 1973). Andrews et al. (1975) demonstrated that this preincubation converted a high Km $[CO_2]$ form of the enzyme to a low Km $[CO_2]$ form in which the activity was sufficient to account for the observed rates of CO_2 fixation in vivo. The kinetics of the activation have been examined by Lorimer et al. (1976) who showed that it was completely reversible, that CO_2 itself rather than HCO_3^- was the active species in the activation, and that this activation involved the formation of a ternary complex of enzyme-CO_2-Mg^{2+}. They also suggested that the activating CO_2 molecule does not take part in the carboxylase reaction. Walter and Lilley (1976) have also concluded that the low Km $[CO_2]$ and high maximal velocity of the reaction are sufficient to account for the in vivo activity of the enzyme.

In 1971 Bowes et al., demonstrated that purified Fraction I protein from soybeans possessed, in addition to its carboxylase activity, an oxygenase function. This converted RuBP to one molecule of phosphoglyceric acid and one molecule of phosphoglycollic acid. The dual catalytic function was later shown to be common to the carboxylases of *Euglena* (McFadden et al. 1975) as well as the anaerobic photo-organotrophic bacterium *Rhodospirillum rubrum* (McFadden 1974). The discovery of the oxygenation reaction immediately raised the possibility that this would constitute the first step in the photorespiratory pathway. As with the carboxylase reaction, it was initially thought the reaction was too slow to support photorespiration (Zelitch 1973) but Andrews et al. (1975) again showed that pre-incubation with CO_2 and Mg^{2+} activated the oxygenase function to a level which is compatible with observed in vivo rates of photorespiration.

Fraction I protein thus is responsible for the initial stages of both photosynthesis and photorespiration. Since the balance between these reactions is a significant factor controlling the productivity of crops, considerable efforts have been directed toward improving that balance by breeding to reduce or eliminate carbon losses caused by photorespiration. Lorimer and Andrews (1979) have offered the interesting suggestion that the oxygenase activity is an inevitable consequence of the increasing oxygen level in the atmosphere during evolution and that both the CO_2 and the oxygen react with the same carbanion intermediate derived from RuDP. The predictions arising from this suggestion are, firstly, that all RuDP carboxylases should also have an oxygenase function and secondly that it is unlikely that the oxygenase function can be bred out of crop plants without a consequent loss of CO_2 fixing ability. So far neither of these predictions has been disproved. In fact the observation of oxygenase activity in RuBP carboxylase from *Rhodospirillum rubrum* (McFadden 1974) which carries out anaerobic photosynthesis, supports the idea that the oxygenase activity may be an unavoidable adjunct to carboxylase activity. *R. rubrum* RuBP carboxylase contains only one subunit (M.W. 5.7×10^4 daltons) which suggests that the two enzymatic activities reside with the large subunit of Fraction I. This idea is supported by the work of Gray and Kekwick (1974a) who found that the RuBP carboxylase activity of Fraction I from *Phaseolus vulgaris* was inhibited by the antiserum to the large subunit but not by the antiserum to the small subunit. In addition, Takabe and Akazawa (1973) showed that the large subunit of Fraction I from *Chromatium* retains up to 14% of the carboxylase activity after separation from the small subunit. Badger and Lorimer (1976) have demonstrated that the activation of the oxygenase function by Mg^{2+} and CO_2 which indicates that the oxygenase and carboxylase functions are tightly coupled. The recent claim by Brändén (1978) that in parsley the carboxylase activity could be separated from the oxygenase activity by gel filtration has not been confirmed by McCurry et al. (1978).

VI. The Occurrence of Fraction I in C_4 Plants

Following the demonstration that some tropical grasses possessed a novel pathway of CO_2 fixation (Slack and Hatch 1967) the existence has been recognized of a large class of plants which depend for their primary CO_2 fixation on PEP carboxylase rather than RuBP carboxylase (Hatch 1976). These plants carry out primary fixation in the mesophyll cells to yield the C_4 dicarboxylic acids-malic acid and aspartic acid. The malate and aspartate are transported to the bundle-sheath cells where they are decarboxylated and the CO_2 is produced refixed by the Calvin cycle via RuBP carboxylase.

Although it soon became apparent that the RuBP carboxylase activity in the leaves of C_4 plants was very low (Slack and Hatch 1967) it was not until 1976 that Huber et al. demonstrated that, whereas the bundle-sheath chloroplasts contained Fraction I, the mesophyll chloroplasts were completely lacking this protein or its subunits. In addition, they found that the ratio of soluble protein to chlorophyll in the mesophyll was considerably lower than that in bundle-sheath cells.

It has been observed (Bottomley, unpubl. results) that, unlike spinach chloroplasts, isolated mesophyll chloroplasts from a number of C_4 plants do not incorporate amino acids into the large subunit of Fraction I. As will be discussed later, the genetic message for the large subunit of Fraction I is contained in the chloroplast DNA and so this lack of expression of the genetic message for the large subunit in the mesophyll chloroplasts may prove a useful tool in the study of factors which control the expression of the genetic information of chloroplasts.

VII. Site of Synthesis of Fraction I Protein

Until relatively recently, one of the main questions in chloroplast biochemistry was "Is the chloroplast an autonomous, self-perpetuating organelle?". Since it was established around 1965 that chloroplasts contained all the necessary components for autonomy, namely DNA, DNA-dependent RNA polymerase, and the complete machinery for protein synthesis (Kirk and Tilney-Bassett 1967) it appeared possible that chloroplast DNA contained the information to code for all chloroplast constituents and that chloroplasts could transcribe and translate this DNA into the unique enzymes and structural proteins which were required for organelle development and for the functioning of the photosynthetic apparatus. Since Fraction I protein was the most abundant of the proteins known to be associated with chloroplasts, and ribulosebisphosphate carboxylase activity was relatively simple to assay, it was not surprising that this was one of the first proteins used to study the site of its synthesis and also the site of the genetic message coding for its structure. The three main approaches to the question of whether Fraction I protein is synthesized within the chloroplast on the 70S ribosomal system or on the 80S ribosomal system of the cytoplasm have been (1) the in vivo use of inhibitors which differentially inhibit the two translating systems, (2) the use of isolated chloroplast preparations to make specific polypeptides in the absence of cytoplasmic factors, and (3) isolation of chloroplast polyribosomes and immunoprecipitation of either the nascent polypeptides attached to the polyribosomes or the in vitro products of translation from these polysomes.

A. Inhibitor Studies

Caution needs to be exercised in the interpretation of the results obtained by the use of inhibitors in plants. Generally the specificity of inhibitors has been established using bacterial and animal systems and it may well be assumed that plants will react in the same way to these inhibitors as the former systems. In addition, it is not always possible to know the inhibitor concentration at the site of action especially when whole plants or tissues are used.

Most authors using *Euglena* (Davis and Merrett 1975), *Chlamydomonas* (Margulies 1971; Armstrong et al. 1971; Iwanij et al. 1975) or higher plants (Graham et al. 1970; Criddle et al. 1970; Owens and Bruening 1975; Beisenherz and Koth 1975) have reported that ribulosebisphosphate carboxylase activity is decreased both by inhibitors such as chloramphenicol, which affect the 70S chloroplast protein synthesizing system, as well as those such as cycloheximide, which inhibit the 80S cytoplasmic system.

Most inhibitors of the two systems did not always affect both systems equally and most observations are consistent with the theory that the large subunit of Fraction I is synthesized in the chloroplast and the small subunit in the cytoplasm, and that there is a degree of coupling or coordination of the synthesis of the two. Even when differential effects were observed on the two subunits there was usually some degree of inhibition of the synthesis of both as would be expected if the stringency of coupling between the synthesis of the two polypeptides varied in different organisms. The two observations that do not fit with this conclusion are the two cases where cycloheximide was found to stimulate the synthesis of Fraction I (Smillie et al. 1971; Givan 1974).

B. Isolated Chloroplast Systems

Although isolated chloroplasts had been demonstrated to have the capacity to incorporate radioactive amino acids into polypeptides as early as 1965 (Spencer) it was not until 1968 that Margulies and Parenti, who investigated the ATP-driven incorporation of amino acids into isolated bean chloroplasts, obtained a peak of radioactivity in the products which corresponded with that of Fraction I on sucrose density gradients. A similar result was obtained by Harris et al. (1973) using *Euglena gracilis*, who found that the radioactive protein synthesized by isolated chloroplasts coincided with RuBP carboxylase activity on gel filtration columns.

Ramirez et al. (1968) observed that chloroplasts of spinach, when isolated in a medium using KCl as osmoticum, incorporated labeled amino acids into the protein fraction using light to generate the necessary ATP via endogeneous photophosphorylation. The products of this reaction in pea chloroplasts were examined by Blair and Ellis (1973) who found the sole soluble radioactive product coincided with the large subunit of Fraction I on SDS-polyacrylamide gels. Its identity with the large subunit was confirmed by peptide mapping. Bottomley et al. (1974) found that similar light-driven amino acid incorporation occurred in spinach chloroplasts which had been isolated in a sorbitol medium. These chloroplasts, unlike those isolated by the Ramirez procedure, were unswollen and were capable of high rates of photosynthetic oxygen evolution. Under these conditions at least four soluble peptides were synthesized, one of which coincided with the large subunit of Fraction I on SDS-polyacrylamide gels. It has been shown (Vasconcelos 1976) that intact chloroplasts of *Euglena*, purified by centrifugation on silica gel gradients, also made the large subunit of Fraction I as the major soluble product when incubated in the light. It is clear from these results on isolated chloroplasts that the mRNA for the large subunit is contained in isolated chloroplasts and that the chloroplast protein-synthesizing system has the capacity to synthesize this polypeptide.

C. Immunoprecipitation of Polyribosome Products

Because of its theoretical specificity, the technique of immunoprecipitation is a potentially powerful tool for isolating a single species of polypeptide from a heterogeneous mixture, but at the same time its use is fraught with difficulties because of the possibility of artifacts arising from nonspecific adsorption.

Several workers have used this procedure to approach the question of the site of synthesis of the Fraction I subunits. The general approach has been to isolate either 70S or 80S polyribosomes from plants actively synthesizing Fraction I, and to determine whether either the nascent peptides on the polyribosomes or the polypeptide chains completed and released in vitro are precipitated by the antiserum made against the large subunit or the small subunit of Fraction I. In 1973 Gooding et al. isolated both membrane-bound and free ribosomes from wheat seedlings and separated the 70S and 80S size classes by sucrose gradients. When the nascent peptides were released by treatment with [³H]-puromycin it was found that antiobodies to the large subunit of Fraction I precipitated peptides only from free 70S ribosomes, whereas both anti-small subunit and anti-large subunit antibodies precipitated peptides from the membrane-bound 80S ribosomes. They later (Roy et al. 1976b) raised the possibility that the latter observation could be due to adsorption artifacts. Gray and Kekwick (1973a, 1974b) examined the polypeptides released from 80S cytoplasmic ribosomes of greening bean leaves following incubation with radioactive amino acids and high speed supernatant enzymes of either bean leaves or rat liver. Of the polypeptides released 30% were precipitated by antiserum to Fraction I small subunit, whereas only 6% were precipitated with large subunit antiserum. The protein precipitated by small subunit antiserum appeared to coincide with the small subunit on gel filtration, while that precipitated by large subunit antiserum also had about 50% small subunits. However, when the polysomes themselves, with their nascent polypeptides attached, were reacted with the two antisera, that to the large subunit precipitated 14% of the polyribosomes, while antiserum to the small subunit precipitated only 4%. They postulated that a pool of free large subunit molecules existed which became attached to the nascent small subunit polypeptides before their release from the polyribosome.

Alscher et al. (1976) carried out similar experiments on polysomes from greening barley leaves using a 100,000 g supernatant from $E.$ $coli$ as a source of accessory components. They found that this system specifically translated the chloroplast polysomes and that immunoprecipitation and peptide mapping of the products indicated that the large subunit of Fraction I was a major product. They concluded that the mRNA for the large subunit was associated with the 70S chloroplast ribosomes.

Roy et al. (1976b) studied the immunoprecipitation of the radioactive polypeptides released from greening wheat leaf polysomes in the presence of a 30,000 g supernatant fraction prepared from wheat germ. They found that chloroplast polyribosomes are inactive in this system, a finding in accord with that of Bottomley et al. (1976). Roy et al. reported that this system was inhibited 80% by $10\mu g/ml$ cycloheximide, whereas that of Gray and Kekwick was only sensitive to the extent of 18% at a concentration of 100 $\mu g/ml$. Precipitation of the product with antiserum to the small subunit of Fraction I followed by SDS-acrylamide gel electrophoresis resulted in a peak of activity coincident with the small subunit. They present arguments, based on the consideration of a number of parameters including tryptic peptide mapping, to conclude that the small subunit is synthesized on the 80S cytoplasmic polyribosomes.

Translation of the poly-A containing fraction of RNA from *Chlamydomonas* in a wheat germ system followed by immunoprecipitation with the antibodies to the small subunit of Fraction I resulted in the isolation of a polypeptide which was about 3500 daltons longer than the small subunits (Dobberstein et al. 1977). This precursor was processed down to the correct size of the small subunit by a polysomal endoprotease from *Chlamydomonas*. They postulate that the extra piece of polypeptide may cause the translating polyribosome complex to attach to the outer chloroplast membrane and facilitate the passage of the small subunit through the membrane where it would then be cut down to size before becoming associated with the large subunits being synthesized within the chloroplast. It may be significant that Roy et al. (1976b) also found a larger polypeptide from the cytoplasmic polysomes of wheat which was precipitated by the antiserum to the small sbubunit.

In spite of the problems associated with immunoprecipitation, it is clear that the evidence from this appraoch is generally consistent with the results of most other techniques in indicating that the large subunit of Fraction I is synthesized in the chloroplast on 70S ribosomes, whereas the small subunit is synthesized in the cytoplasm on 80S ribosomes as a precursor of higher molecular weight than the small subunit itself.

D. The Assembly of the Subunits

We have no direct evidence as yet as to the site or mechanism of assembly of the two subunits into an active RuBP carboxylase complex. It would appear logical that the subunits are assembled within the chloroplasts otherwise it would entail the transport of the large subunit across the chloroplast outer membrane and then later, the transport of the assembled enzyme back into the chloroplast.

The observation of Gooding et al. (1973) that in wheat seedlings the small subunit appears to be synthesized on membrane-bound cytoplasmic polysomes led them to suggest that this membrane can, at some time, become attached to, or part of, the chloroplast outer membrane and so allow the small subunits to pass into the chloroplast either during synthesis or following dissociation from the polyribosomes. This suggestion was supported by the observation of Cobb and Wellburn (1976) that a protein with the properties of the small subunit of Fraction I became associated with the chloroplast envelope fraction of *Avena sativa* during the early greening process. That the chloroplast is the site of assembly of the two subunits is supported by recent work which indicates that the small subunit is synthesized in the cytoplasm as a precursor of about 20,000 daltons. This precursor has been observed in the product from cell-free synthesis on polyribosomes of wheat by Roy et al. (1976b), as well as in the translation products of cytoplasmic mRNA of *Chlamydomonas* (Dobberstein et al. 1977) and peas (Cashmore et al. 1978; Highfield and Ellis 1978). In addition, the precursor has been shown to be processed to the small subunit by an endoprotease associated with polysomes (Dobberstein et al. 1977) or by intact chloroplasts (Highfield and Ellis 1978). The latter authors also found that when the precursor was processed by intact chloroplasts the small subunit was resistant to protease digestion and concluded that it must have passed through the outer membrane into the chloroplast. This strongly suggests that the site of assembly of the subunits must be within the chloroplast envelope.

It has been postulated by Ellis et al. (1978) that protein synthesis within the chloroplast is controlled by the synthesis of nuclear coded polypeptides. This theory is supported by the observation of Feierabend and Wildner (1978) that the small subunit accumulates in rye leaves at 32° C where chloroplast protein synthesis is inhibited due to a deficiency of 70S ribosomes. On the other hand Hallier et al. (1978) have shown that in Fraction I-deficient mutants of *Oenothera*, where the deficiency is caused by a mutation in the plastid DNA, no small subunit could be detected, suggesting that the synthesis of the small subunit is being controlled by the synthesis of the large subunit.

It seems probable, therefore, that the subunits of Fraction I protein are assembled within the plastid, but the mechanism controlling the synthesis and assembly is, as yet, not understood.

VIII. Location of the Structural Gene for Fraction I Protein

To determine the site of the genetic information coding for Fraction I protein, it is not sufficient to establish the site of synthesis of the peptides, or even the location of the messenger RNA which directs the synthesis, because it is always possible that the mRNA itself, in some form, could be exported from the nucleus to the chloroplast, or from the chloroplast to the cytoplasm.

The site of the cistrons coding for both the large and small subunit of Fraction I have been established by Wildman and his collaborators (Chan and Wildman 1972; Kawashima and Wildman 1972). These experiments take advantage of the relative ease with which interspecific hybridization within the genus *Nicotiana* takes place, and also of the fact that the DNA from chloroplasts of this and many other species is only inherited from the maternal parent (cytoplasmic inheritance), whereas the nuclear genome is inherited from both partents (Mendelian inheritance).

The first method used by these authors was to resolve differences in the tryptic peptide composition of the subunits of different species, and to follow the inheritance patterns of these peptides. It was found (Kawashima and Wildman 1972) that the small subunit peptide maps of *Nicotiana tabacum* exhibited one peptide that was not present in that of *N. glutinosa* and two that were absent from *N. glauca*. When reciprocal crosses were made using *N. tabacum* alternatively as the male or female parent, it was found that these peptides appeared in the F_1 progeny when *N. tabacum* was either the male or female parent, showing that the genetic information could be transmitted via the pollen. This demonstrated that the cistrons for the small subunits of Fraction I are contained in the nuclear DNA.

The large subunit exhibits considerably less variability in amino acid composition than the small subunit (Kawashima and Wildman 1970a). No differences were found in the tryptic peptide maps of the large subunits of any *Nicotiana* spp. indigenous to the Western hemisphere but it was discovered that species from Australia contained one peptide not present in *N. tabacum*. The reciprocal crosses demonstrated that this peptide was present only in hybrids when an Australian species was the maternal parent and hence the genetic information for the large subunit must be contained in the chloroplast DNA (Chan and Wildman 1972).

These results have been supported by the examination of the mode of inheritance of differences in the polypeptide composition revealed by isoelectric focusing of the S-carboxymethylated large and small subunits of interspecific hybrids of *Nicotiana* spp. Again it was shown that the information for the composition of the small subunit is inherited in a Mendelian manner, whereas that for the large subunit is inherited maternally (Kung et al. 1974).

Direct evidence that chloroplast DNA contains the large subunit cistron has been obtained by the cell-free transcription and translation of a cloned fragment of corn chloroplast DNA (Coen et al. 1977) and also of total spinach chloroplast DNA (Bottomley et al. 1979).

It seems clear now that the large subunit of Fraction I is coded for by chloroplast DNA and synthesized by the 70S ribosomal system in the chloroplast, whereas the small subunit is coded for by the nuclear DNA and synthesized on 80S ribosomes in the cytoplasm. The major problem remaining on the mechanism of synthesis of this protein is the means by which the synthesis of these two types of subunits is coordinated and how they are assembled into the active enzyme.

IX. The Effect of Light on Fraction I Protein

When etiolated plants are transferred to light, the etioplasts begin to develop and, following the commencement of chlorophyll synthesis, the internal membrane structures differentiate. Since Fraction I protein makes up a large proportion of the soluble (stromal) protein of the majority of mature chloroplasts, it is relevant to ask whether the controls on synthesis of this protein and on chloroplast development are related. In this context it is important to distinguish between changes in Fraction I protein content and changes in ribulosebisphosphate carboxylase activity, since it is possible that the specific activity of the enzyme may change during differentiation of the chloroplast.

In 1956 Lyttleton reported a threefold increase in the Fraction I protein content of etiolated wheat leaves following exposure to 20 h of white light. Later Kupke (1962) observed a correlation between the increase in the Fraction I protein peak in the ultracentrifuge and chlorophyll synthesis when dark-grown bean seedlings were illuminated for several days while Huffaker et al. (1966) found a linear correlation between ribulosebisphosphate carboxylase activity and chlorophyll content following the transfer of etiolated barley seedlings to white light for up to 48 h.

Graham et al. (1968, 1971) showed that, when etiolated pea seedlings were given brief red-light treatments, the activity of ribulosebisphosphate carboxylase in the apices increased by as much as 91–fold in five days, whereas the amount of Fraction I protein increased only 12–fold. Continuous white light resulted in a small but rapid increase in enzyme activity under conditions where there was no increase in the amount of Fraction I present during the first 24 h of illumination. Chen et al. (1967) also observed a rapid increase in ribulosebisphosphate carboxylase activity on illumination of dark-grown corn leaves. From this it appears that the ribulose-bisphosphate carboxylase activity of Fraction I can vary in vivo and also, since the red light effects could be reversed by a subsequent far-red treatment, both the carboxlase activity and Fraction I synthesis are, in some degree, under the control of the phyotochrome system. This has been confirmed by Frosch et al. (1976) using etiolated mustard seedling cotyledons, where it was found that continuous red or far-red light resulted in the same increases in ribulosebisphosphate carboxylase activity as continuous white light. Since the red or far-red light treatments did not cause chlorophyll synthesis or significant differentiation of the chloroplast membrane system, it follows that the synthesis of Fraction I protein and the activation of ribulosebisphosphate carboxylase are not dependent on the development of the chloroplast.

Smith et al. (1974) found that when excised etiolated barley leaves were exposed to continuous white light, there was a lag of about 6–8 h before any increase in ribulosebisphosphate carboxylase activity was observed, whereas the synthesis of Fraction I protein increased without any detectable lag. This, once again, demonstrates that the ribulosebisphosphate carboxylase activity of Fraction I protein can vary without a concomitant change in the amount of the protein. Earlier, Kleinkopf et al. (1970b) had reported that the increase in carboxylase activity was due to de novo synthesis of Fraction I while Kannangara (1969) found that the ratio of Fraction I protein content to ribulosebisphosphate carboxylase activity of barley plants remained constant during the first 24 h of greening of etiolated barley in white light, and Blenkinsop and Dale (1974) also found this ratio remained constant when shading of green barley caused the levels of enzyme to vary.

The effects of light on Fraction I protein are not confined to etiolated plants, since it has been reported (Blackwood and Leaver 1977) that light stimulates both the synthesis of the large subunit and also the assembly of the subunits into the holoprotein in fully greened leaves of Lemna minor.

Siddell and Ellis (1975) isolated the plastids from peas at various stages during greening in continuous white light and examined them for their ability to synthesize proteins using either light or ATP as an energy source. They found that, during the first 48 h of greening, the incorporation of labeled amino acids in either system increased, yielding mainly the large subunit of Fraction I and a membrane protein. However, after 96 h in the light, the ability to make the large subunit disappeared in spite of the claim that these chloroplasts were fully differentiated and that fully differentiated chloroplasts do make the large subunit. The controls operating which distinguish these 96 h greened chloroplasts from mature chloroplasts would seem to be a worthwhile subject for further investigation.

When *Chlamydomonas* are transferred to light, RuBP carboxylase activity increases without any apparent lag phase, as do the synthesis of both the large and small subunits of Fraction I (Iwanij et al. 1975). Under these conditions, chlorophyll synthesis has a lag period of about 4 h.

Although the chloroplasts of the mesophyll cells of C_4 plants may not be typical, the fact that they develop and differentiate without the formation of Fraction I protein (Huber et al. 1976) suggests that this protein is not essential to the functioning of chloroplasts apart from its role in photosynthetic carbon reduction. Conversely, the formation of significant amounts of Fraction I protein under conditions where little or no synthesis of chlorophyll or development of the chloroplast structure takes place points to the fact that differentiation is not essential for Fraction I formation.

X. Messenger RNA for Fraction I Protein

In attempting to determine the factors which control the synthesis of Fraction I protein, it would be clearly an advantage to be able to purify the messenger RNA which directs the synthesis of its subunits, thus simplifying the separation of controls which act at the level of transcription from those acting on translation.

Rosner et al. (1975) reported a sharp increase in the synthesis of a 0.5×10^6 dalton RNA from *Spirodela* about 4 h after its transfer from darkness to light. This RNA had a high template activity in an *E. coli* protein-synthesizing system. They suggest that this is of the size expected for the mRNA carrying the cistron for the large subunit of Fraction I. Sagher et al. (1976), using light-grown *Euglena*, have partially purified a nonpoly A-containing RNA which could be translated in a wheat germ protein-synthesizing system to yield the large subunit of Fraction I. One curious aspect of this is that Bottomley et al. (1976) have reported that the wheat germ system would not translate spinach chloroplast RNA but that an *E. coli* system using the same RNA synthesized the large subunit.

Recently, Howell et al. (1977a, 1977b) and Gelvin and Howell (1977) have reported the isolation of mRNA for the large subunit from *Chlamydomonas* by making use of their finding that the large subunit messenger activity was associated with polyribosomes with only about three ribosomes attached. When the RNA from this polyribosomes fraction was hybridized with chloroplast DNA that had been digested with the restriction endonuclease EcoRI, it hybridized to a specific fragment of M.W. 3.2×10^6 daltons (Gelvin et al. 1977), thus providing a further direct demonstration that the cistron for the large subunit is contained in chloroplast DNA.

While it is probable that the mRNA coding for the large subunit will be isolated in the near future, two major problems which hinder the achievement of this goal are the apparent lack of poly-A sequence on the mRNA (Wheeler and Hartley 1975) and also the similarity of its probable size (0.5×10^6 daltons) to that of the RNA of the small ribosomal subunit (0.56×10^6 daltons).

XI. Conclusion

Fraction I protein is one of the proteins most necessary for life as we know it today, since it is responsible for the initial step in the photosynthetic carbon reduction process on which most of our energy and food requirements depend. In addition, since it forms such a high proportion of the protein of green tissue, it directly provides a significant proportion of the protein of the diet of most domestic animals.

Because of its occurrence in large amounts in green tissue, and because it is a chloroplast-localized enzyme, Fraction I has also provided a convenient object to use in the study of the origin of the genetic message for chloroplast proteins, as well as the controls regulating that synthesis. Since it is now well established that the large subunit of Fraction I is coded for by chloroplast DNA and synthesized within the chloroplast while the small subunit is coded for by nuclear DNA and synthesized in the cytoplasms, the range of possible regulatory mechanisms is complex. While the carboxylase activity of Fraction I protein itself does not require light, light is needed for the in vivo activity since it is dependent on photophosphorylation for a source of ATP. It is not surprising, then, that the synthesis of Fraction I in both etiolated and green plants is responsive to light. The fact that this increase in the synthesis of Fraction I under the influence of light takes place in conditions where little or no differentiation of the chloroplast is observed indicates that the light response is, to some extent, a direct one, and not part of a general response to development of the organelle.

Unfortunately, the identification of the site of action of phytochrome, which appears to have a role in the control of Fraction I synthesis, has not yet been rigorously achieved, but it would be of interest to know whether it was more likely to affect the synthesis of the large subunit or the small subunit. In that way it may be possible to deduce more directly than at present whether the primary control of the synthesis of Fraction I was exerted by the chloroplast system or the nuclear-cytoplasmic. At present it appears that the coupling of the synthesis of the two subunits is very stringent since, although pools of one or other of the subunits have sometimes been invoked to explain results, only one demonstration of such pools has so far been made (Feierabend and Wildner 1978). The only other situation when one subunit appears to be synthesized in the absence of the other is the case of isolated organelles which make the large subunit only. However, these organelles are not in their normal environment and may be regarded as "uncoupled" systems.

The synthesis of both the large subunit (Gooding et al. 1973; Hallier et al. 1978) and of the small subunit (Ellis 1976) have been suggested as the site at which the coordination of the two systems is regulated. It seems logical to suppose that the assembly of the holoprotein takes place in the plastid since, otherwise, it would require the large subunit to be transported through the plastid membrane to the cytoplasm and then back again as the completed protein. The demonstration by Dobberstein et al. (1977) that the small subunit, as synthesized by a wheat germ protein-synthesizing system programmed by poly-A RNA, has an extra piece of polypeptide (F) which can be cleaved off by an endopeptidase, led them to suggest that this peptide may facilitate the transport of the small subunit across the outer membrane of the chloroplast. It is interesting to speculate that this extra polypeptide could be involved in the control of the synthesis of the large subunit. That is, if polypeptide F was cleaved from the precursor polypeptide of the small subunit after transport through the outer membranes of the plastid, and then caused the synthesis of one molecule of the large subunit, strict coupling of the production of the two subunits would be achieved.

That the complexity of the mechanism of the synthesis of Fraction I protein involving two distinct systems may not be unique is suggested by the finding of Mendiola-Morgenthaler et al. (1976), who showed that isolated chloroplasts were capable of synthesizing only three of the five subunits of the chloroplast coupling factor (CF1), and hence that the other two subunits are probably made in the cytoplasm. If, indeed, this very close coordination of the two systems is found to be a common feature of many of the plastid constituents, it will become a major factor to be explained when considering the mechanism of evolution of chloroplasts.

References

Akazawa, T.: The structure and function of Fraction I protein. Regulatory aspects of photosynthetic CO_2-fixation in chloroplasts. In: Progress in Phytochemistry; Reinhold, L., Liwschitz, Y. (eds.); Vol. II, pp. 107-141. London: Interscience 1970

Alscher, R., Smith, M.A., Petersen, L.W., Huffaker, R.C., Criddle, R.S.: In vitro synthesis of the large subunit of ribulose diphosphate carboxylase on 70S ribosomes. Arch. Biochem. Biophys. 174, 216-225 (1976)

Anderson, L.E., Price, G.B., Fuller, R.C.: Molecular diversity of the ribulose-1,5-diphosphate carboxylase from photosynthetic microorganisms. Science 161, 482-484 (1968)

Andrews, T.J., Badger, M.R., Lorimer, G.H.: Factors affecting interconversion between kinetic forms of ribulose diphosphate carboxylase-oxygenase from spinach. Arch. Biochem. Biophys. 171, 93-103 (1975)

Armstrong, J.J., Surzycki, S.J., Moll, B., Levine, R.P.: Genetic transcription and translation specifying chloroplast components in Chlamydomonas reinhardii. Biochemistry 10, 692-701 (1971)

Badger, M.R., Lorimer, G.H.: Activation of ribulose-1,5-bisphosphate oxygenase. The role of Mg^{2+}, CO_2 and pH. Arch. Biochem. Biophys. 175, 723-729 (1976)

Bahr, J.T., Jensen, R.G.: Ribulose diphosphate carboxylase from freshly ruptured spinach chloroplasts having an in vivo Km [CO_2]. Plant Physiol. 53, 39-44 (1974)

Baker, T.S., Eisenberg, D., Eiserling, F.A., Weissman, L.: The structure of form 1 crystals of D-ribulose-1,5-diphosphate carboxylase. J. Mol. Biol. 91, 391-399 (1975)

Baker, T.S., Eisenberg, D., Eiserling, F.: Ribulose bisphosphate carboxylase: A two-layered, square-shaped molecule of symmetry 422. Science 196, 293-295 (1977)

Beisenherz, W.W., Koth, P.: Der Einfluß von Chloramphenicol und Cycloheximid auf die Synthese von Ribulose-1,5-diphosphat-Carboxylase, NADP-abhängiger Glycerinaldehyd-3-phosphat-Dehydrogenase und Chlorophyll während der Leukoplasten-Chloroplasten-Transformation in Gewebekulturen von Nicotiana tabacum. Z. Pflanzenphysiol. 75, 201-210 (1975)

Blackwood, G.C., Leaver, C.J.: The effect of light on protein synthesis in green leaves. In: Acides Nucléiques et Synthèse des Protéins chez les Végétaux; Bogorad, L., Weil, J.H. (eds.); pp. 611-615. Paris: C.N.R.S. 1977

Blair, G.E., Ellis, R.J.: Protein synthesis in chloroplasts. I. Light-driven synthesis of the large subunit of Fraction I protein by isolated pea chloroplasts. Biochim. Biophys. Acta 319, 223-234 (1973)

Blenkinsop, P.G., Dale, J.E.: The effects of shade treatment and light intensity on ribulose-1,5-diphosphate carboxylase activity and Fraction I protein level in the first leaf of barley. J. Exp. Bot. 25, 899-912 (1974)

Bottomley, W., Whitfeld, P.R.: Cell-free transcription and translation of total spinach chloroplast DNA. Eur. J. Biochem. 93, 31-39 (1979)

Bottomley, W., Spencer, D., Whitfeld, P.R.: Protein synthesis in isolated spinach chloroplasts: Comparison of light-driven and ATP-driven synthesis. Arch. Biochem. Biophys. 164, 106-117 (1974)

Bottomley, W., Higgins, T.J.V., Whitfeld, P.R.: Differential recognition of chloroplast and cytoplasmic messenger RNA by 70S and 80S ribosomal systems. FEBS Lett. 63, 120-124 (1976).

Bowes, G., Ogren, W.L., Hageman, R.H.: Phosphoglycolate production catalyzed by ribulose diphosphate carboxylase. Biochem. Biophys. Res. Commun. 45, 716-722 (1971).

Brändén, R.: Ribulose-1,5-diphosphate carboxylase and oxygenase from green plants are two different enzymes. Biochem. Biophys. Res. Commun. 81, 539-546 (1978).

Brown, R.H., Armitage, T.L., Merrett, M.J.: Ribulose diphosphate carboxylase synthesis in Euglena. III. Serological relationships of the intact enzyme and its subunits. Plant Physiol. 58, 773-776 (1976)

Cashmore, A.R., Broadhurst, M.K., Gray, R.E.: Cell-free synthesis of leaf proteins: Identification of an apparent precursor of the small subunit of ribulose-1,5-bisphosphate carboxylase. Proc. Natl. Acad. Sci. USA 75, 655-659 (1978)

Chan, P.H., Wildman, S.G.: Chloroplast DNA codes for the primary structure of the large subunit of Fraction I protein. Biochim. Biophys. Acta 277, 677-680 (1972)

Chan, P.H., Sakano, K., Singh, S., Wildman, S.G.: Crystalline Fraction I protein: Preparation in large yield. Science 176, 1145-1146 (1972)

Chen, K., Johal, S., Wildman, S.G.: Role of chloroplast and nuclear DNA genes during evolution of Fraction I protein. In: Genetics and Biogenesis of Chloroplasts and Mitochondria; Bücher, Th., Neupert, W., Sebald, W., Werner, S. (eds.); pp. 3-11. Amsterdam: North-Holland 1976a

Chen, K., Kung, S.D., Gray, J.C., Wildman, S.G.: Subunit polypeptide composition of Fraction I protein from various plant species. Plant Sci. Lett. 7, 429-434 (1976b)

Chen, S., McMahon, D., Bogorad, L.: Early effects of illumination on the activity of some photosynthetic enzymes. Plant Physiol. 42, 1-5 (1967)

Chollet, R., Anderson, L.L., Hovsepian, L.C.: The absence of tightly bound copper, iron, and flavin nucleotide in crystalline ribulose-1,5-bisphosphate carboxylase oxygenase from tobacco. Biochem. Biophys. Res. Commun. 64, 97-107 (1975)

Chu, D.K., Bassham, J.A.: Activation and inhibition of ribulose 1,5-diphosphate carboxylase by 6-phosphogluconate. Plant Physiol. 52, 373-379 (1973)

Cobb, A.H., Wellburn, A.R.: Polypeptide binding to plastid envelopes during chloroplast development. Planta 129, 127-131 (1976)

Coen, D.M., Bedbrook, J.R., Bogorad, L., Rich, A.: Maize chloroplast DNA fragment encoding the large subunit of ribulosebisphosphate carboxylase. Proc. Natl. Acad. Sci. USA 74, 5487-5491 (1977)

Criddle, R.S., Dau, B., Kleinkopf, G.E., Huffaker, R.C.: Differential synthesis of ribulosediphosphate carboxylase subunits. Biochem. Biophys. Res. Commun. 41, 621-627 (1970)

Davis, B., Merrett, M.J.: The glycolate pathway and photosynthetic competence in Euglena. Plant Physiol. 55, 30-34 (1975)

Dobberstein, B., Blobel, G., Chua, N.H.: In vitro synthesis and processing of a putative precursor for the small subunit of ribulose-1,5-bisphosphate carboxylase of Chlamydomonas reinhardii. Proc. Natl. Acad. Sci. USA 74, 1082-1085 (1977)

Ellis, R.J.: Fraction I Protein. Comm. Plant Sci. 4, 29-38 (1973)

Ellis, R.J.: The search for plant messenger RNA. In: Perspectives in Experimental Biology. Vol. II Botany; Sunderland, N. (ed.); pp. 283-298. Oxford: Pergamon 1976

Ellis, R.J., Highfield, P.E., Silverthorne, J.: The synthesis of chloroplast proteins by subcellular systems. In: Photosynthesis 77; Hall, D.O., Coombs, J., Goodwin, T.W. (eds.); pp. 497-506. Proc. 4th Congr. Photosynth. London: The Biochemical Society 1978

Feierabend, J., Wildner, G.: Formation of the small subunit in the absence of the large subunit of ribulose 1,5-bisphosphate carboxylase in 70S ribosome-deficient rye leaves. Arch. Biochem. Biophys. 186, 283-291 (1978)

Frosch, S., Bergfeld, R., Mohr, H.: Light control of plastogenesis and ribulosebisphosphate carboxylase levels in mustard seedling cotyledons. Planta 133, 53-56 (1976)

Gelvin, S., Howell, S.H.: Identification and precipitation of the polyribosomes in *Chlamydomonas reinhardii* involved in the synthesis of the large subunit of D-ribulose-1,5-bisphosphate carboxylase. Plant Physiol. 59, 471-477 (1977)

Gelvin, S., Heizmann, P., Howell, S.H.: Identification and cloning of the chloroplast gene coding for the large subunit of ribulose-1,5-bisphosphate carboxylase from *Chlamydomonas reinhardii*. Proc. Natl. Acad. Sci. USA 74, 3193-3197 (1977)

Gibson, J.L., Tabita, F.R.: Different molecular forms of D-ribulose-1,5-bisphosphate carboxylase from *Rhodopseudomonas sphaeroides*. J. Biol. Chem. 252, 943-949 (1977)

Givan, A.L.: Ribulose diphosphate carboxylase synthesis in *Chlamydomonas reinhardii:* Inhibition by chloramphenicol and stimulation by cycloheximide. Planta 120, 181-188 (1974)

Givan, A.L., Criddle, R.S.: Ribulosediphosphate carboxylase from *Chlamydomonas reinhardii:* purification, properties and its mode of synthesis in the cell. Arch. Biochem. Biophys. 149, 153-163 (1972)

Goldthwaite, J.J., Bogorad, L.: A one-step method for the isolation and determination of leaf ribulose-1,5-diphosphate carboxylase. Anal. Biochem. 41, 57-66 (1971)

Gooding, L.R., Roy, H., Jagendorf, A.T.: Immunological identification of nascent subunits of wheat ribulose diphosphate carboxylase on ribosomes of both chloroplast and cytoplasmic origin. Arch. Biochem. Biophys. 159, 324-335 (1973)

Graham, D., Grieve, A.M., Smillie, R.M.: Phytochrome as the primary photoregulator of the synthesis of Calvin cycle enzymes in etiolated pea seedlings. Nature (London) 218, 89-90 (1968)

Graham, D., Hatch, M.D., Slack, C.R., Smillie, R.M.: Light-induced formation of enzymes of the C_4-dicarboxylic acid pathway of photosynthesis in detached leaves. Phytochemistry 9, 521-532 (1970)

Graham, D., Grieve, A.M., Smillie, R.M.: Phytochrome-mediated plastid development in etiolated pea stem apices. Phytochemistry 10, 2905-2914 (1971)

Gray, J.C, Kekwick, R.G.O.: Synthesis of the small subunit of ribulose 1,5-diphosphate carboxylase on cytoplasmic ribosomes from greening bean leaves. FEBS Lett. 38, 67-69 (1973a)

Gray, J.C., Kekwick, R.G.O.: A serological investigation of ribulose 1,5-diphosphate carboxylase and its subunits. Biochem. Soc. Trans. , 455-458 (1973b)

Gray, J.C., Kekwick, R.G.O.: An immunological investigation of the structure and function of ribulose 1,5-bisphosphate carboxylase. Eur. J. Biochem. 44, 481-489 (1974a)

Gray, J.C., Kekwick, R.G.O.: The synthesis of the small subunit of ribulose 1,5-bisphosphate carboxylase in the french bean *Phaseolus vulgaris*. Eur. J. Biochem. 44, 491-500 (1974b)

Gray, J.C., Wildman, S.G.: A specific immunoabsorbent for the isolation of Fraction I protein. Plant Sci. Lett. 6, 91-96 (1976)

Hallier, U.W., Schmitt, J.M., Heber, U., Chaianova, S.S., Volodarsky, A.D.: Ribulose-1,5-bisphosphate carboxylase-deficient plastome mutants of *Oenothera*. Biochim. Biophys. Acta 504, 67-83 (1978)

Harris, E.H., Preston, J.F., Eisenstadt, J.M.: Amino acid incorporation and products of protein synthesis in isolated chloroplasts of *Euglena gracilis*. Biochemistry 12, 1227-1234 (1973)

Haslett, B.G., Yarwood, A., Evans, I.M., Boulter, D.: Studies on the small subunit of Fraction I protein from *Pisum sativum* L. and *Vicia faba* L. Biochim. Biophys. Acta 420, 122-132 (1976)

Hatch, M.D.: The C_4 pathway of photosynthesis: mechanism and function. In: Carbon Dioxide Metabolism and Plant Productivity; Burris, R.H., Black, C.C. (eds.); pp. 59-81. Baltimore: University Park Press 1976

Hayaishi, O.: Molecular Mechanisms of Oxygen Activation. London-New York: Academic Press 1974

Highfield, P.E., Ellis, R.J.: Synthesis and transport of the small subunit of ribulose bisphosphate carboxylase. Nature (London) 271, 420-424 (1978)

Hirai, A.: Random assembly of different kinds of small subunit polypeptides during formation of fraction I protein macromolecules. Proc. Natl. Acad. Sci. USA 74, 3443-3445 (1977)

Howell, S.H., Heizmann, P., Gelvin, S.: Properties of the mRNA and localization of the gene coding for the large subunit of ribulose bisphosphate carboxylase in *Chlamydomonas reinhardii*. In: Acids Nucléiques et Synthèse des Protéins chez les Végétaux; Bogorad, L., Weil, J.H. (eds.); pp. 313-318. Paris: C.N.R.S. 1977a

Howell, S.H., Heizmann, P., Gelvin, S., Walker, L.L.: Identification and properties of the messenger RNA activity in *Chlamydomonas reinhardii* coding for the large subunit of D-ribulose-1,5-bisphosphate carboxylase. Plant Physiol. 59, 464-470 (1977b)

Huber, S.C., Hall, T.C., Edwards, G.E.: Differential localization of Fraction I protein between chloroplast types. Plant Physiol. 57, 730-733 (1976)

Huffaker, R.C., Obendorf, R.L., Keller, C.J., Kleinkopf, G.E.: Effects of light intensity on photosynthetic carboxylative phase enzymes and chlorophyll synthesis in greening leaves of *Hordeum vulgare* L. Plant Physiol. 41, 913-918 (1966)

Iwanij, V., Chua, N.H., Siekevitz, P.: Synthesis and turnover of ribulose bisphosphate carboxylase and of its subunits during the cell cycle of *Chlamydomonas reinhardii*. J. Cell Biol. 64, 572-585 (1975)

Kagan-Zur, V., Lips, S.H.: Studies on the intracellular location of enzymes of the photosynthetic carbon-reduction cycle. Eur. J. Biochem. 59, 17-23 (1975)

Kannangara, C.G.: The formation of ribulose diphosphate carboxylase protein during chloroplast development in barley. Plant Physiol. 44, 1533-1537 (1969)

Kawashima, N.: Comparative studies on fraction I protein from spinach and tobacco leaves. Plant Cell Physiol. 10, 31-40 (1969)

Kawashima, N., Wildman, S.G.: Fraction I Protein. Annu. Rev. Plant Physiol. 21, 325-358 (1970a)

Kawashima, N., Wildman, S.G.: A model of the subunit structure of Fraction I protein. Biochem. Biophys. Res. Commun. 41, 1463-1468 (1970b)

Kawashima, N., Wildman, S.G.: Studies on Fraction I protein. I. Effect of crystallization of Fraction I protein from tobacco leaves on ribulose diphosphate carboxylase activity. Biochim. Biophys. Acta 229, 240-249 (1971a)

Kawashima, N., Wildman, S.G.: Studies on Fraction I protein. II. Comparison of physical, chemical, immunological and enzymatic properties between spinach and tobacco Fraction I proteins. Biochim. Biophys. Acta 229, 749-760 (1971b)

Kawashima, N., Wildman, S.G.: Studies on Fraction I protein. IV. Mode of inheritance of primary structure in relation to whether chloroplast or nuclear DNA contains the code for a chloroplast protein. Biochim. Biophys. Acta 262, 42-49 (1972)

Kawashima, N., Kwok, S.Y., Wildman, S.G.: Studies on Fraction I protein. III. Comparison of the primary structure of the large and small subunits obtained from five species of *Nicotiana*. Biochim. Biophys. Acta 236, 578-586 (1971)

Kirk, J.T.O.: The genetic control of plastid formation: recent advances and strategies for the future. Subcell. Biochem. 1, 333-361 (1972)

Kirk, J.T.O., Tilney-Bassett, R.A.E.: The Plastids. Their Chemistry, Structure, Growth and Inheritance. London: Freeman 1967

Kleinkopf, G.E., Huffaker, R.C., Matheson, A.: A simplified purification and some properties of ribulose 1,5-diphosphate carboxylase from barley. Plant Physiol. 46, 204-207 (1970a)

Kleinkopf, G.E., Huffaker, R.C., Matheson, A.: Light-induced *de novo* synthesis of ribulose 1,5-diphosphate carboxylase in greening leaves of barley. Plant Physiol. 46, 416-418 (1970b)

Kung, S.D.: Tobacco Fraction I protein: a unique genetic marker. Science 191, 429-434 (1976)

Kung, S.D., Lee, C.I.: Evolutionary conservation of chloroplast genes coding for the large subunit of Fraction I protein. Plant Physiol. 60, 89-94 (1977)

Kung, S.D., Sakano, K., Wildman, S.G.: Multiple peptide composition of the large and small subunits of *Nicotiana tabacum* Fraction I protein ascertained by fingerprinting and electrofocusing. Biochim. Biophys. Acta 365, 138-147 (1974)

Kupke, D.W.: Correlation of a soluble leaf protein with chlorophyll accumulation. J. Biol. Chem. 237, 3287-3291 (1962)

Kwok, S.Y., Kawashima, N., Wildman, S.G.: Specific effect of ribulose-1,5-diphosphate on the solubility of tobacco Fraction I protein. Biochim. Biophys. Acta 234, 293-296 (1971)

Lorimer, G.H., Andrews, T.J.: Plant photorespiration — an inevitable consequence of the existence of atmospheric oxygen. Nature (London) 243, 359-360 (1973)

Lorimer, G.H., Andrews, T.J., Tolbert, N.E.: Ribulose diphosphate oxygenase. II. Further proof of reaction products and mechanism of action. Biochemistry 12, 18-23 (1973)

Lorimer, G.H., Badger, M.R., Andrews, T.J.: The activation of ribulose-1,5-bisphosphate carboxylase by carbon dioxide and magnesium ions. Equilibria, kinetics, a suggested mechanism, and physiological implications. Biochemistry 15, 529-536 (1976)

Lowe, R.H.: Crystallization of Fraction I protein from tobacco by a simplified procedure. FEBS Lett. 78, 98-100 (1977)

Lyttleton, J.W.: Relationship between photosynthesis and a homogeneous protein component of plant cytoplasm. Nature (London) 177, 283-284 (1956)

Lyttleton, J.W., Ts'o, P.O.P.: The localization of Fraction I protein of green leaves in the chloroplasts. Arch. Biochem. Biophys. 73, 120-126 (1958)

Margulies, M.M.: Concerning the sites of synthesis of proteins of chloroplast ribosomes and of Fraction I protein (Ribulose-1,5-diphosphate carboxylase). Biochem. Biophys. Res. Commun. 44, 539-545 (1971)

Margulies, M.M., Parenti, F.: *In vitro* protein synthesis by plastids of *Phaseolus vulgaris*. III. Formation of lamellar and soluble chloroplast protein. Plant Physiol. 43, 504-514 (1968)

McCurry, S.D., Hall, N.P., Pierce, J., Peach, C., Tolbert, N.E.: Ribulose-1,5-bisphosphate carboxylase/oxygenase from parsley. Biochem. Biophys. Res. Commun. 84, 895-900 (1978)

McFadden, B.A.: Autotrophic CO_2 assimilation and the evolution of ribulose diphosphate carboxylase. Bacteriol. Rev. 37, 289-319 (1973)

McFadden, B.A.: The oxygenase activity of ribulose-1,5-bisphosphate carboxylase from *Rhodospirillum rubrum*. Biochim. Biophys. Res. Commun. 60, 312-317 (1974)

McFadden, B.A., Lord, J.M., Rowe, A., Dilks, S.: Composition, quaternary structure, and catalytic properties of D-ribulose-1,5-bisphosphate carboxylase from *Euglena gracilis*. Eur. J. Biochem. 54, 195-206 (1975)

Mendiola-Morgenthaler, L., Morgenthaler, J.J., Price, C.A.: Synthesis of coupling factor CF_1 protein by isolated spinach chloroplasts. FEBS Lett. 62, 96-100 (1976)

Moon, K.E., Thompson, E.O.P.: Subunits from reduced and S-carboxymethylated ribulose diphosphate carboxylase (Fraction I protein). Aust. J. Biol. Sci. 22, 463-470 (1969)

Murai, T., Akazwawa, T.: Bicarbonate effect on the photophosphorylation catalyzed by chromatophores isolated from *Chromatium* Strain D. XIII. Structure and function of chloroplast proteins. Plant Physiol. 50, 568-571 (1972)

Owens, R.A., Bruening, G.: The pattern of amino acid incorporation into two cowpea mosaic virus proteins in the presence of ribosome-specific protein synthesis inhibitors. Virology 64, 520-530 (1975)

Pistorius, E.K., Axelrod, B.: Iron, an essential component of lipoxygenase. J. Biol. Chem. 249, 3183-3186 (1974)

Pon, N.G., Rabin, B.R., Calvin, M.: Mechanism of the carboxydismutase reaction. I. The effect of preliminary incubation of substrates, metal ion and enzmye on activity. Biochem. Z. 338, 7-19 (1963)

Purohit, K., McFadden, B.A.: Heterogeneity of large subunits of ribulose-1,5-bisphosphate carboxylase from *Hydrogenomonas eutropha*. Biochem. Biophys. Res. Commun. 71, 1220-1227 (1976)

Rabinowitz, H., Reisfeld, A., Sagher, D., Edelman, M.: Ribulose diphosphate carboxylase from autotrophic *Euglena gracilis*. Plant Physiol. 56, 345-350 (1975)

Ramirez, J.M., del Campo, F.F., Arnon, D.I.: Photosynthetic phosphorylation as energy source for protein synthesis and carbon dioxide assimilation by chloroplasts. Proc. Natl Acad. Sci. USA 59, 606-612 (1968)

Rosner, A., Jakob, K.M., Gresel, J., Sagher, D.: The early synthesis and possible function of a 0.5×10^6 M_r RNA after transfer of dark-grown *Spirodela* plants to light. Biochem. Biophys. Res. Commun. 67, 383-391 (1975)

Roy, H., Alvarez, O., Mader, L.: Monomeric behavior of the small subunit of ribulose-1,5-bisphosphate carboxylase. Biochem. Biophys. Res. Commun. 70, 914-919 (1976a)

Roy, H., Patterson, R., Jagendorf, A.T.: Identification of the small subunit of ribulose 1,5-bisphosphate carboxylase as a product of wheat leaf cytoplasmic ribosomes. Arch. Biochem. Biophys. 172, 64-73 (1976b)

Roy, H., Terenna, B., Cheong, L.C.: Synthesis of the small subunit of ribulose-1,5-bisphosphate carboxylase by soluble fraction polyribosomes of pea leaves. Plant Physiol. 60, 532-537 (1977)

Rutner, A.C., Lane, M.D.: Nonidentical subunits of ribulose diphosphate carboxylase. Biochem. Biophys. Res. Commun. 28, 531-537 (1967)

Sagher, D., Grosfeld, H., Edelman, M.: Large subunit ribulosebisphosphate carboxylase messenger RNA from *Euglena* chloroplasts. Proc. Natl. Acad. Sci. USA 73, 722-726 (1976)

Siddell, S.G., Ellis, R.J.: Protein synthesis in chloroplasts. Characteristics and products of protein synthesis *in vitro* in etioplasts and developing chloroplasts from pea leaves. Biochem. J. 146, 675-685 (1975)

Slack, C.R., Hatch, M.D.: Comparative studies on the activity of carboxylase and other enzymes in relation to the new pathway of photosynthetic carbon dioxide fixation in tropical grasses. Biochem. J. 103, 660-665 (1967)

Smillie, R.M., Bishop, D.G., Gibbons, G.C., Graham, D., Grieve, A.M., Raison, J.K., Reger, B.J.: Determination of the sites of synthesis of proteins and lipids of the chloroplast using chloramphenicol and cycloheximide. In: Autonomy and Biogenesis of Mitochondria and Chloroplasts; Boardman, N.K., Linnane, A.W., Smillie, R.M. (eds.); pp. 422-433 (Amsterdam: North-Holland 1971

Smith, M.A., Criddle, R.S., Peterson, L., Huffaker, R.C.: Synthesis and assembly of ribulosebisphosphate carboxylase enzyme during greening of barley plants. Arch. Biochem. Biophys. 165, 494-504 (1974)

Spencer, D.: Protein synthesis by isolated spinach chloroplasts. Arch. Biochem. Biophys. 111, 381-390 (1965)

Steer, M.W., Gunning, B.E.S., Graham, T.A., Carr, D.J.: Isolation, properties and structure of Fraction I protein from *Avena sativa* L. Planta 79, 254-267 (1968)

Strobaek, S., Gibbons, G.C.: Ribulose-1,5-diphosphate carboxylase from barley *(Hordeum vulgare)*. Isolation, characterization and peptide mapping studies of the subunits. Carlsberg Res. Commun. 41, 57-72 (1976)

Sugiyama, T., Akazawa, T.: Subunit structure of spinach leaf ribulose 1,5-diphosphate carboxylase. Biochemistry 9, 4499-4504 (1970)

Sugiyama, T., Nakayama, N., Ogawa, M., Akazawa, T., Oda, T.: Structure and function of chloroplast proteins. II. Effect of p-chloromercuribenzoate treatment on the ribulose 1,5-diphosphate carboxylase activity of spinach leaf Fraction I protein. Arch. Biochem. Biophys. 125, 98-106 (1968)

Sugiyama, T., Ito, T., Akazawa, T.: Subunit structure of ribulose 1,5-diphosphate carboxylase from *Chlorella ellipsoidea*. Biochemistry 10, 3406-3411 (1971)

Tabita, F.R., McFadden, B.A.: Regulation of ribulose-1,5-diphosphate carboxylase by 6-phospho-D-gluconate. Biochem. Biophys. Res. Commun. 48, 1153-1159 (1972)

Tabita, F.R., Stevens, S.E., Jr., Quijano, R.: D-ribulose 1,5-diphosphate carboxylase from blue-green algae. Biochem. Biophys. Res. Commun. 61, 45-52 (1974)

Takabe, T., Akazawa, T.: Catalytic role of subunit A in ribulose-1,5-diphosphate carboxylase from *Chromatium* strain D. Arch. Biochem. Biophys. 157, 303-308 (1973)

Takabe, T., Nishimura, M., Akazawa, T.: Presence of two subunit types in ribulose-1,5-bisphosphate carboxylase from blue-green algae. Biochem. Biophys. Res. Commun. 68, 537-544 (1976)

Trown, P.W.: An improved method for the isolation of carboxydismutase. Probable identity with Fraction I protein and the protein moiety of protochlorophyll holochrome. Biochemistry 4, 908-918 (1965)

Vasconcelos, A.C.: Synthesis of proteins by isolated *Euglena gracilis* chloroplasts. Plant Physiol. 58, 719-721 (1976)

Walker, D.A., Lilley, R.McC.: Ribulose bisphosphate carboxylase — an enigma resolved? In: Perspectives in Experimental Biology. Vol. II Botany; Sunderland, N. (ed.); pp. 189-198. Oxford: Pergamon 1976

Weissbach, A., Horecker, B.L., Hurwitz, J.: The enzymatic formation of phosphoglyceric acid from ribulose diphosphate and carbon dioxide. J. Biol. Chem. 218, 795-810 (1956)

Wheeler, A.M., Hartley, M.R.: Major mRNA species from spinach chloroplasts do not contain poly(A). Nature (London) 257, 66-67 (1975)

Wildman, S.G., Bonner, J.: The proteins of green leaves. I. Isolation, enzymatic properties and auxin content of spinach cytoplasmic proteins. Arch. Biochem. Biophys. 14, 381-413 (1947)

Wildman, S.G., Hongladarom, T., Honda, S.J.: Chloroplasts and mitochondria in living plant cells: cinephotomicrographic studies. Science 138, 434-436 (1962)

Wishnick, M., Lane, M.D., Scrutton, M.C., Mildvan, A.S.: The presence of tightly bound copper in ribulose diphosphate carboxylase from spinach. J. Biol. Chem. 244, 5761-5763 (1969)

Wishnick, M., Lane, M.D., Scrutton, M.C.: The interaction of metal ions with ribulose 1,5-diphosphate carboxylase from spinach. J. Biol. Chem. 245, 4939-4947 (1970)

Zelitch, I.: Plant productivity and the control of photorespiration. Proc. Natl. Acad. Sci. USA 70, 579-584 (1973)

Factors in Chloroplast Differentiation

C. Sundqvist*, L.O. Björn**, and H.I. Virgin*

*Department of Plant Physiology, University of Göteborg
Göteborg, Sweden
**Department of Plant Physiology, University of Lund
Lund, Sweden

I. General Introduction

It is evident that light plays the predominant role in differentiation of the chloroplasts. The development of chloroplast structure is, however, greatly affected by additonal factors, particularly hormones, nutritional state, and water stress. In the present chapter the most important factors and their effects will be discussed. A rather extensive discussion will be devoted to the effects exerted by hormones and nutrient state, because rather few such compilations are found in the literature. Comparative studies of photosynthesis and plastid structure are important to reveal the mechanism of photosynthesis, since differentiation of chloroplast structures can be shown to be well correlated with the ontogenesis of various parts of the photosynthetic apparatus.

II. The Light Factor

A. Introduction—Chlorophyll Formation in Darkness

In plants requiring light for chlorophyll formation only etioplasts are developed in darkness. They are characterized by rather undifferentiated structures, the most evident being the prolamellar body containing the precursor to the chlorophylls, protochlorophyll. This pigment, like the chlorophylls, can exist either in a nonesterified form, protochlorophyllide, or as an esterified form, protochlorophyll in the proper sense. In the light these etioplasts differentiate into fully developed

---

Abbreviations. $P-628$, $P-634$, $P-640$: Different forms of protochlorophyll(ide), the numbers denoting the wavelengths of their maximum absorption in the red. ALA: δ-aminolevulinic acid. P_{fr}: Phytochrome in its far-red absorbing form. P_{total}: Total content of phytochrome. $NADP$: Nicotinamide adenine dinucleotid phosphate. $NADPH$: NADP in reduced form. RNA: Ribonucleic acid. DNA: Deoxyribonucleic acid. GA: Gibberellic acid. ABA: Abscisic acid. IAA: Indoleacetic acid. BA: Benzyl adenine. $2.4\text{-}D$: 2.4-dichlorophenoxyacetic acid. CCC: (2-chloroethyl)-trimethyl-ammoniumchlorid. $AMO\ 1618$: 2-isopropyl-4-dimethylamino-5-methylphenyl-1-piperidene-carboxylate methyl chloride. $B\ 995$: N-dimethylamino succinamic acid. $PS\ I$ and $PS\ II$: Photosystems I and II in photosynthesis. $P-700$: The form of chlorophyll with max absorption at 700 nm, comprising the reaction center in photosystem I.

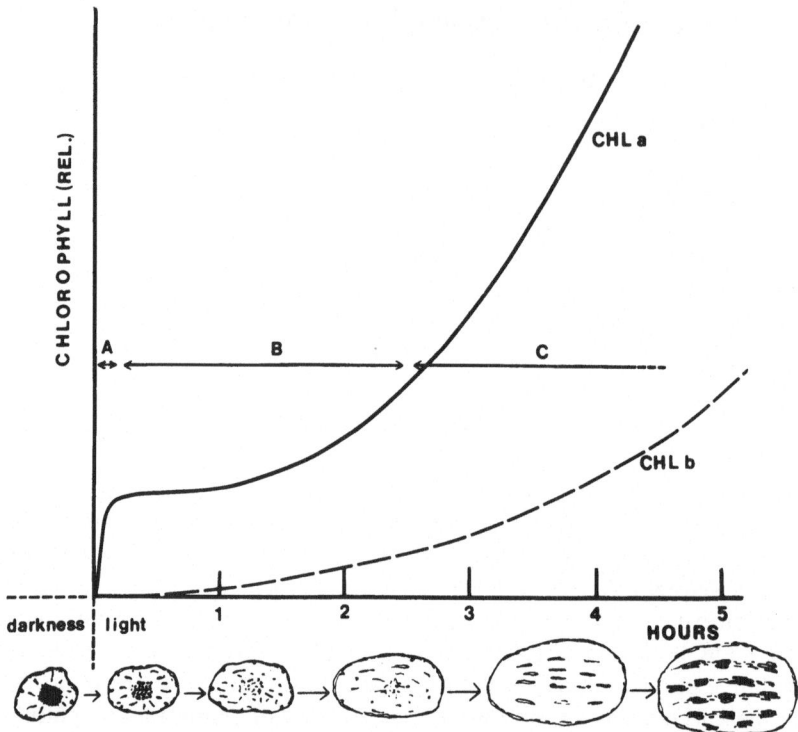

Fig. 1. Scheme of the gradual differentiation of an etioplast from a dark grown barley leaf during continuous irradiation together with the change in chlorophyll content during the greening process. *A* The phototransformation of protochlorophyll to chlorophyll a. *B* The lag phase in chlorophyll formation. *C* The accelerating phase in chlorophyll formation

chloroplasts through a series of stages which are described elsewhere in this volume. The differentiation pattern of an etioplast which in situ is exposed to light is shown in Fig. 1. This differentiation involves, besides ultrastructural changes linked to protein and lipid syntheses of, e.g., chlorophylls and their precursors, carotenes and enzymes, and other substances necessary for the function of the mature chloroplast. Among plants with a dark synthesis of chlorophyll, seedlings of gymnosperms and some algae have been best studied. Some algae require light, others do not, variations occurring within the same species.

The plastids in species not requiring light are often fully differentiated with distinct thylakoids. Dark conversion of protochlorophyll(ide) to chlorophyll(ide) a varies from more or less complete in *Pinus,* intermediate in *Picea,* to rather incomplete in *Larix.* In *Metasequoia glyptostroboides* grana form even in ordinary leaves developed in darkened buds contrary to most conifers. In species with incomplete transformation in darkness the plastids can contain prolamellar bodies in addition to grana. Except for some algae and conifers, the existing information concerning chlorophyll formation in darkness is very scanty and investigations with modern techniques are highly desirable.

Light acts in several different ways, and at least four different pigment systems are involved in the absorption of the active light energy: (1) protochlorophyll(ide), the excitation of which drives the reduction of protochlorophyll(ide) to chlorophyll(ide) a and also indirectly regulates the synthesis of the protochlorophyll(ide) precursor δ-aminolevulinic acid (ALA); (2) phytochrome, which seems to regulate the synthesis of many important chloroplast enzymes and which, indirectly, affects chlorophyll synthesis; (3) (in some plant materials) an unknown pigment absorbing blue and long-wavelength UV light, but not red light.

This chapter will primarily deal with systems (2) and (3). Some effects of strong light on chloroplasts (photodestruction), of UV inhibition with photoreactivation, and of short consecutive light flashes on plastid differentiation will also be mentioned.

B. Photoconversion of Protochlorophyll(ide) to Chlorophyll(ide) a with Protochlorophyll(ide) as the Photoreceptor

Protochlorophyll(ide) itself absorbs the photons active in its photoconversion (phototransformation) to chlorophyll(ide) a. (Koski et al. 1951). In addition to the photoconvertible, aggregated form of protochlorophyllide (P-650), which lacks phytol, leaves contain monomeric forms of protochlorophyll(ide). One of these forms, P-628, is not transformable, usually esterified with phytol (Bovey et al. 1974), and probably not protein-bound. There are perhaps several protochlorophyll(ide) forms having absorption spectra with maxima between these two extremes, but we shall refer to them simply as P-634. Usually P-634 is not directly transformable, but may serve as a precursor for the aggregated and transformable P-650. Aggregation is, however, no prerequisite for transformability. P-650 may be disaggregated by freezing and thawing of the plant material (Butler and Briggs 1966), or by more complicated methods (Nielsen and Kahn 1973) without loss of transformability. In many dark-grown plant materials which are able to form chlorophyll upon illumination, the experimentally determined absorption spectrum shows peaks at about 634 nm (Zeldin et al. 1975; Björn 1976). In some of these cases the P-634 may not itself be phototransformable. Instead, there might be spectroscopically undetectable traces of P-650 present which may undergo phototransformation. As soon as the initial P-650 has disappeared, new P-650 is formed from P-634. In *Euglena* only P-634 is spectroscopically detectable (Zeldin et al. 1975), but action spectra of chlorophyll formation (and other aspects of chloroplast development) show peaks at 650 nm (Egan et al. 1975).

Thus, it is not the state of aggregation per se that determines the phototransformability of protochlorophyll(ide). The important distinction seems to be whether or not the pigment is associated with a special "photoenzyme" (Sironval and Michel 1967), which also carries the reductant (see below). The photoenzyme has never been isolated, and only a few of its properties have been indirectly determined (Sundqvist 1973).

The photoconversion of protochlorophyllide to chlorophyllide a and the subsequent esterification to chlorophyll a are the dominating paths for chlorophyll a synthesis. At the transformation of protochlorophyll(ide) to chlorophyll(ide) a the pigment is reduced (two hydrogen atoms are added per molecule). Protochlorophyll(ide) dissolved in an organic solvent is not transformed into chlorophyll(ide) a by light, except under special conditions (Suboch et al. 1974). From certain plants like bean and barley, it is, however, possible to extract with aqueous buffer a protochlorphyll(ide)–protein complex (protochloro-phyll holochrome, Smith 1952) which is transformed by light to a chlorophyll(ide)–protein complex. The source of the two hydrogen atoms which are bound to or an integral part of the protein is still unknown. The direct source of the two hydrogen atoms is still unknown. However, NADPH can supply hydrogen at this process (Griffiths 1975) in vitro.

The photochemical conversion of protochlorophyll(ide) into chlorophyll(ide) triggers a number of light-independent processes (usually, but somewhat misleadingly, referred to as "dark reactions"; they also take place in light). Although the nature of these processes is not completely understood, they can be followed spectroscopically as shifts of the absorption maximum in the red.

The synthesis of ALA, δ-aminolevulinic acid, (Beale 1971) and thus that of protochlorophyll(ide), is resumed after the plants are illuminated, and can in darkness be seen as an increase in absorption at 650 nm. In dark-grown plants protochlorophyll(ide) synthesis is regulated at the point of ALA formation (Nadler and Granick 1970). Several explanations can be given for this regulation. One possibility is that ALA synthesizing enzyme is inhibited by P-650 (i.e., by the pigment–photoenzyme complex) (Sundqvist 1970). Another is that the photoenzyme is identical to the enzyme synthesizing ALA, and that it is inactive when protochlorophyllide is bound to it.

In *Euglena* light absorption in a substance with an absorption peak in the green region influences chloroplast development (Egan et al. 1975), but no further details are known about this pigment.

C. The Phytochrome System

Effects of red light on the growth and fine structure of chloroplasts have been known for some time (Mego and Jagendorf 1961), and it was evident quite early that they could be attributed to effects of the phytochrome system. Chloroplast formed under irradiation with far-red light [with wavelength components giving small amounts of P_{fr} but not absorbed by protochlorophyll(ide)] have a diameter of about 5 μm (in mustard seedlings), with large paracrystallinic prolamellar bodies, while the corresponding diameter of etioplasts from dark-grown plants is only 1.5 μm. Here also the prolamellar bodies are smaller (Mohr and Kasemir 1975). Even a brief red light impulse—equal to that diminishing the lag phase—increases the proportion of crystalline bodies during the subsequent dark period (Berry and Smith 1971). The formation of thylakoids, however, does not seem to be under the control of phytochrome (Kasemir et al. 1975) but is correlated to the protochlorophyll(ide)–chlorophyll a transformation.

Isolated chloroplasts are also affected by red light. Like plastids in situ, they show an increase in the rate of ultrastructural development compared to controls.

The response shows a clear reversibility between effects by red and far-red irradiation, typical for a phytochrome-governed response (Wellburn and Wellburn 1973). While red light—via the phytochrome system—seems to be of fundamental importance for the differentiation of plastid strucutures, even low intensities of green light have been reported to have effects on the ultrastructure (Possingham et al. 1975). Even the replication of the chloroplasts is stimulated by green light, but at higher intensities. In this respect, it equals white, blue, and red light if supplied at intensities of the order of 50 μW × cm^{-2}. Thus, all parts of the visible spectrum are active with respect to plastid development and differentiation, and one must be very careful when using the so-called green safe light for work in a dark room. In green "safe light" a photostationary state of about 50% P_{fr}/P_{total} will be obtained (Smith 1975). The effect of green light might very well be a phytochrome effect well distinguished from the effects exerted by blue light to be described later in this article.

The phytochrome-governed effects on pigment synthesis and chloroplast development and differentiation in general is mirrored in effects on enzyme systems of great importance for general cell metabolism (Smillie and Scott 1969). A list of 22 enzymes for which an influence by phytochrome in far-red form has been definitely proven and of 13 for which more definite proof is lacking is given by Schopfer (1972). There is no reason to believe that such an influence is lacking for enzyme systems located in the chloroplasts, even if there are indications that the synthesis of enzymes of chlorophyll biosynthesis takes place in the cytoplasm (Beisenherz and Schneider 1974). In fact, it has been found that red light also has a pronounced effect on the enzyme activity in dark-grown plastids (Margulies 1965).

From studies of the effect of inhibitors of nucleic acid and of protein synthesis, it has become increasingly clear that the phytochrome system is also involved in protein synthesis and differentiation of the morphological framework of the chloroplast (Lange and Mohr 1965). Of particular interest in this connection is the phytochrome-governed synthesis of the protein which, together with protochlorophyll(ide), forms the holochrome capable of phototransformation (Sundqvist 1970) and the effects on the esterification with phytol (Liljenberg 1966), on carotenoid synthesis (Cohen and Goodwin 1962), on lipoquinone synthesis (Mitrakos 1961) and on the content of plastid RNA (Scott et al. 1971). An important fact in this connection is that continuous irradiation is not essential for the production of RNA. Even a short light impulse will have a long-lasting effect on RNA synthesis. The synthesis of at least part of the chloroplast RNA seems to be templated by chloroplast DNA (Gibbs 1967). According to newer findings the phytochrome seems to regulate the membrane permeability during chloroplast development (Hampp and Schmidt 1977).

D. Effects on Chloroplast Development Specifically Induced by Blue Light

In some plant materials there is a specific requirement of short wave radiation (near ultraviolet, violet, or blue, henceforth called blue for simplicity) for the development of chloroplasts.

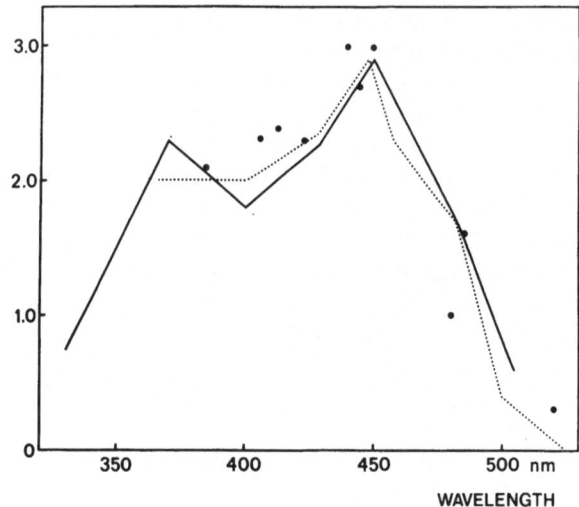

Fig. 2. Action spectra for induction of chloroplast formation in wheat roots (Björn 1967b; *dotted line*), chloroplast rearrangement in *Vaucheria* (Fischer-Arnold 1963; *solid line*), and leaf dropping in *Mimosa* (Fondeville et al. 1967; *dots*)

An absolute blue-light requirement for chloroplast formation (measured as chlorophyll accumulation) was first described by Björn et al. (1963) for excised wheat roots growing in aseptic cultures. Blue light has an inductive effect: three days of blue light is in itself not enough to cause development of chloroplasts, but this pretreatment enables the roots to form chloroplasts under subsequent red light. Conversely, red light preceding the blue increases chlorophyll accumulation during the blue light period (Björn 1965). The action spectrum for the specific blue-light effect (Björn 1967b) is in Fig. 2 compared with action spectra for two other specific blue light effects in plants. These action spectra are very similar to absorption spectra for various carotenoids and flavoproteins present in plants.

Blue light is necessary not only for the synthesis of chlorophyll, but also for the development of the chloroplast as a whole. A typical chloroplast enzyme, D-glyceraldehyde 3-phosphate: NADP oxidoreductase (phosphorylating, EC 1.2.1.13) is formed in excised wheat roots only after irradiation with blue light, and synthesis sets in after the same lag phase as for chlorophyll (Björn 1967c). The underlying cause seems to be a blue light-induced synthesis of specific types of RNA (Dirks and Richter 1973).

Wheat roots are not unique in their absolute blue light requirement for chloroplast development. Roots of cucumber *(Cucumis sativus)* and pea *(Pisum sativum)* show a similar behavior as do cell cultures of *Nicotiana tabacum* (Bergmann and Berger 1966) and dark-adapted potato tubers (Berger and Bergmann 1967).

Among plants showing a quantitative blue light requirement for chloroplast development, fern prothallia and the alga *Acetabularia* have been particularly well studied. Although chloroplasts develop also under red light, they become much larger under blue (Mohr 1956). The effect is reversible: The time for half-way adaptation to new light conditions is about a week at 20° C, but the effect is perceptible already after 5 h (Cran and Dyer 1975). Blue light-induced growth of chloroplasts is dependent upon blue light-induced RNA and protein synthesis (Ohlenroth and Mohr 1963; Bergfeld 1964). There is a difference in protein synthesis between blue light-developed and red light-developed fern chloroplasts (Raghavan and de Maggio 1971).

Chloroplasts of pea plants grown under blue light contain a greater number and larger grana than do those of red light-grown plants (Vlasova et al. 1971). When fern protonemata are transferred from red to blue light (but not when they are transferred from red to white light), prolamellar bodies temporarily appear after about one hour (Cran and Dyer 1975). Apparently prolamellar bodies are formed only when a blue-mediated process (protein synthesis?) runs ahead of a red-mediated one [protochlorophyll(ide) conversion or a phytochrome-dependent reaction]. Blue light (as compared to red light) changes the synthetic functions of chloroplasts, generally speaking in the direction of more nitrogenous assimilates and away from carbohydrate synthesis.

One functional change is probably directly dependent on structural changes in the thylakoid membrane: chloroplasts from plants grown under blue light have a larger electron transport capacity (per unit chlorophyll) than those grown under red light (Appleman and Pyfrom 1955).

Blue light also triggers or greatly favors the development of peroxisomes, and thus enzymes typical for these bodies, such as glycolate: O_2 oxidoreductase, EC 1.1.3.1. In wheat roots the blue light requirement for this process, as for chloroplast development, is absolute (Björn 1967b), while in other cases (Voskresenskaya et al. 1970) it is quantitative. Although peroxisomes are organelles closely cooperating with chloroplasts, the light effect on peroxisome development can be demonstrated under conditions when chloroplast development does not take place (Feierabend and Beevers 1972). The blue light effect on peroxisome development thus seems to be independent of that on chloroplast development.

Finally blue light may have a stimulating effect on chloroplast division (replication). In *Acetabularia mediterranea* the effects on division and on protein synthesis are in balance, so that the amount of protein per chloroplast remains unchanged (Schmid and Clauss 1974).

E. Effects of Ultraviolet Radiation

Ultraviolet radiation in the wavelength range 200–300 nm has two main effects on chloroplasts: (1) Inhibition of photosynthetic electron transport at a site close to photosystem II, probably by affecting the plastoquinone (Bishop 1961) or by disrupting the structural integrity of the photosynthetic membranes (Mantai and Bishop 1970). (2) Inactivation of plastid DNA, resulting in inhibition of plastid replication. UV affects plastid replication with a main peak at 260 nm and a minor at 280 nm (Schiff and Epstein 1966), probably due to the effects on a plastid DNA-protein complex. The inactivation kinetics agrees with target theory better than most UV effects on DNA (Lyman et al. 1961).

The inactivation of the replication process is completely reactivated if a sufficient dose of reactivating light is administered immediately after the inactivating UV. The action spectrum for reactivation has a broad band extending from below 330 nm to about 470 nm, with a maximum near 370 nm (Schiff et al. 1961). UV radiation may delay chlorophyll break-down during senescence by inactivating certain enzymes of importance to this process (Skokut et al. 1975).

F. Special Light Effects—Photodestruction and Effects of Flashing Light

The differentiation of the proplastid into chloroplasts with distinct grana and stroma regions is greatly influenced by, in addition to light quality, light intensity and the time sequence in which the light is administered. At an intensity just high enough to transform the protochlorophyll(ide) initially present in the proplastids also causing a destruction of the primary prolamellar bodies (Virgin et al. 1963), grana formation eventually starts. After about 4 h of irradiation, however, a new formation of prolamellar bodies takes place (Henningsen and Boynton 1970). This formation begins after the end of the lag phase in chlorophyll formation and is thus correlated with an increased rate of protochlorophyll(ide) formation. The effect of light intensity on the later steps of differentiation is mainly on the thylakoid formation, i.e., the higher the intensity, the more thylakoid layers.

This relationship is valid up to a certain intensity. High intensity light, at least when given in the earlier phases of chloroplast differentiation, gives rise to photooxidations, and a certain destruction of the plastid morphology can be observed. The different forms of protochlorophyll(ide) and chlorophyll(ide) formed during the so-called Shibata shift (Shibata 1957) show different sensitivities to light destruction (Axelsson 1974, 1976). This implies that the degree of sensitivity decreases as chloroplast differentiation proceeds. During the lag phase in chlorophyll formation light sensitivity is very high, and there is evidence that the lag phase itself is partly the result of a certain pigment destruction at the start of irradiation (Virgin 1972).

A normal structure of the chloroplast with distinct grana and stroma is only obtained in comparatively long regimes of irradiation. If the light is given in short periods with darkness in between or the etioplasts are irradiated with photoflashes, they will show a quite different morphology. Layers of parallel elongated double membranes are formed very much like the structures found in the bundle-sheath chloroplasts in C_4-plants. These membranes have sometimes been called primary thylakoids (Sironval et al. 1968). There are no grana stacks characteristic of mature chloroplasts, but only an occasional stacking of two or three thylakoids. The plastids thus obtained do not exhibit any photosystem II activity (Remy 1973), and the ratio chlorophyll a/chlorophyll b can reach very high values (20–30; Akoyunoglou et al. 1966).

III. Chemical Factors

A. Introduction

Numerous chemicals influence chloroplast differentiation. They can arbitrarily be divided into five main groups: (1) metabolites which can function as precursors or energy-providing compounds; (2) substances which can function as growth regulators; (3) metabolic inhibitors; (4) substances functioning as herbicides or pollutants; (5) mineral nutrients. As mentioned in the general introduction, most of the discussion will be devoted to groups 2 and 5. As to effects of herbicides on chloroplast differentiation, few reports are available (review see Anderson and Thomson 1973). A study of this influence might, however, give an insight as to the function of these compounds.

Effects of several metabolites on the chloroplast metabolic flow have been shown, e.g., a diminished lag phase with sucrose supply (Wolf and Price 1960) and an increased production of lipoquinones with shikimic acid and phenylalanin (Lichtenthaler 1973).

The chloroplast differentiation is thus dependent on an exchange of substances between the chloroplast and other parts of the cell. Factors altering the exchange, e.g., an increased supply of metabolites, can provoke an abnormal pattern of differentiation. The rate of formation of protochlorophyllide and chlorophyll can be increased with ALA, the precursor of porphyrins, but simultaneously the sensitivity to photobleaching is increased, and thus the stability of the pigments is lost (Granick 1959; Axelsson 1974). This shows the necessity of a precise combination between different metabolites, a combination which is probably dependent on the presence of suitable mineral nutrients and regulated by various plant hormones.

B. Effects of Hormones

The hormonal regulation of chloroplast maintenance seems at present to be very complex. At least four exogenously supplied growth regulators have a direct influence on the chloroplast (cytokinins, GA, ABA, IAA) and some of them can also be found inside the chloroplast. From the data available in the literature it can be anticipated that the phytohormones control not only the biosynthesis of thylakoids, but also their rate of turnover (Lichtenthaler and Grumbach 1974). Furthermore, there are several indications that growth regulators also influence chloroplast differentiation by promoting or inhibiting production of different substances in the cytoplasm. A very little-explored field in connection with this problem is the survival and/or differentiation of etioplast/chloroplast preparations in media containing various growth regulators. At present this seems to be one of the more promising ways to distinguish a real chloroplast effect from that induced via the cytoplasm. It is necessary to solve this question before a clear picture of the hormonal control of chloroplast differentiation can be obtained.

1. Cytokinins

The most well-known influence of cytokinins on chloroplasts is the preservation of chlorophyll during senescence (Chibnall 1939; Kende 1971; Hall 1973). Also the formation of chlorophyll in dark-grown leaves exposed to light is affected (Banerji and Laloraya 1967; Shlyk et al. 1970). Incubation in benzyl-adenine (BA) before irradiation stimulates chlorophyll accumulation in cucumber. Treatment with BA

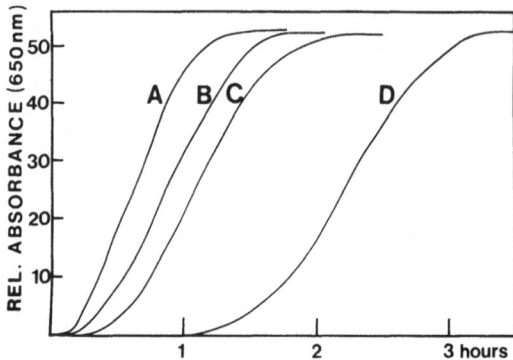

Fig. 3. The influence of age and cytokinin on the regeneration of absorbance at 650 nm in dark-grown leaves of wheat after a light-flash treatment. Dark-grown wheat leaves were irradiated with a light flash, which converted the protochlorophyllide to chlorophyllide. The regeneration of protochlorophyllide was measured in vivo as increase in absorbance at 650 nm. *A* 6-day-old material; *B* 7-day-old material; *C* 6-day-old material floated 20 h on a 10^{-4} M kinetin solution; *D* 6-day-old material floated 20 h on a phosphate buffer solution (pH = 6.6)

eliminates the lag phase in chlorophyll formation and in the production of ALA after competitive inhibition with levulinic acid (Fletcher et al. 1973). The effect of BA on chlorophyll production is mediated by an increased protein synthesis. The effect of BA on leaves resembles the effect of pre-irradiation with red light (Beevers et al. 1970). ALA-dehydratase is supposed to be one of the enzymes affected (Sundqvist et al. 1975), but an effect on the permeability of the chloroplast membranes is also highly probable.

In aging dark-grown seedlings, cytokinins counteract the loss of protochlorophyll(ide), thereby increasing protochlorophyll(ide) (P-634) formation and protochlorophyll(ide) (P-650) regeneration, and eventually chlorophyll formation ability (Stobart et al. 1972). It seems likely that chloroplast differentiation is influenced by the cytokinins in such a way that aging is prevented. In Fig. 3 it is shown that the decrease in ability to generate protochlorophyll(ide) was delayed by kinetin treatment, but it was never restored to the level existing before excision of the leaves. The same was found for chlorophyll formation ability. In dark-grown material or young developing seedlings, cytokinin application probably keeps the cells in an immature state, but cytokinin is seldom able to increase the activity of any special biosynthetic pathway over the value of the initial materials. Besides the chlorophyll pigments, several other chloroplast components are influenced by cytokinins, e.g., plastoquinone-9, α-tocopherol and carotenoids (Straub and Lichtenthaler 1973a). It seems likely that kinetin promotes the formation and incorporation of prenyl chains C-45 into plastoquinone-9 and of C-20 into chlorophyll leaving less isoprenoid precursors for formation of carotenoids and α-tocopherol. It is also evident that kinetin in *Raphanus* seedlings increases the number of photosynthetic reaction centers and reduces the chlorophyll antenna, which enables the plant to achieve higher photosynthetic rates and growth (Buschmann and Lichtenthaler 1974).

In rye seedlings cytokinin increases the photosynthetic enzyme activities, whereby NADP-glyceraldehyde-3-phosphate dehydrogenase and ribulose-bisphosphate carboxylase are more influenced than are mitochondrial or cytoplasmic enzymes (Feierabend 1969). A disturbance of the normal cytokinin supply caused by removing the roots from the seedlings also lowers the activity of ribulosebisphosphate carboxylase. The original activity is restored if cytokinins are supplied. Kinetin application to green leaves results in an increased amount of linolenic acid (Kull and Büxenstein 1974). In greening cotyledons, cytokinin accelerates ultrastructural differentiation. Grana and thylakoids are formed more abundantly and prolamellar bodies become larger (Mlodzianowski and Gezela 1974), even if conversion of the prolamellar bodies to thylakoid material seems to be facilitated. In nongreen root callus of *Melilotus alba*, cytokinins can induce chlorophyll production (El Hinnawy 1974). Rapid chlorophyll synthesis often takes place when growth of cultivated tissues becomes more or less stationary (Laetsch and Stetler 1965). By decreasing the growth rate of tobacco callus, formation of prolamellar bodies can be induced (Stetler and Laetsch 1968).

2. Gibberellins

Plants treated with gibberellic acids (GA) are more or less pale green in appearance. The first interpretation of this effect was that the increase in cell volume caused by GA was not correlated with an increase in chlorophyll content, and that thus a "dilution" of the chlorophyll content of the leaves was obtained. This might be an explanation, but it is also known that GA treatment affects the chlorophyll content not only per unit leaf area, but also per whole cotyledon or whole leaf. In GA-treated seedlings of *Raphanus* and *Hordeum* chlorophyll a showed a greater decrease than chlorophyll b (Szalai 1968). Even other chloroplast constituents show a response to GA treatment, such as carotenes (Szalai 1969; Tietz and Dörffling 1969; Straub and Lichtenthaler 1973a), xanthophylls (Szalai 1969), isoprenoid benzoquinons, prenyl lipids, and α-tocopherol (Straub and Lichtenthaler 1973a). These effects are difficult to interpret as both increases and decreases occur. The gibberellins have, like the cytokinins, a retarding effect on chloroplast senescence. The GA effect is partly counteracted by abscisic acid (Back and Richmond 1971). In some cases, when kinetin had no effect, GA was found to prevent the breaking down of chlorophyll (Bata and Neskovic 1974).

3. Abscisic Acid

ABA is present in the chloroplast (Railton et al. 1974). Further addition affects chloroplast differentiation by decreasing chlorophyll, carotenoids, and vitamin K_1 in greening barley seedlings (Lichtenthaler and Becker 1970) and in maize (Mercer and Pughe 1969). It has been suggested that ABA depresses the light-induced gene activation, leading to decreased chlorophyll and thylakoid formation (Lichtenthaler and Becker 1970). ABA inhibits the activity of glutamate dehydrogenase (Huber and Sankhla 1974), which is interesting as the biosynthetic pathway for ALA seems to include glutamate and α-ketoglutarate (Beale and Castelfranco 1974). A more direct influence than via the genes can thus be presumed. Recently it was also found that ABA is released from etioplasts during water stress (Loveys 1977).

ABA accelerates senescence, shown as an increased chlorophyll break-down (El-Antably et al. 1967). It also accelerates ultrastructural changes in detached leaves incubated in darkness (Mittelheuser and Van Stevenink 1971a). The grana and chloroplast ribosomes disappear, and osmiophilic globules are formed. In the light, ABA inhibits the accumulation of starch (Mittelheuser and Van Stevenink 1971b).

4. Auxins

The influence of auxins on chloroplast differentiation is found primarily in callus cultures. An increase of the indole acetic acid (IAA) concentration in the medium promotes starch formation in the plastids. The shape of the plastid changes from ellipsoid to spherical. 2,4-D in high concentrations inhibits the development of thylakoids and induces formation to amyloplasts (Sunderland and Wells 1968). In *Raphanus* cotyledons, IAA-enhanced synthesis of chlorophylls, carotenoids, and plastoquinone-9 is coupled to an increase in thylakoid formation. IAA enhances (as does kinetin, but less pronouncedly) the light-induced change of benzoquinone synthesis in etioplasts (more plastoquinone-9, less α-tocopherol formation) (Straub and Lichtenthaler 1973a,b). The concentration of chlorophyll P-700 and the Hill-activity rates are much higher in the IAA-treated plants, which points to an increase in the number of photosynthetic reaction centers (Buschmann and Lichtenthaler 1974).

5. Growth Retardants

CCC, AMO-1618, B995 and coumarin inhibit chlorophyll biosynthesis in greening cotyledons of cucumber. They also delay senescence seen as degradation of chlorophyll, starch, protein and RNA (Harada 1966; Beevers and Guersney 1967; Mishra and Misra 1968; Knypl 1969a).

C. Effects of Mineral Nutrients

1. Introduction

Most mineral deficiencies alter the pattern of chloroplast development and affect the ultrastructure and activity. Indeed, lack of chlorophyll (chlorosis) is one of the most conspicious features of nutrient deficiency. We know very little about how these alterations take place. In only a few cases a more precise function for the mineral in question has been clearly elucidated. The following survey gives a short orientation on the possible effects of different mineral nutrients on chloroplast development.

Table 1. The influence of mineral deficiency on the content of different chloroplast constituents and on chloroplast activity. Values given as % of control, except for the chlorophyll a/b ratio

Parameter measured	Content or activity at different mineral deficiencies					
	N	P	K	Ca	S	Mg
Chl[a]	17	105	62	233	67	26
mg/g dry wt						
Ratio *Chl* a/b[a]	2.01	2.83	2.67	2.63	2.45	2.55
Plastoquinone[a]	36	70	150	76	51	34
Vitamin K[a]	96	134	100	99	17	24
α-Tocopherol[a]	21	122	23	207	83	37
CO_2 fixation						
CO_2/min/g f wt[b]	22	38	71	93	30	47
CO_2/min/mg *Chl*[b]	55	41	41	74	54	70
PS I activity						
μmol acceptor reduced/mg						
Chl/h						
$TMPD \rightarrow MV$[c]	100	100	100	91	80	80
PS II activity						
μmol acceptor reduced/mg						
Chl/h						
$H_2O \rightarrow DCIP$[c]	241	79	98	97	171	100
$DPC \rightarrow DCIP$[c]	245	80	94	79	167	97
μmol acceptor reduced/mg						
Chl/h						
mg chloroplast protein/	439	166	136	181	153	152
mg *Chl*[c]						

Values recalculated after: [a] Barr et al. (1971), [b] Bottrill et al. (1970), [c] Baszynski et al. (1972) Abbreviations: *DCIP* 2,6'-dichlorophenol indophenol; *DPC* diphenylcarbazide; *TMPD* N, N, N', N-tetramethyl-p-phenylene diamine; *Chl* chlorophyll

2. Macronutrients (N, P, K, Ca, S, Mg)

Decreased chlorophyll formation during nitrogen deficiency is a well-known phenomenon. Nitrogen deficiency also affects the formation of carotenoids in higher plants with a strong correlation between carotenoid and chlorophyll biosynthesis. The inhibition of carotenoid biosynthesis seems to be at the dimerising step of geranyl-geranyl-pyrophosphate. The isoprenoid plastid quinones, the naphto-quinone vitamin K_1 and the benzoquinone derivatives plastoquinone-9, α-tocopherol and α-tocoquinone are, in turn, less affected. Relative to chlorophyll their synthesis is even increased (Barr et al. 1971; Verbeek and Lichtenthaler 1973). If is assumed that the reduced rate of protein and aromatic acid formation increases the flow of aromatic precursors towards vitamin K_1 and the lipophilic benzoquinones (Lichtenthaler 1973).

Nitrogen deficiency can reduce the chloroplast to about one-half of the normal length. At the same time there is an increase in the size of osmiophilic plastoglobuli (Tevini 1971a; Hall et al. 1972) containing the isoprenoid plastid quinones (Lichtenthaler and Weinert 1970) and other excess lipids. Nitrogen deficiency in

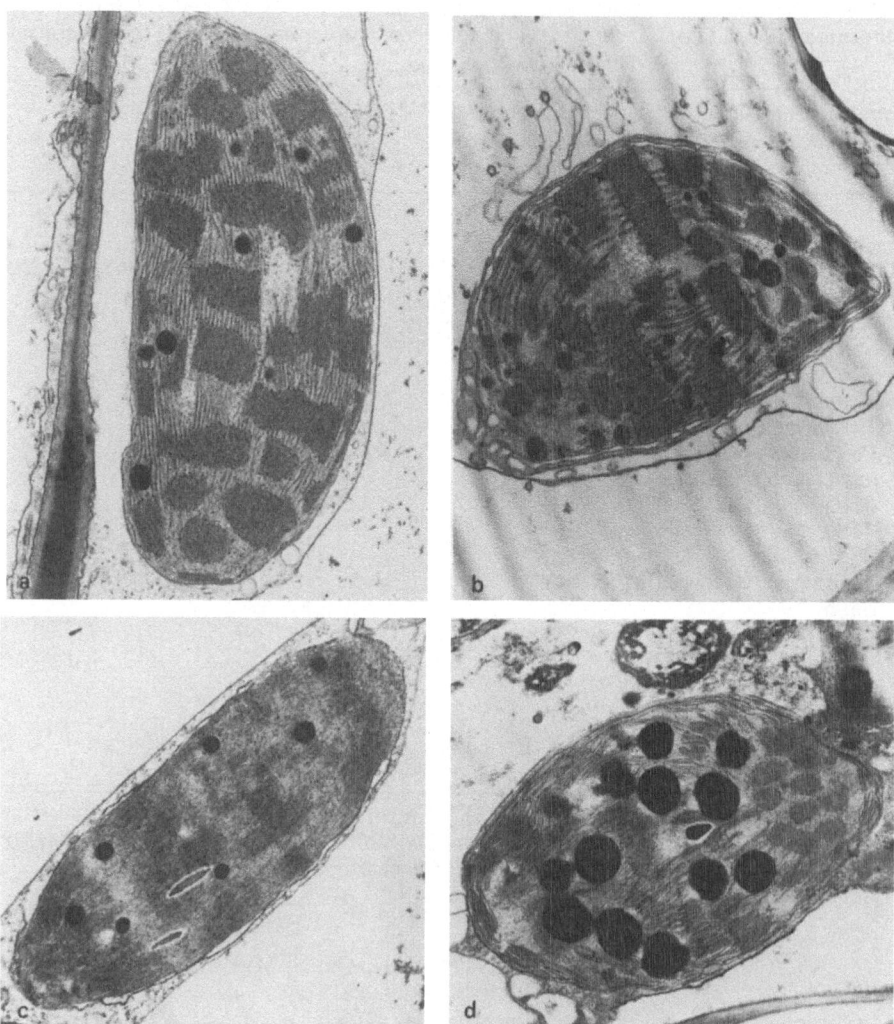

Fig. 4. a Chloroplast from full-nutrient plant. Osmiophilic globules, well developed grana and stroma lamellae, grana stacks, and peripheral reticulum are present. **b** Chloroplast from sulfur-deficient plant. Large grana stacks are randomly oriented. There is a marked reduction in stroma lamellae. Grana stacks are large and well-developed. Numerous small osmiophilic globules are present. **c** Chloroplast from potassium-deficient plant. The chloroplast is filled with a lamellar system having poor contrast. Grana stacks are difficult to distinguish. Starch grains and osmiophilic globules are visible. **d** Chloroplast from magnesium-deficient plant. Numerous large osmiophilic globules occur in the stroma. The electron micrographs, showing mesophyll chloroplasts of maize, are reproduced by kind permission of Prof. J.D. Hall (Hall et al. 1972)

Impatiens reduces the rate of chlorophyll and of glyco- and phospholipid formation, the latter being the major group of thylakoid lipids (Tevini 1971a).

The nitrogen deficiency-induced decline in photosynthesis is strongly correlated with the decline in ultrastructural order (Table 1). The effect is seen as lower CO_2 uptake per unit leaf area (Bouma 1970) and per chlorophyll basis (Bottrill et al.

1970). PS II activity increases while PS I activity shows a smaller change as counted on a chlorophyll basis. This can probably be attributed to a decreased amount of antenna chlorophyll with an unchanged amount of reaction centers (Baszynski et al. 1975). Effects of deficiency of nitrogen, phosphorus, potassium, calcium, sulphur and magnesium on the chloroplast content of chlorophyll and lipoquinones, as well as on chloroplast function, are assembled in Table 1.

The development of the ultrastructure can be more or less normal although macronutrients are missing, but, for example, calcium deficiency causes the lamellae to break down totally (Das 1973). The appearance of magnesium, potassium, and sulphur deficient chloroplasts is shown in Fig. 4. Phosphorus deficiency decreases the levels of chlorophylls monogalactosyldiglyceride and phospholipids; while there is a pronounced increase in the concentration of digalactosyldiglyceride and the chloroplast sulpholipid (Tevini 1971b). At the same time the size of osmiophilic plastoglobuli shows an increase (Tevini 1971a). As a constituent of the chlorophyll molecule, magnesium obviously has a major role in chloroplast development. Magnesium deficiency has a pronounced influence on the ultrastructure of chloroplasts, Fig. 4 (Whatley 1971; Hall et al. 1972), concentrations of lipoquinones (Barr et al. 1971), chloroplast enzyme activity (Evans and Sorger 1966) and rates of photosynthesis and respiration (Bottrill et al. 1970).

3. Micronutrients (Fe, Mn, Cu, B, Mo, Zn)

Lack of iron inhibits chlorophyll synthesis (Price 1968). In higher plants, at least two metabolic steps in the biosynthesis of chlorophyll are influenced by iron: (1) the formation of ALA and (2) the conversion of coproporphyrinogen to protoporphyrinogen (Marsh et al. 1963a; Stephan and Machold 1969). Iron deficiency is manifested by decreased incorporation of ^{14}C precursors into ALA and a subsequent restoration of incorporation when iron is added (Marsh et al. 1963b; Hsu and Miller 1965; Machold and Stephan 1969). In iron-deficient leaves, incorporation of ^{14}C from α-ketoglutarate decreased more than incorporation from succinate compared to normal leaves (Marsh et al. 1963b), in accordance with the proposed synthesis of ALA from glutamate or α-ketoglutarate (Beale et al. 1975). Iron deficiency causes the grana lamellae to swell (Whatley 1971). Upon severe deficiency, there is a decrease in the number and dimensions of grana and a dissolution of the discs in the grana and intergranal thylakoids (Lamprecht 1961). CO_2 fixation as expressed per unit fresh weight shows a decrease in iron-deficient leaves, but when expressed on a per chlorophyll basis it is not influenced (Bottrill 1970). Photophosphorylation seems to be stimulated in chlorotic tissue when calculated per unit chlorophyll (Ostrovskaya and Zaitseva 1967).

Manganese is known to be associated with PS II activity in the chloroplast (Eyster 1958). PS II activity is found to be proportional to the content of manganese within the chloroplast (Possingham and Spencer 1962). Removal of 2/3 of the normal Mn pool (ca. 3 Mn per 200 chlorophyll molecules in system II) by extraction abolished the capacity for O_2 evolution. Addition of manganese to deficient cells caused reactivation of PS II (Cheniae and Martin 1970). The final 1/3 of the manganese pool is very tightly bound to the chloroplast lamellae and has been suggested to have a structural role in the stacking of the lamellae (Possingham et al. 1964). The structural manganese has recently been shown to bind to a structural protein consisting of a peptid chain of M.W. of 25,000 (Lagoutte and Duranton 1975). This agrees with the observation that mild deficiency has no influence on the ultrastructure of chloroplasts (Cheniae and Martin 1968). Upon severe deficiency, however, the lamellar system is broken down (Mercer et al. 1962).

Copper is necessary micronutrient for the chloroplast as a part of plastocyanin, but other micronutrients also influence the ultrastructure and activity of chloroplasts. Lack of Cu, B, Zn, and Mo decrease CO_2 fixation measured per minute and gram fresh weight. Measurement on a chlorophyll basis gives a decrease only for plants grown with Mo deficiency. The chlorophyll content is strongly decreased in Mo-deficient plants as well as in those lacking Zn, but is slightly increased in Cu- and B-deficient plants (Bottrill et al. 1970). A reduced number of grana and intergrana connections was obtained in Zn-deficient plants and Cu deficiency caused swelling of the grana. For all micronutrient deficiencies the stroma showed a relative increase (Vesk et al. 1966).

IV. Effects of Water Stress

Chlorophyll formation, as well as structural and functional development of chloroplasts, is influenced by water stress in the plant (reviewed by Hsiao 1973). Chloroplasts of dark-grown wheat leaves, irradiated for 48 h (10,000 lux) under water stress (-14 bars), showed a reduced number of grana per plastid and a reduced number of thylakoids per grana compared to non-stressed chloroplasts. Chlorophyll formation was decreased by 25% (Duysen and Freeman 1974). At still stronger stress conditions (about -19 bars), the decrease can be up to 80% (Virgin 1965). The plastids in stressed leaves showed a decreased length and a dilation of the membranes in grana and stroma thylakoids, evident after 24 h of irradiation, as well as loss of chloroplast ribosomes. Plastoglobuli and fibrillar structures also seem to be influenced by the stress condition. The difference between stressed and non-stressed plants was first visible after 12 h of greening, indicating that as long as the etioplast can rely on reserve material, water stress is of minor importance.

Another conspicuous feature of the plant cell under water stress is the accumulation of free proline (cf. Hsiao 1973). The proline accumulated can be regarded as a reserve material, and when water stress is released it can be metabolized. It has been proposed that proline could act as a precursor for chlorophyll biosynthesis (Breyhan et al. 1959). This is hardly probable, however, as the incorporation of radioactivity from ^{14}C-proline into ALA under normal conditions is very low (Beale et al. 1975). It could be possible that during stress conditions proline can function as reserve material for a renewed chlorophyll synthesis when the stress is relieved.

A lowering of the water potential around the chloroplasts decreases their electron transport measured as reduction of dichlorophenol indophenol (Potter and Boyer 1973). From the viewpoint of differentiation, the findings that desiccation in vivo has a greater effect than exposing isolated chloroplasts to low osmotic potential is especially interesting.

Down to – 11 bars leaf water potential, electron transport in the chloroplast limits photosynthesis (Keck and Boyer 1974). At a more severe desiccation (– 17 bars), photosynthesis is limited by photophosphorylation. CO_2 fixation, too, seems to be affected by the water potential, and 69% inhibition can be obtained by isolated chloroplasts in a sorbitol medium with an osmotic potential of – 16 bars (Plaut and Bravdo 1973). This effect might be attributed to the effect of water stress on the activity of ribulose-5-phosphate kinase and ribulose-bisphosphate carboxylase (Plaut 1971).

V. Conclusions

It is obvious from the foregoing that the differentiation of the chloroplast is subject to many influences; its development is thus an intricate process controlled by external factors such as light and temperature and also by internal factors such as hormones and mineral nutrients.

Our present knowledge may seem comprehensive, but in spite of a vast literature many fundamental questions are still unsolved. There is a tremendous difference between the etioplast in the dark-grown plant and the fully differentiated functioning chloroplast. We can passably describe the changes taking place upon illumination, but confusion exists about the degree of participation of different photoreceptors such as phytochrome, protochlorophyll, chlorophyll, and possible blue light receptors in these changes. The way of transmission of the message from the photoreceptor is also unknown. Likewise our knowledge concerning the differentiation of the photosynthetic membrane and the building up of its complicated molecular organization is still largely hypothetical even if, for example, electron microscopic evidence has revealed a pattern which makes it possible to construct plausible membrane models of the system.

The fact that internal factors such as hormones can have effects similar to those of light indicates a chain of reactions whereby hormone metabolism may play an important role in mediating the light effect. The mineral nutrition and the water potential of the chloroplast-containing cell has long been known to influence chloroplast development. Concerning the mechanism behind this influence, good progress has recently been made, but at the molecular level the real function of only a few elements, e.g., nitrogen and iron, has been fairly well investigated. In this field there is much to be done, and as our knowledge on the mode of action of the various hormones and mineral nutrients is gradually increased, a clearer insight into the mechanism of chloroplast differentiation will be obtained. As the chloroplasts have their own DNA there are also strong reasons to believe that some hormones may, for example, act directly on the gene mechanism independently of the nucleus of the cell in which the chloroplasts are situated.

Note Added in Proof. The manuscript for this article was completet in 1976.

Acknowledgments. The authors are grateful to Mr. Michael Coveney for linguistic revision of the manuscript.

References

Akoyunoglou, G, Argyroudi-Akoyunoglou, J.H., Michel-Wolwertz, M.R., Sironval, C.: Effect of intermittent and continuous light on chlorophyll formation in etiolated plants. Physiol. Plant. 19, 1101-1104 (1966)

Anderson, J.L., Thomson, E.W.: The effects of herbicides on the ultrastructure of plant cells. Residue Rev. 47, 167-187 (1973)

Appleman, D., Pyfrom, H.T.: Changes in catalase activity and other responses induced in plants by red and blue light. Plant Physiol. 30, 543-549 (1955)

Axelsson, L.: In vivo measurements of the photo-oxidation of chlorophyll pigments in dark grown wheat leaves after treatment with δ-aminolevulinic acid. Physiol. Plant. 31, 77-85 (1974)

Axelsson, L.: The photostability of different chlorophyll forms in dark grown leaves of wheat. I. Stability to high intensity red light of forms appearing after photoreduction of protochlorophyllide. Physiol. Plant. 38, 327-332 (1976)

Back, A., Richmond, A.E.: Interrelations between gibberellic acid, cytokinins and abscisic acid in retarding leaf senescence. Physiol. Plant. 24, 76-79 (1971)

Banerji, D., Laloraya, M.M.: Chlorophyll formation in isolated pumpkin cotyledons in the presence of kinetin and chloramphenicol. Plant Cell Physiol. 8, 263-268 (1967)

Barr, R., Hall, J.D, Crane, F.L.: Lipophilic quinones in mineral-deficient maize leaves. Indiana Acad. Sci. 80, 130-139 (1971)

Baszynski, T., Brand, J., Barr, R., Krogmann, D.W., Crane, F.L.: Some biochemical characteristics of chloroplasts from mineral-deficient maize. Plant Physiol. 50, 410-411 (1972)

Baszynski, T., Panczyk, B., Krol, M., Krupa, Z.: The effect of nitrogen deficiency on some aspects of photosynthesis in maize leaves. Z. Pflanzenphysiol. 74, 200-207 (1975)

Bata, J., Neškovic, M.: The effect of gibberellic acid and kinetin on chlorophyll retention in Lemna trisulca L. Z. Pflanzenphysiol. 73, 86-88 (1974)

Beale, S.I.: Studies on the biosynthesis and metabolism of δ-aminolevulinic acid in Chlorella. Plant Physiol. 48, 316-319 (1971)

Beale, S.I., Castelfranco, P.A.: The biosynthesis of δ-aminolevulinic acid in higher plants. Plant Physiol. 53, 291-296 (1974)

Beale, S.I., Gough, S.P., Granick, S.: Biosynthesis of δ-aminolevulinic acid from the intact carbon skeleton of glutamic acid in greening barley. Proc. Natl. Acad. Sci. USA 72, 2719-2723 (1975)

Beevers, L., Guernsey, F.S.: Interaction of growth regulators in the senescence of Nasturtium leaf discs. Nature (London) 214, 941-942 (1967)

Beevers, L., Loveys, B., Pearson, J.A., Wareing, P.F.: Phytochrome and hormonal control of expansion and greening of etiolated wheat leaves. Planta 90, 286-294 (1970)

Beisenherz, W.W., Schneider, H.A.W.: Syntheseort von Enzymen der Chlorophyllbiosynthese. Ber. Dtsch. Bot. Ges. 87, 161-166 (1974)

Berger, Ch., Bergmann, L.: Farblicht und Plastidendifferenzierung in Speichergewebe von Solanum tuberosum L. Z. Pflanzenphysiol. 56, 439-445 (1967)

Bergfeld, R.: Der Einfluss roter und blauer Strahlung auf die Ausbildung der Chloroplasten bei gehemmter Proteinsynthese. Z. Naturforsch. 19b, 1076-1078 (1964)

Bergmann, L., Bälz, A.: Der Einfluss von Farblicht auf Wachstum und Zusammensetzung pflanzlicher Gewebekulturen. I. Nicotiana tabacum var. Samsun. Planta 70, 285-303 (1966)

Bergmann, L., Berger, Ch.: Farblicht und Plastidendifferenzierung in Zellkulturen von Nicotiana tabacum var. Samsun. Planta 69, 58-69 (1966)

Berry, D.R., Smith, H.: Red-light stimulation of prolamellar body recrystallization and thylakoid formation in barley etioplasts. J. Cell Sci. 8, 185-200 (1971)

Bishop, N.I.: The possible role of plastoquinone (Q-254) in the electron transport system of photosynthesis. In: CIBA Found. Symp. Quinones in Electron Transport; Wolstenholme, G.E.W., O'Connor, C.M. (eds.). London: Churchill Press 1961

Björn, L.O.: Chlorophyll formation in excised wheat roots. Physiol. Plant. 18, 1130-1142 (1965)

Björn, L.O.: Some effects of light on excised wheat roots with special reference to peroxide metabolism. Physiol. Plant. 20, 149-170 (1967a)

Björn, L.O.: The light requirement for different steps in the development of chloroplasts in excised wheat roots. Physiol. Plant. 20, 483-499 (1967b)

Björn, L.O.: The effect of blue and red light on NADP-linked glyceraldehydephosphate dehydrogenases in excised roots. Physiol. Plant. 20, 519-527 (1967c)

Björn, L.O.: The state of protochlorophyll and chlorophyll in corn roots. Physiol. Plant. 37, 183-184 (1976)

Björn, L.O., Suzuki, Y., Nilsson, J.: Influence of wavelength on the light response of excised wheat roots. Physiol. Plant. 16, 132-141 (1963)

Bottrill, D.E., Possingham, J.V., Kriedemann, P.E.: The effect of nutrient deficiencies on photosynthesis and respiration in spinach. Plant Soil 32, 424-438 (1970)

Bouma, D.: Effects of Nitrogen Nutrition on leaf expansion and photosynthesis of *Trifolium subterraneum* L. Comparison between different levels of nitrogen supply. Ann. Bot. 34, 1131-1142 (1970)

Bovey, F., Ogawa, T., Shibata, K.: Photoconvertible and non-photoconvertible forms of protochlorophyll(ide) in etiolated bean leaves. Plant Cell Physiol. 15, 1133-1137 (1974)

Breyhan, T., Heiliger, F., Fischnich, O.: Über das Vorkommen und die Bedeutung des Prolins in der Kartoffel. Landwirtsch. Forsch. 12, 293-295 (1959)

Buschmann, C., Lichtenthaler, H.K.: Hill reaction of chloroplasts from *Raphanus* seedlings grown with β-indole-acetic acid and kinetin. In: Proc. 3rd Int. Congr. Photosynth.; Avron, M. (ed.). Amsterdam: Elsevier 1974

Butler, W.L., Briggs, W.R.: The relation between structure and pigments during first stages of protoplastid greening. Biochim. Biophys. Acta 112, 45-53 (1966)

Cheniae, G.M., Martin, I.F.: Studies on the function of manganese in photosynthesis. Brookhaven Symp. Biol. 19, 406-417 (1966)

Cheniae, G.M., Martin, I.F.: Site of manganese function in photosynthesis. Biochim. Biophys. Acta 153, 819-837 (1968)

Cheniae, G.M., Martin, I.F.: Sites of function of manganese within photosystem II. Roles in O_2 evolution and system II. Biochim. Biophys. Acta 197, 219-239 (1970)

Chibnall, A.C.: Protein Metabolism in the Plant. New Haven: Yale Univ. 1939

Cohen, R., Goodwin, T.W.: The effect of red and far red light on carotenoid synthesis by etiolated maize seedlings. Phytochemistry 1, 67-72 (1972)

Cran, D., Dyer, A.F.: The effect of a change in light quality on plastids of protonemata of *Dryopteris borreri*. Plant Sci. Lett. 5, 57-65 (1975)

Das, G.: Influence of calcium on the metabolism of chlorophyll carotene, nucleic acid, and protein in *Scenedesmus*. Can. J. Bot. 51, 121-125 (1973)

Dirks, W., Richter, G.: Die Wirkung des Cytostaticum "Proresid" auf das Wachstum und die Blaulicht-induzierte Chloroplastendifferenzierung isolierter and normaler Keimlingswurzeln von *Pisum sativum*. Planta 112, 101-120 (1973)

Duysen, M.E., Freeman, T.P.: Effects of moderate water deficit (stress) on wheat seedling growth and plastid pigment development. Physiol. Plant. 31, 262-266 (1974)

Egan, Jr., J.M., Dorsky, D., Schiff, J.A.: Events surrounding the early development of *Euglena* chloroplasts VI. Action spectra for the formation of chlorophyll, lag elimination in chlorophyll synthesis, and appearance of TPN-dependent triose phosphate dehydrogenase and alkaline DNAse activities. Plant Physiol. 56, 318-323 (1975)

El-Antably, H.M.M., Wareing, P.F., Hillman, J.: Some physiological responses to d,l-abscisin (dormin). Planta 73, 74-90 (1967)

El Hinnawy, E.: Effect of some growth regulating substances and carbohydrates on chlorophyll production in *Melilotus alba* (Desr.) Callus tissue cultures. Z. Pflanzenphysiol. 74, 95-105 (1974)

Evans, H.J., Sorger, G.J.: Role of mineral elements with emphasis on the univalent cations. Annu. Rev. Plant Physiol. 17, 47-76 (1966)

Eyster, C., Brown, T.E., Tanner, H.A., Hood, S.L.: Manganese requirement with respect to growth, Hill reaction and photosynthesis. Plant Physiol. 33, 235-241 (1958)

Feierabend, J.: Der Einfluss von Cytokininen auf die Bildung von Photosyntheseenzymen in Roggenkeimlingen. Planta 84, 11-29 (1969)

Feierabend, J., Beevers, H.: Developmental studies on microbodies in wheat leaves. I. Conditions influencing enzyme development. Plant Physiol. 49, 28-32 (1972)

Fisher-Arnold, G.: Untersuchungen über die Chloroplastenbewegung bei *Vaucheria sessilis*. Protoplasma 56, 495-520 (1963)

Fletcher, R.A., Teo, C., Ali, A.: Stimulation of chlorophyll synthesis in cucumber cotyledons by benzyladenine. Can. J. Bot. 51, 937-939 (1973)

Fondeville, J.C., Schneider, M.J., Borthwick, H.A., Hendricks, S.B.: Photocontrol of *Mimosa pudica* L. leaf movement. Planta 75, 228-238 (1967)

Gibbs, S.P.: Synthesis of chloroplast RNA at the site of chloroplast DNA. Biochem. Biophys. Res. Commun. 28, 653-657 (1967)

Graham, D., Grieve, A.M., Smillie, R.M.: Phytochrome-mediated plastid development in etiolated pea stem apices. Phytochemistry 10, 2905-2914 (1971)

Granick, S.: Magnesium porphyrin formed by barley seedlings treated with aminolevulinic acid. Plant Physiol. 34, XVIII (1959)

Griffiths, W.T.: Characterization of the terminal stages of chlorophyll(ide) synthesis in etioplast membrane preparations. Biochem. J. 152, 623-635 (1975)

Hall, J.D., Barr, R., Al-Abbas, A.H., Crane, F.L.: The ultrastructure of chloroplasts in mineral-deficient maize leaves. Plant Physiol. 50, 404-409 (1972)

Hall, R.H.: Cytokinins as a probe of developmental processes. Annu. Rev. Plant Physiol. 24, 415-444 (1973)

Hampp, R., Schmidt, H.W.: Regulation of membrane properties of mitochondria and plastids during chloroplast development. I. The action of phytochrome in situ. Z. Pflanzenphysiol. 82, 68-77 (1977)

Harada, H.: Retardation of the senescence of Rumex obtusifolius L. leaves by growth retardants. Plant Cell Physiol. 7, 701-703 (1966)

Henningsen, K.W., Boynton, J.E.: Macromolecular physiology of plastids. VIII. Pigment and membrane formation in plastids of barley greening under low light intensity. J. Cell Biol. 44, 290-304 (1970)

Hsiao, T.C.: Plant response to water stress. Annu. Rev. Plant Physiol. 24, 519-570 (1973)

Hsu, W.-P., Miller, G.W.: Chlorophyll and porphyrin synthesis in relation to iron in *Nicotiana tabacum*, L. Biochim. Biophys. Acta 111, 393-402 (1965)

Huber, W., Sankhla, N.: Abscisic acid-kinetin interaction in growth and activities of enzymes of amino-acid metabolism in *Pennisetum typhoides* seedlings. Z. Pflanzenphysiol. 73, 160-166 (1974)

Kasemir, H., Bergfeld, R., Mohr, H.: Phytochrome-mediated control of prolamellar body reorganization and plastid size in mustard cotyledons. Photochem. Photobiol. 21, 111-120 (1975)

Keck, R.W., Boyer, J.S.: Chloroplast response to low leaf water potentials. III. Differing inhibition of electron transport and photophosphorylation. Plant Physiol. 53, 474-479 (1974)

Kende, H.: The cytokinins. Int. Rev. Cytol. 31, 301-338 (1971)

Knypl, J.S.: The control of RNA, protein and chlorophyll synthesis in senescing leaf tissue of Kale by coumarin and growth retardants. Flora Abt. A 160, 217-233 (1969a)

Koski, V.M., French, C.S., Smith, J.H.C.: The action spectrum for the transformation of protochlorophyll to chlorophyll a in normal and albino corn seedlings. Arch. Biochem. Biophys. 31, 1-17 (1951)

Kull, U., Büxenstein, R.: Effect of cytokinins on the lipid fatty acids of leaves. Phytochemistry 13, 39-44 (1974)

Laetsch, W.M., Stetler, D.A.: Chloroplast structure and function in cultured tobacco tissue. Am. J. Bot. 52, 798-804 (1965)

Lagoutte, B., Duranton, J.: A manganese protein complex within the chloroplast structures. FEBS Lett. 51, 1 (1975)

Lamprecht, I.: Die Feinstruktur der Plastiden von Tradescantia albiflora (Kth.) bei Eisenmangelchlorose. Protoplasma 53, 162-199 (1961)

Lange, H., Mohr, H.: Die Hemmung der Phytochrominduzierten Anthocyansynthese durch Actinomycin D und Puromycin. Planta 67, 107-121 (1965)

Lichtenthaler, H.K.: Regulation der Lipochinonsynthese in Chloroplasten. Ber. Dtsch. Bot. Ges. 86, 313-329 (1973)

Lichtenthaler, H.K., Becker, K.: Inhibition of the light-induced vitamin K_1 and pigment synthesis by abscisic acid. Phytochemistry 9, 2109-2113 (1970)

Lichtenthaler, H.K., Grumbach, K.H.: Observations on the turnover of thylakoids and their prenyl lipids in *Hordeum vulgare* L. In: Proc. 3rd. Int. Congr. Photosynth.; Avron, M. (ed.). Amsterdam: Elsevier 1974

Lichtenthaler, H.K., Weinert, H.: Die Beziehungen zwischen Lipochinonsynthese und Plastoglobulibildung in den Chloroplasten von *Ficus elastica* Roxb. Z. Naturforsch. 25b, 619-623 (1970)

Liljenberg, C.: The effect of light on the phytolization of chlorophyllide a and the spectral dependence of the process. Physiol. Plant. 19, 848-853 (1966)

Loveys, B.R.: The intracellular location of abscisic acid in stressed and non-stressed leaf tissue. Physiol. Plant. 40, 6-10 (1977)

Lyman, M., Epstein, H.T., Schiff, J.A.: Studies of chloroplast development in *Euglena* I. Inactivation of green colony formation by UV light. Biochim. Biophys. Acta 50, 301-309 (1961)

Machold, O., Stephan, U.W.: The function of iron in porphyrin and chlorophyll biosynthesis. Phytochemistry 8, 2189-2192 (1969)

Mantai, K.E., Wong, J., Bishop, N.I.: Comparison studies on the effects of ultraviolet irradiation on photosynthesis. Biochim. Biophys. Acta 197, 257-266 (1970)

Margulies, M.M.: Relationship between red light mediated glyceraldehyde-phosphate dehydrogenase formation and light dependent development of photosynthesis. Plant Physiol. 40, 57-61 (1965)

Marsh, H.V., Evans, H.J., Matrone, G.: Investigations of the role of iron in chlorophyll metabolism. I. Effect of iron deficiency on chlorophyll and heme content and on the activities of certain enzymes in leaves. Plant Physiol. 38, 632-638 (1963a)

Marsh, H.V., Evans, H.J., Matrone, G.: Investigations of the role of iron in chlorophyll metabolism. II. Effect of iron deficiency on chlorophyll synthesis. Plant Physiol. 38, 638-642 (1963b)

Mego, J.I., Jagendorf, A.: Effect of light on growth of Black Valentine bean plastids. Biochim. Biophys. Acta 53, 237-254 (1961)

Mercer, E.I., Pughe, J.E.: The effect of abscisic acid on the biosynthesis of isoprenoid compounds in maize. Phytochemistry 8, 115-122 (1969)

Mercer, F.V., Nittim, M., Possingham, J.V.: The effect of manganese deficiency on the structure of spinach chloroplasts. J. Cell. Biol. 15, 379-381 (1962)

Mishra, D., Misra, B.: Effect of growth regulating chemicals on degradation of chlorophyll and starch in detached leaves of corn plants. Z. Pflanzenphysiol. 58, 193-206 (1968)

Mitrakos, K.: The participation of the red far-red reaction system in chlorophyll metabolism. Physiol. Plant. 14, 497-503 (1961)

Mittelheuser, C.J., Steveninck, R.F.M. van: The ultrastructure of wheat leaves. I. Changes due to natural senescence and the effects of kinetin and ABA on detached leaves incubated in the dark. Protoplasma 73, 239-252 (1971a)

Mittelheuser, C.J., Steveninck, R.F.M. van: The ultrastructure of wheat leaves. II. The effects of kinetin and ABA on detached leaves incubated in the light. Protoplasma 73, 253-262 (1971b)

Mlodzianowski, F., Gezela, E.: Effect of kinetin and chloramphenicol on chlorophyll synthesis and chloroplast development in detached lupin cotyledons under low light intensity. Acta Soc. Bot. Pol. XLIII, 149-168 (1974)

Mohr, H.: Die Abhängigkeit des Protonemawachstums und der Protonemapolarität bei Farnen vom Licht. Planta 47, 127-158 (1956)

Mohr, H., Kasemir, H.: Control of plastid development and chlorophyll synthesis by phytochrome. Proc. Indian Natl. Sci. Acad. 41B, 503-525 (1975)

Nadler, K., Granick, S.: Controls of chlorophyll synthesis in barley. Plant Physiol. 46, 240-246 (1970)

Nielsen, O.F., Kahn, A.: Kinetics and quantum yield of photoconversion of protochlorophyll(ide) to chlorophyll(ide) a. Biochim. Biophys. Acta 292, 117-129 (1973)

Ohlenroth, K., Mohr, H.: Die Steuerung der Proteinsynthese und der Morphogenese bei Farnvorkeimen durch Licht. Planta 59, 427-441 (1963)

Ostrovskaya, L.K., Zaitseva, N.A.: Photochemical activity of the chloroplasts in the presence of an iron deficiency in plants. Dokl. Akad. Nauk SSSR 176, 1178-1180 (1967)

Plaut, Z.: Inhibition of photosynthetic carbon dioxide fixation in isolated spinach chloroplast exposed to reduced osmotic potentials. Plant Physiol. 48, 591-595 (1971)

Plaut, Z., Bravdo, B.: Response of carbon dioxide fixation to water stress. Parallel measurements on isolated chloroplasts and intact spinach leaves. Plant Physiol. 52, 28-32 (1973)

Possingham, J.V., Spencer, D.: Manganese as a functional component of chloroplasts. Aust. J. Biol. Sci. 15, 58-68 (1962)

Possingham, J.V., Vesk, M., Mercer, F.V.: The fine structure of leaf cells of manganese-deficient spinach. J. Ultrastruct. Res. 11, 68-83 (1964)

Possingham, J.V., Cran, D.G., Rose, R.J., Loveys, B.R.: Effects of green light on the chloroplasts of spinach leaf discs. J. Exp. Bot. 26, 33-42 (1975)

Potter, J.R., Boyer, J.S.: Chloroplast response to low leaf water potentials. II. Role of osmotic potential. Plant Physiol. 51, 993-997 (1973)

Price, C.A.: Iron compounds and plant nutrition. Annu. Rev. Plant Physiol. 19, 239-248 (1968)

Raghavan, V., Maggio, A.E. de: Enhancement of protein synthesis in isolated chloroplasts by irradiation with blue light. Plant Physiol. 48, 82-85 (1971)

Railton, I.D., Reid, D.M., Gaskin, P., MacMillan, J.: Characterization of abscisic acid in chloroplasts of Pisum sativum L. cv. Alaska by combined gas-chromatography-mass spectrometry. Planta 117, 179-182 (1974)

Remy, R.: Pre-existence of chloroplast lamellar proteins in wheat etioplasts. Functional and protein changes during greening under continuous or intermittent light. FEBS Lett. 31, 308-312 (1973)

Schiff, J.A., Epstein, H.T.: The replicative aspect of chloroplast continuity in Euglena. In: Biochemistry of Chloroplasts, Vol. I; Goodwin, T.W. (ed.), pp. 341-353. London-New York: Academic Press 1966

Schiff, J.A., Lyman, H., Epstein, H.T.: Studies of chloroplast development in Euglena II. Photoreversal of the UV inhibition of green colony formation. Biochim. Biophys. Acta 50, 310-318 (1961)

Schmid, R., Clauss, H.: Die Vermehrung der Chloroplasten von Acetabularia im Rot- und Blaulicht. Protoplasma 82, 283-287 (1974)

Schopfer, P.: Phytochrome and the control of enzyme activity. In: Phytochrome; Mitrakos, K., Shropshire, W., Jr. (eds.), pp. 486-514. London-New York: Academic Press 1972

Scott, N.S., Nair, H., Smillie, R.M.: The effect of red irradiation on plastid ribosomal RNA synthesis in dark-grown pea seedlings. Plant Physiol. 47, 385-388 (1971)

Shibata, K.: Spectroscopic studies on chlorophyll formation in intact leaves. J. Biochem. (Tokyo) 44, 147-173 (1957)

Shlyk, A.A., Walter, G., Averina, N.G., Savchenko, G.E.: Influence of kinetin on the formation of active protochlorophyllide in green and post-etiolated leaves of wheat. Dokl. Bot. Sci. 111-114, 193-195 (1970)

Sironval, C., Michel, J.-M.: On a "photoenzyme" or, the mechanism of the protochlorophyllide—chlorophyllide photoconversion. Europ. Photobiol. Symp. Hvar. Yugoslavia, 19th-22nd Sept. 1967. Book of Abstr. Zagreb: Yugoslav Acad. Sci. 1967

Sironval, C., Bronchard, R., Michel, J.M., Brouers, M., Kuyper, Y.: Structure macromoléculaire et activités photochimiques des lamelles plastidiales (essais). Bull. Soc. Fr. Physiol. Veg. 14, 195-225 (1968)

Skokut, T.A., Wu, J.H., Daniel, R.S.: Chloroplast ultrastructure in relation to the preservation of green color in leaves irradiated with high dose of ultraviolet light. Plant Physiol. 56, (suppl.) 64 (1975)

Smillie, R.M., Scott, N.S.: Organelle biogenesis. In: Progress in Molecular and Subcellular Biology, Vol. 1; Hahn, F.E. (ed.), p. 173. Berlin-Heidelberg-New York: Springer 1969

Smith, H.: Phytochrome and Photomorphogenesis: An Introduction to the Photocontrol of Plant Development. London: McGraw Hill 1975

Smith, J.H.C.: Factors affecting the transformation of protochlorophyll to chlorophyll. Carnegie Inst. Wash. Yearb. 51, 151-153 (1952)

Stephan, U.W., Machold, O.: Über die Funktion des Eisens bei der Porphyrin- und Chlorophyll-Biosynthese. Z. Pflanzenphysiol. 61, 98-113 (1969)

Stetler, D.A., Laetsch, W.M.: Prolamellar body-like structures in tobacco tissue cultures on a medium lacking kinetin. Am. J. Bot. 55, 709 (1968)

Stobart, A.K., Shewry, P.R., Thomas, D.R.: The effect of kinetin on chlorophyll synthesis in ageing etiolated barley leaves exposed to light. Phytochemistry 11, 571-577 (1972)

Straub, V., Lichtenthaler, H.K.: Die Wirkung von Gibberellinsäure A_3 und Kinetin auf die Bildung der Photosynthesepigmente, Lipochinone und Anthocyane in *Raphanus*-Keimlingen. Z. Pflanzenphysiol. 4, 308-321 (1973a)

Straub, V., Lichtenthaler, H.K.: Die Wirkung von β-Indolessigsäure auf die Bildung der Chloroplastenpigmente, Plastidenchinone, und Anthocyane in *Raphanus*-Keimlingen. Z. Pflanzenphysiol. 70, 34-45 (1973b)

Suboch, V.P., Losev, A.P., Gurinovitch, G.P.: Photoreduction of protochlorophyll and its derivatives. Photochem. Photobiol. 20, 183-190 (1974)

Sunderland, N., Wells, B.: Plastid structure and development in green callus tissues of *Oxalis dispar*. Ann. Bot. (London) 32, 327-346 (1968)

Sundqvist, C.: The conversion of protochlorophyllide$_{636}$ to protochlorophyllide$_{650}$ in leaves treated with δ-aminolevulinic acid. Physiol. Plant. 23, 412-424 (1970)

Sundqvist, C.: The relationship between chlorophyllide accumulation, the amount of protochlorophyllide$_{636}$ and protochlorophyllide$_{650}$ in dark-grown wheat leaves treated with δ-aminolevulinic acid. Physiol. Plant. 28, 464-470 (1973)

Sundqvist, C., Odengård, B., Persson, G.: Light-stimulated accumulation of protochlorophyllide in leaves of different ages treated with δ-aminolevulinic acid. Plant Sci. Lett. 4, 89-96 (1975)

Szalai, I.: Gibberellinsäure und Chlorophyllgehalt des Blattes von *Phaseolus vulgaris* L. Planta 83, 161-165 (1968)

Szalei, I.: Relation between the chlorophyll content and paleness of gibberellic acid-treated leaves. Physiol. Plant. 22, 587-593 (1969)

Tevini, M.: Der Einfluss von Phosphat- und Nitratmangel auf die Synthese der Phospho- und Glykolipide bei *Impatiens balsamina*. Ber. Dtsch. Bot. Ges. 84, 595-606 (1971a)

Tevini, M.: Der Einfluss von Phosphat-Mangel-Ernährung auf die Synthese der Phospho- und Glykolipide bei *Impatiens*. Z. Pflanzenphysiol. 66, 64-72 (1971b)

Tietz, A., Dörffling, K.: Veränderungen im Gehalt von Abscisinsäure und Indol-3-Essigsäure sowie der Chloroplastenfarbstoffe in Pisumkeimlingen durch Gibberellinsäure. Planta 85, 118-125 (1969)

Verbeek, L., Lichtenthaler, H.K.: Der Einfluss von Stickstoffmangel auf die Lipochinon- und Isoprenoid-Synthese der Chloroplasten von *Hordeum vulgare* L. Z. Pflanzenphysiol. 70, 245-258 (1973)

Vesk, M., Possingham, J.V., Mercer, F.V.: The effect of mineral nutrient deficiencies on the structure of the leaf cells of tomato, spinach, and maize. Aust. J. Bot. 14, 1-18 (1966)

Virgin, H.I.: Chlorophyll formation and water deficit. Physiol. Plant. 18, 994-1000 (1965)

Virgin, H.I.: Chlorophyll biosynthesis and phytochrome action. In: Phytochrome; Mitrakos, K., Shropshire, W., Jr. (eds.), pp. 372-404. London-New York: Academic Press 1972

Virgin, H.I., Kahn, A., Wettstein, D. von: The physiology of chlorophyll formation in relation to structural changes in chloroplasts. Photochem. Photobiol. 2, 83-91 (1963)

Vlasova, M.P., Drozdova, I.S., Voskresenskaya, N.P.: Changes in fine structure of the chloroplasts in pea plants greening under blue and red light. Fiziol. Rast. 18, 5-11 (1971)

Voskresenskaya, N.P., Grishina, G.S., Chmora, S.N., Poyarkova, N.M.: The influence of red and blue light on the rate of photosynthesis and the CO_2 compensation point at various oxygen concentrations. Can. J. Bot. 48, 1251-1257 (1970)

Wellburn, F.A.M., Wellburn, A.R.: Response of etioplasts in situ and in isolated suspensions to pre-illumiation with various combination of red, far-red, and blue light. New Phytol. 72, 55-60 (1973)

Whatley, J.M.: Ultrastructural changes in chloroplasts of *Phaseolus vulgaris* during development under conditions of nutrient deficiency. New Phytol. 70, 725-742 (1971)

Wolff, J.B., Price, L.: The effect of sugars on chlorophyll biosynthesis in higher plants. J. Biol. Chem. 235, 1603-1609 (1960)

Zeldin, M.H., Cohen, C.E., Ben-Shaul, Y., Schiff, J.A.: Measurement in vivo of light-induced spectroscopic changes of protochlorophyll(ide) and chlorophyll(ide) in *Euglena*. Plant Physiol. 56, Suppl. 33 (1975)

The Survival, Division and Differentiation of Higher Plant Plastids Outside the Leaf Cell

R.M. LEECH

Department of Biology, University of York, York, Great Britain

I. Introduction

Advance in any scientific field of endeavour can be measured as readily by the questions currently being subjected to experimental examination as by the body of relevant information available. The characteristics and control of plastid differentiation are no exception, and since the 1920's when the non-Mendelian inheritance of the determinants of plastid phenotypes was established by classical genetical analysis, questions have been posed about the interaction between plastids and other components of the leaf cells which they inhabit. Currently such questions concern the inter-molecular and intra-molecular interactions between and within the organelles and demand an understanding of the molecular biochemistry and function of the variety of cellular nucleic acids involved in the control of the synthesis of the proteins of both plastid and cytoplasm. The development of a plastid into a fully functional photosynthetic chloroplast is essentially a cellular process: both nuclear and plastid genomes are involved in the coding of chloroplast proteins and both chloroplast and cytoplasmic ribosomes in their synthesis. (For details see the reviews of Ellis et al. 1973).

One approach which can be expected to become of increasing value in the study of plastid differentiation is the isolation of plastids from leaves and the continuation of their development in an artificial environment outside the cell. So far such studies have been limited to the examination of the survival of differentiated chloroplasts and also to studies of plastids from dark-grown leaves. This second type of plastid, the etioplast, continues to synthesize chlorophyll and proliferate internal membranes in vitro when it is illuminated. The investigations carried out so far suggest that limited plastid culture in liquid media may be a realistic possibility and have already led to the discovery that partially differentiated plastids with small grana have the ability to divide in vitro.

This review is largely restricted to a consideration of studies of leaf cell plastids from higher plants, since only chloroplasts from a restricted range of higher plants are suitable for detailed quantitative biochemical analysis (Walker 1970).

II. The Survival of Fully Differentiated Chloroplasts
Outside the Cell

While the existence of a complex interplay of plastid and cytoplasmic components during cellular-plastid development was clearly established in early genetical studies, the status of the mature, functional, photosynthesising chloroplast vis-a-vis the rest of the cell was until recently more questionable. By the mid-1960's it was already known that the chloroplasts possessed the biochemical elements of the classical genetical system, i.e., DNA, RNA and ribosomes, and were probably also able to synthesise some nucleic acids and proteins (Kirk 1971). In addition, mature isolated chloroplasts had been shown to fix carbon dioxide photosynthetically into a wide range of small carbon compounds (Jensen and Bassham 1966), and evidence was also accumulating that plastids could synthesise small molecules such as amino acids and fatty acids (Kirk 1970), as well as possessing the capacity for the synthesis of some macromolecules. The rates of CO_2 fixation were as high in isolated plastids as in leaf tissue, establishing the chloroplast as the complete photosynthetic unit (Walker 1970). It seemed legitimate at that time to ask whether fully developed chloroplasts were capable of survival as biochemically functional organelles in the absence of other cellular components and to this end the ability of photosynthesising chloroplasts to remain active for periods of hours or days in artificial environments under physiological conditions was examined (Ridley and Leech 1969). Stimulation for these studies came with the discovery that chloroplasts from siphonaceous green algae such as *Caulerpa* or *Codium,* when ingested by marine moluscs such as *Elysia* and *Tridachia,* could remain photochemically functional in the cytoplasm of the gastric mucosal cells of the animals for periods of up to six weeks (Trench et al. 1969; Taylor 1970; Trench and Smith 1970; Smith 1976). Spinach chloroplasts were also reported to continue to fix carbon dioxide after injection into cultured mouse fibroblast cells (Nass 1969; Giles and Sarafis 1974).

It was established, by careful attention to the conditions of incubation, that the structural integrity of chloroplasts could be preserved at 20° C for periods of 48 h but that during this time the photosynthetic characteristics of the plastids progressively decayed (Ridley and Leech 1969; Leech 1972). CO_2-fixing capacity was lost within the first hour, activity of photosystem II over the first 6 h, but photosystem I activity dependent on ferricyanide reduction continued at 50% of its initial rate 24 h after the plastids were isolated. The progressive decay of photofunction follows in reverse the pattern in which the systems apparently develop in greening etioplasts (Bradbeer 1976). The artificial environment therefore merely provides a means of maintaining limited plastid function.

Neither protein nor nucleic acid synthesis has so far been demonstrated to occur in chloroplasts already kept for several hours outside the cell. Such systems, however, offer a very useful means of studying long-term biochemical interactions on a physiological time-scale and have already proved useful in analysing the effects of plant hormones, growth substances and herbicides on plastid funtion.

There is now, however, a good deal of evidence that the synthetic reactions occurring inside a functional chloroplast require a continuous supply of small molecules and ions from the cell cytoplasm (Heber 1974; Walker 1974; Leech and

Murphy 1976), and so the maintenance of activity in already developed plastids in vitro would depend on continuous and probably modulated provision of these materials in the artificial medium. Clearly, as further requirements for chloroplast function are demonstrated, the artificial media will need to be modified, but it can also be expected that further interactions between chloroplast and cytoplasm will be discovered by the development and study of such in vitro systems.

III. Plastid Division Outside the Cell

The ability of chloroplasts with fully developed grana to divide both in vivo and in vitro is now established and the division process has been followed visually in isolated chloroplasts. The current state of knowledge of plastid division inside and outside the higher plant leaf cells has been recently reviewed (Leech 1976). Earlier light microscopic observation of leaf cells had shown the presence of dumb-bell-shaped chloroplasts and more recent electron microscopic studies have confirmed and extended these observations. The dumb-bell profiles of spinach *(Spinacia oleracea)*, tobacco *(Nicotiana tabacum)* and broad bean *(Vicia faba)* chloroplasts are of similar size and internal membrane complexity (Boasson et al. 1972; Cran and Possingham 1972; Leech et al. 1973). If these profiles are indeed of dividing chloroplasts, then division seems to occur while the grana are still small and have not more than three to five thylakoids. The membrane changes associated with chloroplast division have not yet been explained in detail. However, the observation of dumb-bell-shaped plastids alone does not establish that they are dividing unless a subsequent increase in plastid number per cell can be demonstrated. Until recently, proplastids were thought to be the only type of plastid which regularly divided in leaf cells. It is now clear that in spinach (Possingham and Saurer 1969) and maize (Leech and Nice, unpublished) the plastid number per cell more than doubles at the time of chloroplast division, i.e., that the majority of chloroplasts in each mature spinach leaf cell are derived by division of pre-existing chloroplasts.

The division of chloroplasts in isolation has been described in some detail (Ridley and Leech 1970; Kameya and Takahashi 1971). The process is a binary fission in which a constriction zone appears in the center of the plastid and a separation of the two daughter plastids formed occurs after about 6 h. Similar division profiles have been found in leaf cells containing young chloroplasts in every species in which they have been carefully sought. Because of the transient nature of the division process, many samples have to be taken at different times of the day to establish that division is occurring. Diurnal changes may also be important, since it has been proposed that light may initiate plastid division (Possingham and Saurer 1969; Possingham 1973). In those cells in which the chloroplasts are themselves dividing, cell division has generally ceased (Kameya 1972). In spinach leaves, for example, dividing chloroplasts are found in young mesophyll cells which are still rapidly growing but have ceased to divide. A similar sequence of cell and chloroplast division can be followed in some monocotyledonous leaves such as corn *(Zea mays)* and crab grass *(Digitaria sanguinalis)* in which the cells are in linear array with the youngest near the base of the leaf and the oldest near the tip (Leech et al. 1973). In such leaves, a discrete band of cells can be identified in which initially all the plastids are in the process of division in cells which have ceased to divide.

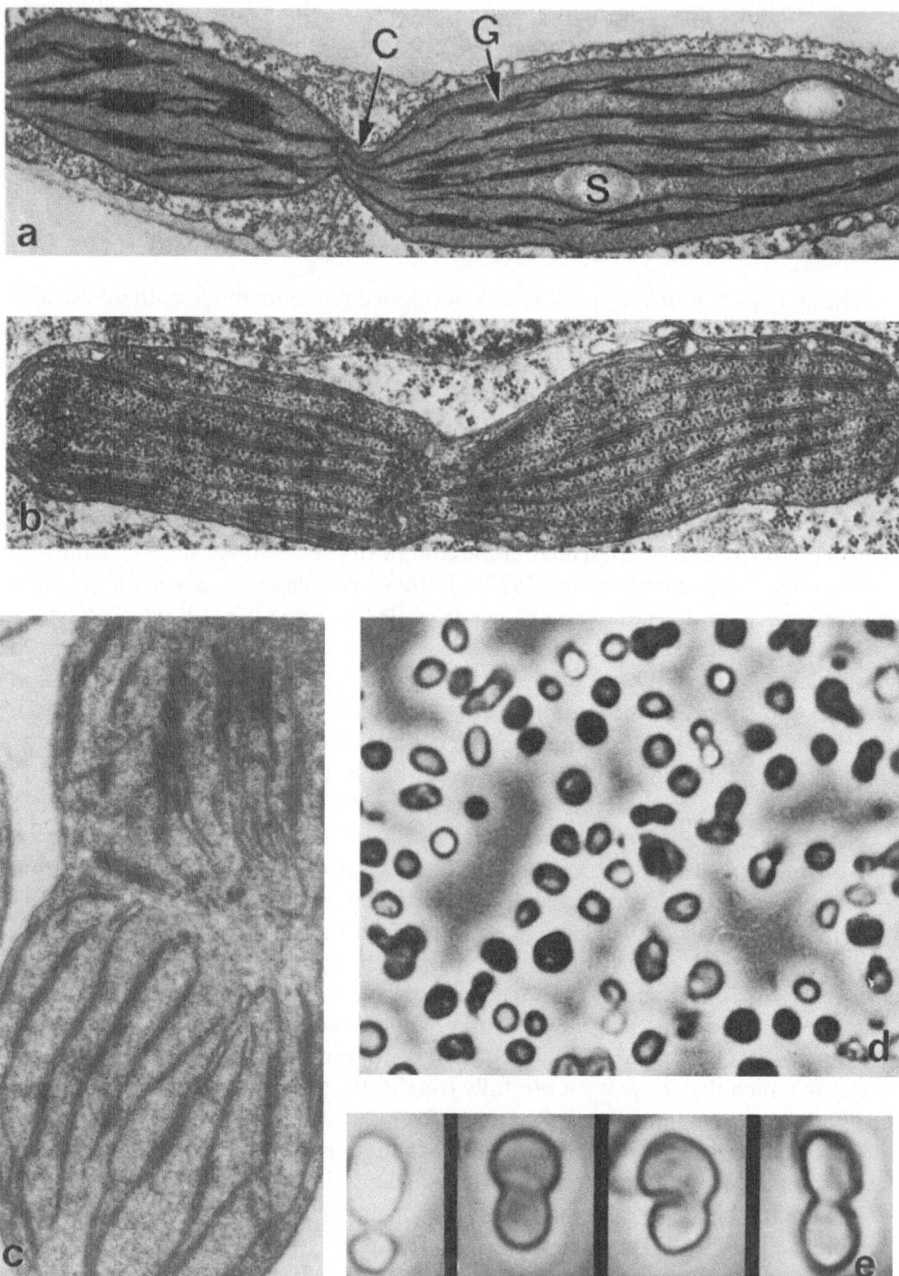

Fig. 1. a Dividing young chloroplast in an etiolated tobacco leaf disc after exposure to white light for 40 h. *C* constriction zone, *G* granum, *S* starch. (Reproduced from Boasson et al. 1972). **b** Dividing young chloroplast in a normally grown green maize leaf. The cell is 3 cm from the base of a leaf of a seven-day old seedling (× 32,000). **c** Dividing young chloroplast after isolation from the base of a maize leaf (Leech, unpublished). **d,e** Light micrographs of chloroplasts dividing 96 h after isolation from *Vicia faba* leaves (Ridley and Leech 1970)

At the time they divide, the plastids are large enough to be observed in the light microscope, so the process of division can be followed visually, but investigation of the details of the membrane changes requires electron microscopic examination. The state of development of the membranes of the young plastids undergoing division is remarkably similar in all the cases observed. Some examples of dividing chloroplasts are shown in Fig. 1. Dividing chloroplasts in leaves of the monocotyledons and dicotyledons, plastids induced to divide by kinetin treatment, and plastids dividing after isolation are all of similar size and complexity. The mean profile length of these plastids in 2–3 μm and they are 1–2 μm wide: each plastid profile has two to six rudimentary grana containing one to three thylakoids. Chloroplasts with more highly developed membrane systems containing grana with large numbers of thylakoids have not so far been observed to divide, so it seems likely that they lose the ability to replicate after a certain stage of maturity. From the studies of leaf discs there is evidence suggesting that light (Possingham and Smith 1972), a supply of kinetin (Boasson et al. 1972; Laetsch and Boasson 1971) and the state of development of the cell (Boasson et al. 1972; Kameya 1972; Possingham and Smith 1972) may each or all trigger the division process, but details of the molecular and genetic changes which initiate and control chloroplast division are unknown. Replication continues in the absence of plastid DNA synthesis immediately prior to the experimental observations, so presumably the plastid already has sufficient copies of DNA to support one division (Boasson and Laetsch 1969). The establishment of an artificial system in which division in many plastids can be followed simultaneously will enable the effects of specific compounds and physical treatments on the division process to be studied. Comparative studies on changes occurring in the tissue at the time of plastid division would clearly indicate which of the effectors functional in vitro have physiological roles.

Although such studies are presently only in their infancy the ultimate goal of in vitro studies is to synchronise the division process so that the nature of the controls may be identified quantitatively and qualitatively.

IV. The Value of Plastid "Culture" in a Study of Plastid Differentiation

So far plastids have been shown to survive with their structure and function partially intact for periods of hours and are still able to divide outside the cell, but the prospect of studying some of the stages of plastid development in artificial environments is an even more exciting one. To date it has only been possible to induce short phases in the developmental process to occur in vitro and only etioplasts, i.e., plastids from plants grown entirely in the dark, have been used for this purpose. Normal chloroplasts develop in plants grown in a diurnal light regime, not from etioplasts, but from undifferentiated proplastids.

The isolation of structurally intact etioplasts from etiolated tissue grown entirely in the dark is relatively straightforward. Methods based on differential centrifugation (Leese et al. 1972), and column chromatography in the dark (Wellburn and Wellburn 1971a) both provide suspensions in which virtually all the

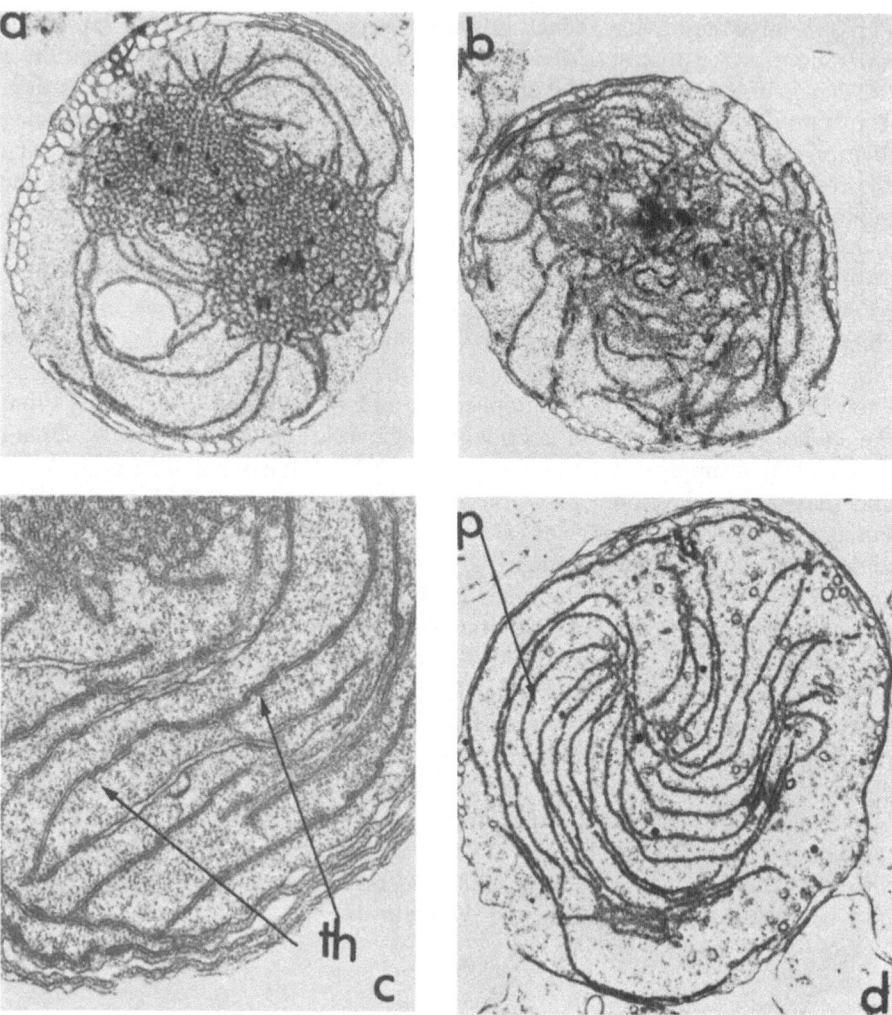

Fig. 2a-d. Electron micrographs of isolated etioplasts from maize after illumination (1.6×10^3 erg cm^2 s^{-1}) for 5 h. **a** An etioplast with partially transformed prolamellar bodies and numerous lamellar perforation ($\times 15,000$); **b** an etioplast showing dispersed prolamellar body and perforated lamellae ($\times 15,000$); **c** part of an etioplast at high magnification ($\times 40,000$) showing extensive bithyalkoid formation *(th)*; **d** an etioplast showing only peripheral transformation of the prolamellar body but extensive bithyalkoid formation ($\times 15,000$) (*p* perforated lamellae). (From Leech 1976)

etioplasts have complete envelopes and are free from cytoplasmic contamination (Wellburn and Wellburn 1971a, 1973; Horton and Leech 1975a,b). Etioplasts completely lack chlorophyll and each has a large central regular lattice of membranes, the prolamellar body (Fig. 2). Illumination of etiolated leaves initiates chlorophyll biosynthesis and the sequential assembly of the light-harvesting systems and reaction centers functional in the photoreactions in photosynthesis follows, and a concomitant dispersal of the prolamellar body takes place. The linear extent of the membrane system of the prolamellar body can be as great as the extent of the chloroplast lamellae into which it finally develops. Other components essential for photosynthetic function, including all the components of the electron transport system and the carboxylating enzyme, ribulose bis-phosphate carboxylase are present in sufficient quantity in some etioplasts before illumination (Plesnicar and Bendall 1973). The consensus of evidence suggests that a photoconversion of protochlorophyllide to chlorophyllide initiates the assembly of the photosynthetic units in developing illuminated etioplasts.

If isolated etioplasts maintained in a suitable artificial environment are illuminated, changes in the prolamellar body, similar to those observed in the leaf, are initiated, and development of the etioplasts can be followed over periods of several hours (Wellburn and Wellburn 1971a,b, 1973; Horton and Leech 1975a,b; Leech 1976). Chlorophyll synthesis and its assembly into membranes, as judged by the occurrence of the Shibata shift, also occurs during this time period (Horton and Leech 1975b). The addition of gibberellin to the media had no effect on the development of Fraction I protein but ATP (Horton and Leech 1975b) and also NADPH (Griffiths 1975) have both been shown to effect control on the membrane and pigment changes. Development of the etioplasts only continues to the bithylakoid stage of membrane assembly and there is no evidence that the etioplasts grow after isolation. The kind of membrane rearrangements which take place in vitro are illustrated in Fig. 2. In the media which have so far been used, etioplast development was arrested after approximately 5 h, but the potential value of the use of both etioplast and proplastid populations for the study of long-term developmental processes is probably considerable. Current problems requiring solution include lack of synchrony in plastid development, bacterial contamination and etioplast decay. Greater care in sampling the leaf material from which the plastids are isolated can be expected to improve the synchrony since in the monocotyledonous leaves from which the etioplasts are generally isolated, several stages of plastid development are present in the same leaf (Robertson and Laetsch 1974). Bacterial contamination may prove more difficult to eradicate, since bacteria are present in the intercellular spaces within leaves. The isolation of plastids from protoplasts may be one method by which this problem can be successfully tackled. The use of antibiotics is not to be generally recommended, since many are known inhibitors of protein synthesis in 70S chloroplast ribosomes, but penicillin is probably a safe antibiotic for the elimination of the Gram-positive bacteria. Improvements and elaboration of alternative media should lead to prevention of etioplast decay and the preservation of their structure and function.

Because chlorophyll synthesis can be initiated and controlled in etiolated tissue, its biosynthesis in etioplasts has been extensively studied. It should be remembered, however, that under natural conditions etioplasts are rarely encountered, and in several respects have been shown to have reached a much more advanced stage of development than proplastids. Normally chloroplasts are derived from proplastids. Proplastids are about 1 μm in diameter and have very few internal membranes. During development their volume increases about tenfold in some species and 30–50-fold in others. There is no information about how the complex reaction centers and electron transport components develop in proplastids, nor the means by which they become fully photosynthetic chloroplasts. Methods are now available for the isolation of such proplastids from leaf meristematic cells (Leese et al. 1972), but no attempts have been made so far to initiate or follow through their development outside the cell. This is a useful project for the future, but plastid development is a cellular function and a complex array of additives will certainly be needed to simulate the changing developmental conditions in the cell.

If proplastids are to be successfully cultured, it will be necessary not only to establish conditions in which repeated divisions occur but also conditions suitable for plastid growth. In the relatively simple defined media so far employed, no plastid growth has been demonstrated. These media contain a variety of mineral salts, bicarbonate, nitrate, sulphate, an osmoticum and a component designed to increase the viscosity (Ridley and Leech 1970). Far more complex additions will be needed to support plastid development since the DNA of the chloroplast contains insufficient genetic information to code for the full complement of chloroplast proteins. From studies of plastid development in leaves, information should increasingly become available about the components which will be needed in more sophisticated artificial environments and these may be expected to include purified molecular species of nucleic acids, in particular messenger RNAs, and probably also cellular organelles other than plastids. In a complementary manner, if it proves possible to initiate and follow in vitro a few steps in the development of plastids over a period of hours or even days, the possibility of identifying specific physical and chemical effectors involved in plastid development in vivo would be considerably increased.

V. Some Characteristics of Plastid Protein Synthesis

Since the immediate cause of development of the plastid resides in the specificities of its own spectrum of proteins and those of the cell containing it, studies of the control and characteristics of protein synthesis during plastid development are central to the understanding of the process and complementary to new attempts to culture plastids. A short description of some recent findings on the synthesis of the most abundant soluble plastid protein, Fraction I protein (see also Sundqvist et al., this vol.) illustrates the problems involved.

The investigation of the synthesis of the two subunits of Fraction I protein is the first example of the rigorous application of the concepts and techniques of molecular biology to problems of plastid biochemistry and illustrates the type of approach which, when applied to the whole complement of plastid proteins, may be expected eventually to provide a quantitative picture of plastid ontogeny. Genetical studies in which the amino acid sequences of the polypeptides of the large and small subunits of Fraction I from several tobacco varieties were analysed, indicated that a mutation in the large subunit (M.W. approx. $5.2–6.0 \times 10^4$ daltons) is inherited via chloroplast DNA, but a mutation in the smaller subunit (M.W. approx. $1.2–2.4 \times 10^4$ daltons) is inherited via nuclear DNA (Chan and Wildman 1972; Kawashima and Wildman 1972; Sakano et al. 1974). A messenger RNA component for the large subunit of Fraction I protein has been isolated from pea leaves (Hartley et al. 1975) and its activity in a reconstituted chloroplast system is being investigated. The synthesis of the two subunits of Fraction I protein coccur in different locations in the cell. The large subunit is synthesised on isolated 70S chloroplast ribosomes and the inhibition of this synthesis by chloramphenicol suggests that the protein also accumulates in chloroplasts in vivo. For a review of the evidence, see Ellis (1976). In contrast, the small subunit whose synthesis is inhibited by cycloheximide is synthesised on 80S cytoplasmic ribosomes. A current model suggests that encoding and translation of the large subunit occurs within the chloroplast, whilst a small subunit is encoded in the nucleus and synthesised in the cytoplasm. A picture is thus emerging of the complex interplay involved in the synthesis of the complete Fraction I protein molecule and presumably the smaller protein must cross the chloroplast envelope for the association between the two subunits to take place.

If, as seems likely, similar collaborative syntheses are involved in the formation of the other plastid proteins, the problems facing the scientist attempting to culture plastids outside the cell may be considerable. The interorganelle reactions will have to be reproduced in vitro if plastid development is to be simulated in an artificial environment.

VI. Conclusions

Isolated chloroplasts can survive in defined simple artificial media for several hours. Their structure is well preserved over this period but only limited biochemical function remains. Etioplasts can be maintained in similar conditions: prolamellar body transformation and dispersion can be induced by illumination, and photoreduction of protochlorophyllide to chlorophyllide takes place. More detailed study of the functional chraracteristics of chloroplasts over periods of hours can be expected to give insight into organelle interaction on a physiological time-scale, but the development of more sophisticated media will be needed. The discovery that partially differentiated chloroplasts can divide after isolation from the cell suggests that the artificial culture of developing plastids as a replicating population may be a realistic possibility. Since chloroplast DNA contains insufficient information to code for the complete spectrum of chloroplast proteins, and the subunits of at least one chloroplast protein are synthesised in different cellular locations, it may be expected that any successful culture system will be a highly sophisticated one including purified macromolecules and cellular membrane and organelles in addition to the plastids.

Note Added in Proof. The literature survey for this review was completed in October 1976.

References

Boasson, R., Laetsch, W.M.: Chloroplast replication and growth in tobacco. Science 166, 749-751 (1969)

Boasson, R., Bonner, J.J., Laetsch, W.M.: Induction and regulation of chloroplast replication in mature tobacco leaf tissue. Plant Physiol. 49, 97-101 (1972a)

Boasson, R., Laetsch, W.M., Price, I.: The etioplast-chloroplast transformation in tobacco: correlation of ultrastructure, replication and chlorophyll synthesis. Am. J. Bot. 59, 217-223 (1972b)

Bradbeer, J.W.: Chloroplast development in greening leaves. In: Perspectives in Experimental Biology; Sunderland, N. (ed.); Vol. II, Botany, pp. 131-144. Oxford-London: Pergamon 1976

Chan, P., Wildman, S.G.: Chloroplast DNA codes for the primary structure of the large subunit of Fraction I proteins. Biochem. Biophys. Acta 277, 677-680 (1972)

Cran, D.G., Possingham, J.V.: Two forms of division profile in spinach chloroplasts. Nature (New Biol.) 235, 57, 142 (1972)

Ellis, R.J.: The search for plant messenger RNA. In: Perspectives in Experimental Biology; Sunderland, N. (ed.); Vol. II, pp. 283-298. Oxford-New York: Pergamon 1976

Ellis, R.J., Blair, G.E., Hartley, M.R.: The nature and function of chloroplast protein synthesis. Biochem. Soc. Symp. 38, 137-162 (1973)

Giles, K.L., Sarafis, V.: On the survival and reproduction of chloroplasts outside the cell. Cytobios. 4, 61-74 (1974a)

Giles, K.L., Sarafis, V.: Implications of rigescent integuments as a new structural feature of some algal chloroplasts. Nature (London) 248, 512-513 (1974b)

Griffiths, W.T.: Characterisation of the terminal stages of chlorophyll(ide) synthesis in etioplast membrane preparations. Biochem. J. 152, 623-625 (1975)

Hartley, M.R., Wheeler, A.M., Ellis, R.J.: Protein synthesis in chloroplasts. V. Translation of messenger RNA for the large subunit of Fraction I protein in a heterologous cell-free system. J. Mol. Biol. 91, 67-77 (1975)

Heber, U.: Metabolite exchange between chloroplasts and cytoplasm. Annu. Rev. Plant Physiol. 25, 393-421 (1974)

Horton, P., Leech, R.M.: The effect of ATP on the Shibata shift and on associated structural changes in the conformation of the prolamellar body in isolated maize etioplasts. Plant Physiol. 55, 393-400 (1975a)

Horton, P., Leech, R.M.: The effect of ATP on the photoconversion of protochlorophyllide in isolated etioplasts of Zea mays. Plant Physiol. 56, 113-120 (1975b)

Jensen, R.G., Bassham, J.A.: Photosynthesis by isolated chloroplasts. Proc. Natl. Acad. Sci. USA 56, 1095-1101 (1966)

Kameya, T.: Cell elongation and division of chloroplasts. J. Exp. Bot. 23, 62-64 (1972)

Kameya, T., Takahashi, N.: Division of chloroplasts in vitro. Jpn. J. Genet. 46, 153-157 (1971)

Kawashima, N., Wildman, S.G.: Studies on Fraction I protein IV mode of inheritance of 1° structure in relation to whether chloroplast or nuclear DNA contains the code for a chloroplast protein. Biochem. Biophys. Acta 262, 42-49 (1972)

Kirk, J.T.O.: Biochemical aspects of chloroplast development. Annu. Rev. Plant Physiol. 21, 11-42 (1970)

Kirk, J.T.O.: Chloroplast structure and biogenesis. Annu. Rev. Biochem. 40, 161-196 (1971)

Laetsch, W.M., Boasson, R.: Effect of growth regulators on organelle development. In: Hormonal Regulation in Plant Growth and Development; Kaldeway, H., Vardar, Y. (eds.); pp. 453-465. Proc. Advan. Study Inst. Izmir. Weinheim: Verlag Chemie 1972

Leech, R.M.: The behaviour of plastids in artificial environments. In: Biology and Radiobiology of Annucleate Systems. II. Plant Cells; Bonotto, S., Goutier, R., Kirschmann, R., Maisin, J.-R. (eds.); pp. 27-49. London-New York: Acadmemic Press 1972

Leech, R.M.: The replication of plastids in higher plants. In: Cell Division in Higher Plants, Yeoman, M. (ed.); pp. 135-159. London-New York: Academic Press 1975

Leech, R.M.: Plastid development in isolated etiochloroplasts and isolated etioplasts. In: Perspectives in Experimental Biology, Sunderland, N. (ed.); Vol. II, Botany, pp. 145-162. Oxford-New York: Pergamon 1976

Leech, R.M., Murphy, D.J.: The cooperative function of chloroplasts in the biosynthesis of small molecules. In: The Intact Chloroplast; Barber, J. (ed.); pp. 365-401. Amsterdam-New York-Oxford: Elsevier/North-Holland 1976

Leech, R.M., Rumsby, M.G., Thomson, W.W.: Plastid differentiation acyl lipid and fatty acid changes in developing green maize leaves. Plant Physiol. 52, 240-245 (1973)

Leese, B.M., Leech, R.M., Thomson, W.W.: Isolation of plastids from different regions of developing maize leaves. In: 2nd Int. Congr. Photosynth., Vol. III, pp. 1485-1494. The Hague: Junk 1972

Nass, M.M.K.: Uptake of isolated chloroplasts by mammalian cells. Science 165, 1128-1131 (1969)

Plesnicar, M., Bendall, D.S.: The photochemical activities and electron carrier of developing barley leaves. Biochem. J. 136, 803-812 (1973)

Possingham, J.V.: Effect of light quality on chloroplast replication in spinach. J. Exp. Bot. 24, 1247-1258 (1973)

Possingham, J.V., Saurer, W.: Changes in chloroplast number per cell during leaf development in spinach. Planta 86, 186-194 (1969)

Possingham, J.V., Smith, J.W.: Factors affecting chloroplast replication in spinach. J. Exp. Bot. 23, 1050-1059 (1972)

Ridley, S.M., Leech, R.M.: Chloroplast survival in *in vitro*. Prog. Photo. Res. 1, 229-244 (1969)

Ridley, S.M., Leech, R.M.: Division of chloroplasts in an artificial environment. Nature (London) 227, 463-465 (1970)

Robertson, D., Laetsch, W.M.: Structure and function of developing barley plastids. Plant Physiol. 54, 148-159 (1974)

Sakano, K., Kung, S.D., Wildman, S.G.: Identification of several chloroplast DNA genes which code for the large subunit of *Nicotiana* Fraction I proteins. Mol. Gen. Genet. 130, 91-97 (1974)

Smith, D.C.: Autotrophic endosymbions of invertebrates. In: Perspectives in Experimental Biology; Sunderland, N. (ed.); Vol. II, pp. 199-206. Oxford-New York: Pergamon 1976

Taylor, D.L.: Chloroplasts as symbiotic organelles. Int. Rev. Cytol. 27, 29-64 (1970)

Trench, R.K., Smith, D.C.: Synthesis of pigment in symbiotic chloroplasts. Nature (London) 227, 196-197 (1970)

Trench, R.K., Greene, R.W., Bystrom, B.G.: Chloroplasts as functional organelles in animal tissues. J. Cell Biol. 42, 404-417 (1969)

Walker, D.A.: Three phases of chloroplast research. Nature (London) 226, 1204-1208 (1970)

Walker, D.A.: Chloroplast and cell—the movement of certain key substances etc. across the chloroplast envelope. In: Med. Tech. Publ. Int. Rev. Biochem.; Northcote, D.H. (ed.); Vol. II, pp. 1-49. London: Butterworth 1974

Wellburn, A.R., Wellburn, F.A.M.: A new method for the isolation of etioplasts with intact envelopes. J. Exptl. Bot. 22, 972-979 (1971a)

Wellburn, A.R., Wellburn, F.A.M.: Developmental changes of etioplasts in isolated suspensions and *in situ*. Ann. Bot. NS 37, 11-19 (1973)

Wellburn, F.A.M., Wellburn, A.R.: Chlorophyll synthesis isolated intact etioplasts. Biochem. Biophys. Res. Commun. 45, 747-750 (1971b)

Subject Index

Results and Problems in Cell Differentiation

A Series of Topical Volumes in Developmental Biology

Editors: W. Beermann, W.J. Gehring, J.B. Gurdon, F.C. Kafatos, J. Reinert

Volume 1
The Stability
of the Differentiated State

Editor: H. Ursprung
1968. 56 figures. XI, 144 pages
ISBN 3-540-04315-2

"This is the first of a series of volumes that will review a few of the central issues in Cell and Developmental Biology, with each author writing on the stated theme with a personal slant. This approach is very successful. The articles have been written for the worker in the field rather than the general reader, and cover a wide variety of cell types, vertebrate and invertebrate, both *in vivo* and *in vitro*. Many of the questions raised are discussed by more than one author – for instance the relationship of differentiation to cessation of DNA synthesis. This gives the reader the advantage of seeing more than one selection of appropriate evidence. On other topics, the authors are pleasingly unanimous – such as the meaningfulness of dedifferentiation. The book is attractively presented, and the editor and publisher should be congratulated."
Quaterly Journal of experimental Physiology

Volume 2
Origin and Continuity of
Cell Organelles

Editors: J. Reinert, H. Ursprung
1971. 135 figures. XIII, 342 pages
ISBN 3-540-05239-9

"It is refreshing to pick up a book entitled *Origin and Continuity of Cell Organelles* and find something more than a discussion of mitochondrial and plastid semi-autonomy... Overall, the volume certainly is well worth reading for many different types of biologists." *Science*

Volume 3
Nucleic Acid Hybridization
in the Study of
Cell Differentiation

Editor: H. Ursprung
1972. 29 figures. XI, 76 pages
ISBN 3-540-05742-0

"... This book is a comprehensive source of knowledge, so that it will be of value to beginners and/or experienced research workers in biochemistry, biology, genetics and experts in other areas of research for gaining more information about new hybridization techniques and interpretation of results." *Neoplasma*

Volume 4
Developmental Studies on Giant
Chromosomes

Editor: W. Beermann
1972. 110 figures. XV, 227 pages
ISBN 3-540-05748-X

"This is the fourth volume of a successful series of highly topical volumes containing articles and short monographs on central issues in cell and developmental biology... The present book is dedicated to the memory of Jack Schultz, one of the pioneers in the study of polytene chromosomes after their "rediscovery" in 1933. Since then, so much material on the morphological, genetic, physiological, developmental, and biochemical aspects of giant chromosomes and puffing has been assembled that a coherent presentation is fully justified...
This excellent volume should be on the bookshelf of every developmental geneticist and of many others interested in the role of chromomeres as sites of complex, integrated regulatory and informative genetic activities." *Genetica*

Volume 5

The Biology of Imaginal Disks

Editors: H. Ursprung, R. Nöthiger
1972. 56 figures. XVII, 172 pages
ISBN 3-540-05785-4

"...The main appeal of the book is that it is very much alive, squarely set in the present time and conveying the excitement in the field. Even if you do not read it until next year it will be more up to date than most books published in 1973. Imaginal disks are a model system for developmental biologists who should find the book fascinating and challenging..." *Nature*

Volume 6
W.J. Dickinson, D.T. Sullivan

Gene-Enzyme Systems in Drosophila

1975. 32 figures. XI, 163 pages
ISBN 3-540-06977-1

"This concise volume summarizes a large and diverse body of literature in an extremely useful format. The authors document the methods used, problems attacked, and results obtained for each gene-enzyme system which has been studied in Drosophila to date (1974)... In all, the authors correlate a large amount of material in an easily readable style focusing on information rather than speculation. This timely book will be useful to cell biologists, geneticists, and especially, developmental biologists."
American Scientist

Volume 7

Cell Cycle and Cell Differentiation

Editors: J. Reinert, H. Holtzer
1975. 92 figures. XI, 331 pages
ISBN 3-540-07069-9

"...The views of Holtzer and his collegues have clearly had a profound influence in this field, and it is therefore appropriate that the book should open with a statement of their position...Lawrence highlights the merits of insects in a critical and penetrating review...I found the book both stimulating and highly instructive...

I consider it a must for libraries of all institutions with an interest in this very fundamental aspect of biology."
Trends in Biochemical Sciences

Volume 8

Biochemical Differentiation in Insect Glands

Editor: W. Beermann

With contributions by numerous experts
1977. 110 figures, 24 tables. XII, 215 pages
ISBN 3-540-08286-7

"...ofters a timely review of three developmental systems in insects...The second article, by F.C. Kafatos and coworkers, deals with the chorions and choriogenesis in the silk moth and in Drosophila and forms the main, and easily the outstanding, part of this volume...this volume offers a number of useful review. The article by F.C. Kafatos and coworkers, in particular, is highly recommended to all who are interested in developmental systems."
Theoretical and Applied Genetics

Volume 9

Genetic Mosaics and Cell Differentiation

Editor: W.J. Gehring
With contributions by numerous experts
1978. 75 figures, 19 tables. XI, 315 pages
ISBN 3-540-08882-2

"...This book, by several eminent workers, presents a detailed appraisal of a number of aspects of new research using experimentally produced mosaics and chimeras of this type..."
Eugenies Society Bulletin

Springer-Verlag
Berlin Heidelberg New York